# AGRICULTURE IN THE TWENTY-FIRST CENTURY

*This book is based on a symposium sponsored by The Colgate Darden Graduate School of Business Administration of the University of Virginia and funded by Philip Morris Incorporated.*

*Books in this series of Twenty-First Century programs include*

Working in the Twenty-First Century
Communications in the Twenty-First Century
Agriculture in the Twenty-First Century

# AGRICULTURE IN THE TWENTY-FIRST CENTURY

**Edited by**

## JOHN W. ROSENBLUM
The Colgate Darden Graduate School of Business Administration
University of Virginia

A WILEY-INTERSCIENCE PUBLICATION

**JOHN WILEY & SONS**

New York    Chichester    Brisbane    Toronto    Singapore

**Library of Congress Cataloging in Publication Data:**

Main entry under title:
Agriculture in the twenty-first century.

Based on a symposium held in Richmond, Va.,
spring of 1983, and sponsored by The Colgate Darden
Graduate School of Business Administration, University of
Virginia.
"A Wiley-Interscience publication."
Includes bibliographical references and index.
1. Agriculture—Congresses. 2. Agriculture—Fore-
casting—Congresses. 3. Twenty-first century—Forecasts—
Congresses. I. Rosenblum, John W. II. Colgate Darden
Graduate School of Business Administration.

S401.A46 1983      630      83–19847
ISBN 0–471–88538-X

Printed in the United States of America

10 9 8 7 6 5 4 3 2 1

# CONTRIBUTORS

RICHARD BARROWS, Ph.D., Professor, Agricultural Economics, University of Wisconsin–Madison

SANDRA S. BATIE, Ph.D., Associate Professor, Agricultural Economics, Virginia Polytechnic Institute and State University

MICHAEL BOEHLJE, Ph.D., Professor, Economics, Iowa State University

JAMES B. BOILLOT, Director of Agriculture, State of Missouri

ERNEST J. BRISKEY, Ph.D., Dean of Agriculture, Oregon State University

JOHN T. CALDWELL, Ph.D., Professor, Political Science, North Carolina State University

W. RONNIE COFFMAN, Ph.D., Professor, Plant Breeding and International Agriculture, New York State College of Agriculture and Life Sciences, Cornell University

ROBERT B. DELANO, President, American Farm Bureau Federation

MARVIN R. DUNCAN, Ph.D., Vice President and Economist, Federal Reserve Bank of Kansas City

J. M. ELLIOT, Ph.D., Professor, Animal Science, New York State College of Agriculture and Life Sciences, Cornell University

KENNETH R. FARRELL, Ph.D., Senior Fellow and Director, Food and Agricultural Policy Program, Resources for the Future

RALPH W. F. HARDY, Ph.D., Director, Life Sciences, Central Research and Development Department, E. I. du Pont de Nemours & Company

HAROLD M. HARRIS, Jr., Ph.D., Extension Agricultural Economist, College of Agricultural Sciences, Clemson University

VIRGIL W. HAYS, Ph.D., Chairman and Professor, Department of Animal Sciences, University of Kentucky

KENZO HEMMI, Ph.D., Professor, College of Agriculture, University of Tokyo

JAMES C. HITE, Ph.D., Alumni Professor, College of Agricultural Sciences, Clemson University

DEAN W. HUGHES, Ph.D., Economist, Federal Reserve Bank of Kansas City

GEORGE HYATT, Jr., Ph.D., Emeritus Director and Professor, North Carolina State University

D. GALE JOHNSON, Ph.D., Eliakim Hastings Moore Distinguished Service Professor of Economics and Chairman, Department of Economics, The University of Chicago

KENNETH R. KELLER, Ph.D., Managing Director, Bright Belt Warehouse Association

W. DAVID KLEMPERER, Ph.D., Associate Professor, Forest Economics, Virginia Polytechnic Institute and State University

RICHARD KRUMME, Editor, *Successful Farming*

MAX LENNON, Ph.D., Vice President for Agricultural Administration, The Ohio State University

W. ARTHUR LEWIS, Ph.D., James Madison Professor of Political Economy, Woodrow Wilson School of Public and International Affairs, Princeton University

ROBERT D. MacELROY, Ph.D., Research Scientist, CELSS Program Manager, NASA Ames Research Center

JOHN T. MARVEL, Ph.D., General Manager, Research Division, Monsanto Agricultural Products Company

JOHN L. MERRILL, Director, Ranch Management Program, Texas Christian University

WALTER W. MINGER, Senior Vice President, Bank of America N.T. & S.A.

ROGER L. MITCHELL, Ph.D., Dean, College of Agriculture, University of Missouri–Columbia

MICHAEL NEUSHUL, Ph.D., Professor, Marine Botany, University of California, Santa Barbara

RICHARD D. ROBBINS, Ph.D., Professor and Chairman, Department of Agricultural Economics and Rural Sociology, North Carolina A&T State University

B. H. ROBINSON, Ph.D., Professor and Head, Department of Agricultural Economics and Rural Sociology, Clemson University

JOHN R. SCHMIDT, Ph.D., Director, North Central Computer Institute; Professor, Agricultural Economics, University of Wisconsin

WAYNE H. SMITH, Ph.D., Director, Center for Biomass Energy Systems, Institute of Food and Agricultural Sciences, University of Florida

STEVEN T. SONKA, Ph.D., Associate Professor, Agricultural Economics, University of Illinois

ARTURO R. TANCO, Jr., Minister of Agriculture, Republic of the Philippines

THOMAS L. THOMPSON, Ph.D., Professor, Agricultural Engineering, University of Nebraska

JO ANN VOGEL, First Vice President, American Agri-Women

JOHN N. WALKER, Ph.D., Associate Dean, College of Agriculture, University of Kentucky

RICHARD O. WHEELER, Ph.D., President and Chief Executive Officer, Winrock International

LARRY R. WHITING, Ph.D., Chairman, Department of Information and Publications, University of Maryland

SYLVAN H. WITTWER, Ph.D., Director Emeritus, Agricultural Experiment Station, College of Agriculture and Natural Resources, Michigan State University

DEMETRIOS M. YERMANOS, Ph.D., Oil Crops Project Leader, University of California, Riverside

# FOREWORD

Agricultural change in this century has been swift and complex. It will be more so in the next century, a scant 17 years away. Everyone who grows food will have to factor economics, technology, demographics, and geopolitics into the agricultural equation: So will those who finance, process, and distribute agricultural products—and those who make agricultural policy. With no less a task at hand than feeding the world, agriculture is on the way to becoming an integrated system to match any we know today.

At a two-day symposium on "Agriculture in the Twenty-First Century" in the Spring of 1983, government, academic, and business leaders from around the world explored the latest available knowledge about this evolving system. We of Philip Morris—with our interest in tobacco, citrus, grains, hops, lumber, and other products of the land—shared the interest and enthusiasm of the conferees as the future of agriculture was put into a manageable perspective. This book is a product of that symposium, held at the Philip Morris Operations Complex in Richmond, Virginia, and sponsored by The Colgate Darden Graduate School of Business Administration of the University of Virginia in association with a consortium of 11 distinguished schools of agriculture.

"Agriculture in the Twenty-First Century" was the third examination of life in the next century sponsored by The Colgate Darden School with help from Philip Morris. The conference series began in 1979 with "Working in the Twenty-First Century," cosponsored by The Wharton School of the University of Pennsylvania. In 1981 we looked at "Communications in the Twenty-First Century," assisted by The Annenberg Schools of Communications of the University of Pennsylvania and the University of Southern California. These meetings produced the first two books in this series.

Philip Morris has participated in these symposia because we believe the future of a corporation cannot be separated from the future of the society in which it lives and does business. We look at the future less to divine what will happen than to get useful bearings as we enter unfamiliar terrain.

The act of looking ahead serves to remind us of the job we face and affirms our confidence that, by carefully and conscientiously preparing for it, we can make a difference in how the future takes shape.

George Weissman
Chairman of the Board
Philip Morris Incorporated

# PREFACE

*Agriculture in the Twenty-First Century* brings together experts from diverse disciplines to examine the issues and possibilities of agriculture around the world. In opening remarks at the symposium on which this book is based, Ross R. Millhiser, Vice Chairman of the Board of Philip Morris Incorporated, said, "Agriculture is one of the oldest and most essential of human pursuits. It is also one of the most complex and confounding. A basic question is what can be done to bring the world's food supply and its population into balance."

As the world's population approaches 6 billion in the year 2000 the productivity and imagination of agriculturists will be taxed as never before. Although there is more food being produced today than ever before, 70 to 80 percent of the earth's inhabitants exist on substandard diets, and 10 percent are near starvation.

One cannot speak of agriculture without confronting the aspirations of the developing nations and the formidable problems they face. Among them are dependence on a narrow base of crops and the need for external supplies of feed and grains and the volatility of commodity prices. Always before us is the immediate problem of widespread hunger. Arturo R. Tanco, Jr., Minister of Agriculture of the Philippines, aptly stated in his keynote address, "We cannot have even the approaching drums of war deafen us to the cries of the hungry, for hunger has killed more people than all the wars put together."

The United States, blessed with a wealth of rich land, a market system that spurs production, and excellent research and development activities, could theoretically satisfy the food requirements of the world. But this country also has an important role to play in the education of farmers in the developing nations and in the transfer of knowledge and technology to those who would like to elevate their own standards of living.

Although the United States has the most advanced and productive agricultural system in the world, we cannot become complacent and neglect the research that will build the agricultural technologies of the future. We know that we must make prudent use of natural resources, including petroleum and its by-products. We also know that agriculture itself can lead to new sources of energy.

Agricultural history is an account of our control over nature. Because of emergent biotechnologies, including genetic engineering, we will have greater control and experience new record-breaking yields. In science and technology lies the answer to the problems of dwindling resources. And agriculture in the sea and in space has yet to be fully explored.

This is a hopeful book, for much on the record suggests good reason for hope. The farmers of the future will be infinitely more sophisticated than their predecessors. With management skills and computerized information systems, they will be alert to the problems of finance and risk. They will be better able to control the variables of nature and society that affect their destiny.

As we move toward the next century we do so with our greatest resource, our ingenuity. This ingenuity produced a promising agricultural system in the past and will surely succeed in the future.

JOHN W. ROSENBLUM

*Charlottesville, Virginia*
*September 1983*

# ACKNOWLEDGMENTS

As with most books, many people contributed importantly to this volume. First among them are the speakers at the "Agriculture in the Twenty-First Century" conference; their insights and expertise are the foundation of this work. My thanks for their participation and support which have made this book possible.

I am also deeply indebted to the conference Advisory Board, my academic colleagues who contributed richly to the content of the symposium program: Charles O. Meiburg, Ph.D., Associate Dean for Academic Affairs, The Colgate Darden School; Luther P. Anderson, Ph.D., Dean, College of Agricultural Sciences, Clemson University; B. B. Archer, Ph.D., Dean, School of Agriculture and Applied Sciences, Virginia State University; W. W. Armistead, Ph.D., Vice President for Agriculture, Institute of Agriculture, University of Tennessee; Charles E. Barnhart, Ph.D., Dean, College of Agriculture, University of Kentucky; David L. Call, Ph.D., Dean, New York State College of Agriculture and Life Sciences, Cornell University; William P. Flatt, Ph.D., Dean, College of Agriculture, University of Georgia; J. E. Legates, Ph.D., Dean, School of Agriculture and Life Sciences, North Carolina State University; Max Lennon, Ph.D., formerly Dean, College of Agriculture, University of Missouri–Columbia—now Vice President for Agricultural Administration, The Ohio State University; James R. Nichols, Ph.D., Dean, College of Agriculture and Life Sciences, Virginia Polytechnic Institute and State University; Irwin W. Sherman, Ph.D., Dean, College of Natural and Agricultural Sciences, University of California, Riverside; Leo M. Walsh, Ph.D., Dean, College of Agriculture and Life Sciences, University of Wisconsin–Madison; also conference consultant Dale E. Hathaway, Ph.D., Principal and Vice President, the Consultants International Group, Inc.

To Philip Morris Incorporated go special thanks for funding the symposium and providing the excellent facilities of the Philip Morris Operations Complex in Richmond as the conference site.

I am specifically grateful to the many Philip Morris executives who gave their counsel as members of the Conference Committee: George Weissman, Chairman of the Board; Ross R. Millhiser, Vice Chairman of the Board; Clifford H. Goldsmith, President; Hugh Cullman, Group Executive Vice President and Chairman and Chief Executive Officer, Philip Morris U.S.A.; James C. Bowling, Senior Vice President, Assistant to the Chairman of the Board, and Director of Corporate Affairs; Shepard Pollack, Vice President and President and Chief Operating Officer, Philip Morris U.S.A.; James A. Remington, Vice President and Executive Vice President, Operations, Philip Morris U.S.A.; William G. Longest, Vice President, Leaf Operations, Philip Morris U.S.A.; O. Witcher Dudley, Vice President, Leaf, Philip Morris U.S.A.; Vincent R. Clephas, Director, Corporate Public Affairs; Emily Leonard, Assistant to the Senior Vice President, Corporate Affairs; Robert J. Moore, Director, Community Relations, Philip Morris U.S.A.; Larry M. Sykes, Ph.D., Director, Agricultural Programs, Philip Morris U.S.A.; Carl P. Johnson, Manager, Government Relations–Kentucky, Philip Morris U.S.A.; and Bernard J. Kosakowski, Manager, Administrative Services, Philip Morris U.S.A.; also William Ruder, William Ruder Inc. Joan Mebane, Philip Morris Manager of Communications Research, and Gina Gallovich, Communications Research Administrator, organized the conference and coordinated the editorial production of this book. Editorial assistance was also provided by Communications Research staff members, Sherryl Post, Sandra Pecan, and Cynthia Hawkins.

Finally, my thanks to David Maxey and Charles McLaughlin, experienced editors, who worked with our authors to fashion the manuscript.

J.W.R.

# CONTENTS

# AGRICULTURE IN THE TWENTY-FIRST CENTURY

Harvested rice is stacked into water buffalo-drawn cart. (Source: World Bank Photo by Edwin G. Huffman 1975.)

# PROLOGUE

# THE FIRST OF ALL IMPERATIVES

*Banish Hunger in Our Time*

ARTURO R. TANCO, JR.

In no other area is there more need for a restructuring of the old international economic order than in that of food and agriculture. In no other area is it more compelling to succeed. In no other area is it more important to look forward into the twenty-first century to see even dimly what faces us.

At this time, when the world's attention is riveted on the continuing worldwide recession, on the explosive conflicts between nations, on the problems of energy and inflation, we must reassert the primacy of the problem of hunger. We cannot allow the approaching drums of war to deafen us to the cries of the hungry, for hunger has killed more people than all the wars put together. Truly the problem of food towers above all of humankind's problems.

As I and others before me have said, the right to food is the first of human rights. We cannot hope for peace in a world where nearly one-fourth of humanity goes hungry every day, where one-eighth of the world's population owns four-fifths of the world's wealth, where the rich eat and the poor go without.

Despite increased efforts to combat hunger and malnutrition about 1 billion human beings continue to be hungry, 455 million of them severely malnourished, and almost all of them living in the poorest countries of the developing world. This appalling situation persists because, basically, the developing countries as a whole have not been able to accelerate their food production growth rates enough to keep up with their rapidly increasing populations.

In the developing countries of Asia, Africa, and Latin America food production grew more slowly in the decade of the 1970s than it did in the 1960s. The latest estimates of the Food and Agriculture Organization (FAO) show that food production in these countries increased by only 2.6 percent per year in the 1970s, less than the 2.9 percent growth achieved in the 1960s, and well below the 4 percent target called for by the World Food Conference of 1974.

In the 43 food-priority or food-short countries identified by the World Food Council, food production has moved even more slowly at only 2 percent per year—far less than their population growth rate and much lower than their 2.5 percent food production growth rate in the previous decade. As a result, per capita food production, which had been barely positive in the 1960s, was negative in the 1970s at the rate of −0.5 percent per year. Simply put, there is less food per person in these food-priority countries than there was 10 years ago.

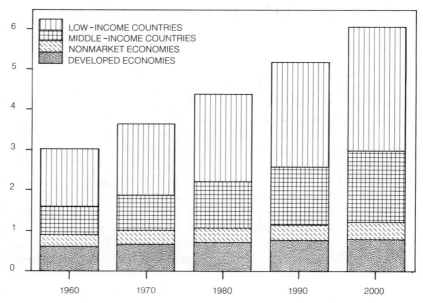

Distribution of the world population, 1960–2000 (*Source:* Adapted from *World Development Report 1982,* The World Bank.)

Africa was the hardest hit of the continents. The yearly rate of growth of agricultural output in Africa declined against a rapid acceleration of population growth from 2.7 percent in the 1960s to only 1.3 percent in the 1970s. Thus, per capita output grew at only 0.2 percent per year during the 1960s and then fell by −1.4 percent per year in the 1970s. Put simply, Africans had less food per person by the end of the 1970s than they had in the previous decade.

Because of these growing shortfalls in production, the food-priority countries have become more and more heavily dependent on grain imports, which have swollen from 19 million tons in 1960 to 85 million tons in 1980 for all developing countries. A study by the International Food Policy Research Institute projects that these imports could rise to between 125 and 145 million tons by 1990. And this burden falls heaviest on the low-income countries which cannot even afford to pay for the food they need to import.

Especially alarming for the future is that current and projected levels of investment in food production and distribution fall short of assuring that the food needs of the world's growing populations can be met. External assistance to developing countries for food production currently amounts to $8.6 billion in 1980 prices, still far short of the $13 billion needed by food-deficient countries to hit the required 4 percent growth rate that would enable them to catch up with their growing population.

At the same time, here in the United States, excess production and reduced world demand have brought prices of farm products down sharply—some say to their lowest levels since the 1930 depression years. This has resulted in the unprecedented retirement of about 65 million acres of farmland from production.

In the face of these grim statistics we must face the reality that the world hunger problem is getting worse rather than better and that a major crisis lies ahead unless we act quickly to forestall it. In the face of this bleak situation might those who advocate despair and doom not be right after all? On balance I believe not.

I believe that the world community has the capacity to surmount whatever obstacles are in the way of abolishing hunger by the beginning of the twenty-first century, if only because the welfare and prosperity of any country increasingly depends on the existence of growth and stability in other nations. The current world economic recession has served to impress upon us once again that the futures of our countries—rich and poor alike—are inextricably intertwined. We are all part of an interlocking whole, each dependent on the other for food, raw materials, and manufactured goods and markets, and lately, each endangered by hunger, economic upheaval, and a wasteful arms race. The time has come for rich and poor countries together to arrive at durable solutions to problems of mutual interest.

I believe, along with others, that the goal of eliminating hunger by the turn of the century is attainable and that the main constraints are neither technological nor financial, but rather political, social, and organizational. And because all of us—rich and poor countries alike—have a stake in this endeavor, the responsibility for increasing food production must be shared by all. I base my optimism on the fact that some progress has been made in the past few years. More important, I am optimistic because the political commitment among world leaders to banish the scourge of hunger from the earth has spread and grown stronger in the last few years. From my travels around the world as president of the World Food Council I can bear personal witness to this heightened commitment.

Now more and more, Third World leaders are giving the highest priority to their food problems by allocating more resources and managerial expertise to the food sector. This growing political commitment at the national level has been a major catalyst in the food production successes achieved in many developing countries.

In many developed countries the problem of hunger in the world has become a major public issue. The U.S. Presidential Commission on Hunger, in its preliminary report submitted in June 1980, urged that ". . . the United States make the elimination of hunger the primary focus of relations with developing countries." It is our hope that this primary focus on food in U.S. foreign relations will continue to be the centerpiece of the Reagan Administration and of all others that succeed it.

Although the supply of resources continues to be much less than what is needed, it is encouraging to note that development assistance going to the food sector in developing countries has in fact doubled since 1974, and concessionary aid for food production could reach the 1974 target of $6.5 billion sometime this year. International financing institutions have played a crucial role in these increases. The World Bank has tripled its lending to food and agriculture in the past few years. Regional development banks—the Asian Development Bank, the African Development Bank, and the Inter-American Development Bank—are likewise channeling more investment to food production.

The $1-billion International Fund for Agricultural Development (IFAD), which was established with the help of the World Food Council in 1977, is well on its way toward making an impact on food production in developing countries. The Fund has already received the replenishment it needed and deserved, so its impact should grow to even greater proportions in the future. This new Fund is especially significant for two reasons: it is the first and only fund devoted solely to food and agricultural production, and it is the first international fund to which OPEC countries contributed a substantial amount.

In past years we have taken a few tentative steps toward building a good security network, including the establishment of a new Food Aid Convention, separate from the International Wheat Convention, which raised the level of guaranteed food aid to 7.6 million tons—almost twice the level guaranteed by the previous Convention. Food aid flows have improved.

Especially encouraging, thanks to the world of international research centers in the past 20 years, is the fact that humankind now possesses much of the technology to double or even triple the yields in tropical and subtropical areas where hunger is most widespread. The Consultative Group on International Agricultural Research (CGIAR), another creation of the World Food Conference, has contributed quietly and superbly to the mobilization of increased resources for such research.

The new technology, additional funds, and determined political commitment have enabled many developing countries to dramatically increase their food production in the last decade. China and India in Asia, Colombia and Brazil in Latin America, the Ivory Coast and Kenya in Africa, among others, have proven that self-sufficiency in basic staples can be attained through a combination of actions within a very short period of time.

These steps have brought us a little closer to our cherished vision of a world without hunger. But the challenge of this vision is far from being met. If we are to banish hunger from our lives, the international community must act even more quickly with far greater determination and with a much higher level of resources transferred to hungry nations.

The United States, in particular, has a crucial role in this global food endeavor. It is, after all, the largest supplier of food to the world and the largest contributor of financial resources for food production in developing countries. Consider the following:

- The United States alone supplies more than 50 percent of the grain imports of the Third World and about 60 percent of total food aid.
- The United States is the largest single contributor to the World Bank, which in turn is the largest single financier of food production in the Third World.
- Half of the OECD contribution of $400 million to IFAD, or $200 million, was contributed by the United States.
- The United States is the major contributor to the CGIAR, which supports all the international agricultural research centers in the world, some 13 of them now.

Before I go into the main elements of a global action program on food, let me look forward to the twenty-first century. Various scenarios have been painted for the year 2000. The FAO has one such scenario drawn on the assumption that all past trends continue. Under this scenario we would see the following events:

- Population growth of developing countries would rise by 2.4 percent per year, or a staggering 1.6 billion between 1980 and the year 2000, as against only 1.2 billion in the preceding two decades. This means that during the 1990s alone the world would have to feed an additional 9.4 million people every year. That is equivalent to feeding an additional Bangladesh every year.
- Food demand would go up by 2.9 percent per year—against a 2.4 percent population growth rate—but food production would only go up by 2.9 percent per year. This means that there would be declining food levels per capita. In Africa it would be worse, with a 3.4 percent projected demand for food per year against a projected production increase of only 2.6 percent.
- The present calorie gap between the average 3475 calories consumed per capita in rich nations and the 2175 calories in poor nations would widen, with countries in Africa and Asia being the most severely hit areas.
- Agricultural imports of the least developed countries would increase by 250 percent by the year 2000.
- Worldwide there would continue to be surpluses of grain in developed countries of 215 million tons as against a projected deficit of 165 million tons in developing countries; however, there would be worldwide deficits of livestock products.
- Developing countries would produce so much excess sugar, citrus, and vegetable oils and oil seeds as to outstrip import demand of the developed nations.
- There would be about 150 million more malnourished people in the developing world by the beginning of the twenty-first century. World Bank projections on "absolute poverty" show that even if per capita food production and income growth rates go back to the relatively good levels of the 1960s there would still be 600 million people living in extreme poverty and therefore in hunger and malnourishment by the year 2000.

This is the FAO scenario on the basis of past trends. It does not seem far from the present picture. While the United States and other developed countries should help, in the final analysis it is only the developing countries themselves who can solve their own food problems by transferring a greater portion of their internal resources to food production and nutrition. Since their resources are limited, however, they will need external assistance to supplement internal efforts.

Proceeding from this basic premise, while I was president of the World Food Council we evolved and initiated the "food strategy approach" in 1979 as the central mechanism for solving the world hunger problem. The idea was to get each developing country to determine and adopt through its own food strategy the precise mix of policies and incentives needed to overcome the most critical constraints arising out of its specific circumstances. Once adopted at the highest political level, a food strategy would have the effect of drawing greater internal attention and more financial and managerial resources to a developing country's food sector. In this way, donor countries and international agencies could be better convinced to give increased assistance to those countries that demonstrate through a food strategy that they are willing to help themselves. A food strategy would thus provide a rational focus for both internal efforts and external assistance.

The food strategy approach has gained widespread international support. So far, 50 developing countries are now in various stages of adopting, designing, or implementing food strategies. The majority of these countries are in Africa, where food shortages and malnutrition are most severe.

Donor countries such as Canada, France, the Federal Republic of Germany, the Netherlands, and the United States are providing assistance bilaterally. Multilateral agencies such as the World Bank, IFAD, and the regional banks have also extended financial and technical assistance for the preparation of food strategies or the implementation of specific projects arising out of these food strategies.

If the world hunger problem is to be stemmed before it worsens further, now is the time for developing countries which have not done so to adopt food strategies and thereby produce more food within their own borders. Now is the time for the rich nations to help them.

As developing countries demonstrate their willingness and capacity to increase food production, using as much of their own resources as possible, the richer nations must shoulder their fair share of this burden. Clearly, the investments required are too enormous for developing countries to bear on their own. This is especially true in the case of the low-income countries whose already scarce foreign exchange resources have continued to be drained by rising costs while export earnings have slumped due to the shrinking world market brought about by the economic recession.

In the short run, however, it is unrealistic to expect substantially increased investments from developed countries while the recession continues. External investments, which doubled in real terms between 1973 and 1979, have in fact been slowing down significantly during the last few years in real terms.

For the time being at least, the feasible approach may be to redirect current levels of bilateral assistance to the food strategies of the 31 low-income countries identified by the World Food Council as having "extremely severe" food deficits or "very severe" food deficits. Since such a redirection would mean that some amounts of concessionary assistance would be diverted away from the so-called "middle-income" countries, these countries could in turn be compensated by replacing every dollar of bilateral aid diverted away from them with two dollars of concessionary financing from the World Bank.

In the longer run, as the recession lifts, external assistance should be stepped up to finance the food strategies not only of the 31 hungriest countries but of the middle-income countries as well. At present, external investments in the food sector of developing countries still fall short of the amounts needed to push food production to the 4 percent growth rate required to keep pace with population growth. In this respect it has been pointed out that a mere 7.5 percent reduction in the $550 billion spent annually on armaments would free enough funds to solve the world's food problems in six months.

Apart from bilateral investments in food strategies, there is also a need to increase the resources of international financing agencies which have been assisting the food production efforts of developing countries. The World Bank, in particular, needs funding for its expanding role in world food and agricultural production. It is significant to note that the Bank more than doubled its lending to agricultural and rural development from $1.6 billion in 1976 to about $3.5 billion in 1980. Agriculture accounted for 33 percent of its total lending in 1980 compared to 21 percent in the 1970–1974 period and only 10 percent in the mid-1960s.

The Asian Development Bank, the African Development Bank, and the Inter-American Development Bank have likewise been giving more and more priority to food and agriculture. They should be given increased resources to support their investments in the food production efforts of countries within their respective regions.

As we move toward the twenty-first century research will continue to be a vital effort, perhaps the most critical one, as it has been in the past. As swelling populations put increasing pressure on limited land and other finite resources, the world will have to rely more and more on science and technology to produce higher yields per hectare. Therefore, it is essential that developed countries lend increased support to the work of international research centers which have been developing new technologies for tropical lending.

Since its creation in 1971, the Consultative Group for International Agricultural Research (CGIAR) has done an admirable job in mobilizing more funds for the 13 international agricultural research centers it supports all over the world. I understand that this year CGIAR has raised about $160 million to fund the research programs of such centers as the International Rice Research Institute (IRRI) in the Philippines, the International Crops Research Institute for the Semi-Arid Tropics (ICRISAT) in India, the International Maize and Wheat Improvement Center or CIMMYT in Mexico, and the International Institute for Tropical Agriculture (IITA) in Nigeria, among others.

But I also understand that these stepped-up contributions to CGIAR have fallen short of the amounts needed to expand research in real terms due to the economic recession in the major donor countries. The United States merits commendation in this respect because it raised its commitment by $50 million last year. Other OECD donors, however, have yet to follow suit.

One way of increasing the resources available for international research is to expand the membership of the CGIAR to include not only traditional OECD donors but a greater number of the OPEC countries as well as middle-income countries. The Philippines takes pride in having been the first non-OPEC developing country to have become a CGIAR donor, swiftly followed by India, Mexico, Brazil, and a few others.

While developing countries must share the burden of support, the developed countries will still have to provide the bulk of increased funding for international research institutes through CGIAR. No more worthy enterprise is deserving of funds; no other investment is more profitable insofar as food production is concerned. It is crystal-clear that research *pays!* There is no better proof than the quiet technological revolution in tropical agriculture spawned by the work of international research institutes over the past decade, boosting yields twofold or even threefold throughout Asia, Africa, and Latin America.

While developed countries should step up funding for international research, they have much to contribute as well by way of transferring research from their own laboratories to farmers' fields in the developing world. This is an area where the United States, specifically its land-grant universities and private agribusiness corporations, can play a vital role. One exciting field of cooperation would be applied microbiology and genetic engineering, an area where land-grant universities and private laboratories have achieved phenomenal breakthroughs. Research could be undertaken jointly with institutions in developing countries to engineer crop strains that can thrive under the environmental and soil stresses of the tropics. Research into recombinant DNA, nitrogen-fixing bacteria, tissue culture, and other scientific frontiers could go a long way toward advancing agricultural development in the tropical areas. Such transfers of technology are the key to the rapid development of food production in developing countries.

Food security and food trade are two vital concerns that need concerted action by both developed and developing countries in order to establish a balanced and rational global food system that can serve the needs of all by the twenty-first century. It is unfortunate that despite recurring cycles of boom and bust, which have occurred since biblical times, we still have not established an adequate system of world food security that will ensure enough food in times of scarcity at prices that hungry countries can afford.

The conclusion of a new International Wheat Trade Convention would have provided the core of such a system. The proposed Convention would have set aside a buffer stock, financed and coordinated internationally, with agreed floor and ceiling prices to trigger stock accumulation and release. In late 1980, after strenuous negotiations between exporting and importing countries over a period of 5 years, substantial agreement had been reached on most of the elements of the proposed Convention, thus clearing the way for a possible compromise agreement. In June 1981, however, the United States, which supplies more than 50 percent of the world's traded wheat, announced that it would not participate in the Convention. Negotiations were therefore postponed indefinitely.

One of the results of our failure to agree is the present glut in wheat in the United States. In view of our failure to conclude a new binding Convention to replace the present consultative arrangement, the World Food Council has put forward a less ambitious, interim proposal for a modest, developing-country-owned reserve for 9 million tons to be made available at an agreed price. This reserve would ensure that the essential import requirements of some 72 low-income countries, over and above food-aid shipments, can be met in the event of a global shortage.

Some form of concessionary assistance will be needed by developing countries to finance stock acquisition. The reserve would be accumulated over a 3-year period and would be incorporated within a broader Wheat Trade Convention when the conclusion of such an agreement becomes possible.

I believe that proposal merits our study and support. It is essential that all countries concerned view the question of food security beyond economic self-interest. The large wheat-exporting countries, in particular, should take the lead in the shift to a more humanitarian view of food security.

The long-term solution to the problem of hunger and malnutrition in developing countries is intimately linked to their ability to expand agricultural exports. More liberal access to developed-country markets would enable hungry nations to pay for their growing food import needs and invest more funds in food production. In the longer run this would reduce the cost of food security arrangements for both developed and developing countries. Moreover, the relaxation of trade barriers would expand trade among all. A recent study by the International Food Policy Research Institute shows that a 50 percent decrease in the level of protection by OECD countries—affecting some 99 agricultural commodities—would increase agricultural exports from some 56 developing countries by about 11 percent, equivalent to $30 billion of additional sales at 1977 prices.

But the persisting protectionist policies of some developed countries, both in terms of import tariffs and export subsidies, have long been barriers to increased trade earnings by developing countries. This is not new. What is alarming is that such protectionism is increasing, thus aggravating the instability of world supplies and prices and, more importantly, impeding the development efforts of developing countries. In the light of these facts it is up to the rich nations to temper their hard-nosed protectionism and to adopt more liberal trade policies that will enhance rather than hamper the food security and economic well-being of the developing world.

While it is clear that some progress has been made on several fronts in the efforts to vanquish hunger, it is also clear that much remains to be done. Some say that the cost of effective action is too great. I believe that the cost of feeble and ineffective action will be prohibitively greater over time. To stand still in this case, as in others, is to move backward. From 4.5 billion people today we will climb to more than 6 billion by the year 2000. We must run fast if we are not to move backward in the common task of banishing hunger from this earth.

We all have a share in this task. Let us, like Socrates, think of ourselves neither as Athenians nor Greeks but as citizens of the world. If we are to give substance to our vision of a world where peace and human dignity reign, we must recognize anew that each person's grief is our own and that the first of all imperatives is to banish hunger in our time.

There is a stirring passage in the writings of Mahatma Gandhi in which he says "the hungry see God in the form of bread." Our sacred task—large nations and small, scientists and businesspeople, scholars and politicians—is to band together and act so that all the hungry in the world are finally able to see God in the form of bread.

# I

# NATURAL RESOURCES
# AND AGRICULTURE

# 1

# CRITICAL CHOICES
# FOR NATURAL RESOURCES

KENNETH R. FARRELL

*Man's relationship to the natural environment, and nature's influence upon the course and quality of human life, are among the oldest topics of speculation of which we are aware. Myth, folktale, and fable; custom, institutions, and law; philosophy, science, and technology—all, as far back as records extend, attest to an abiding interest in these concerns.*

From *Scarcity and Growth,*
by Harold J. Barnett and Chandler Morse

Two years ago cropland in agricultural production in the United States reached a record high of 391 million acres—about 60 million acres more than in the early 1970s. But this year 82 million acres are being withdrawn from production under government programs, and cropland in agricultural production may be near the lowest level of the century, which was about 325 million acres in 1910. Ten years ago world stocks of grain were dangerously low, and the prices of energy, agricultural commodities, and land were soaring. Now world grain stocks are at very high levels, and prices of commodities, land, and energy have moved sharply lower. No more than 2 years ago the media were featuring impending shortages of cropland and water because of projected high rates of growth in U.S. agricultural exports. Now, agricultural exports are falling, and we resort to a variety of government actions to prevent them from falling further. The Malthusians "held court" in the late 1970s; in the early 1980s the Cornucopians have staged a comeback!

These vagaries of U.S. agriculture illustrate its dynamic, unstable nature and confirm the hazards of forecasting its future. Projections vary according to the selection and use of evidence, and especially its interpretation. Is the glass half full or half empty?

Nevertheless, we shall continue to speculate about the future, and such speculation is beneficial if it causes us to consider more fully our goals, the range of possible outcomes, and the choices by which the best might be achieved or the worst avoided. Exploring the future of natural resources and the natural environment is particularly appropriate, for planning horizons are long and choices made today may be costly or impossible to reverse in the future.

Two major interrelated issues involving the future of agriculture and natural resources seem to warrant our attention. The first is whether the availability of natural resources will become a serious constraint on the development of U.S. agriculture. The second is whether the future development of agriculture will result in serious deterioration in the quality of the natural environment. Neither issue is new, but each may pose critical choices for us, individually and collectively, on our path to the year 2000 and beyond.

## NATURAL RESOURCE AVAILABILITY

Long historical trends suggest there will be no major constraint on food production deriving from the availability of natural resources until well into the twenty-first century. For a century real costs of nearly all resource services and major agricultural commodities have gone irregularly downward, trends that are scarcely indicative of chronic scarcities of either natural resources or food. Of course, the future need not replicate the past. There is no guarantee that the complex interaction of resources, technology, economics, institutions, and human capital underlying the decline will yield the same results in the next century. The Malthusians contend that increases in world population to more than 6 billion in the year 2000, to more than 8 billion in 2025, and to a stationary population of perhaps 11 billion toward the end of the twenty-first century would change resource/population ratios so fundamentally that projections based upon long historical trends are misleading, if not irrelevant. Perhaps the point to be drawn from the century-long decline in the costs of resources and food is simply that humans seeking to improve their quality of life can stimulate change in the availability of the resources they need.

Three major sets of relationships are relevant to the issue of the adequacy of the U.S. natural resource base to sustain agricultural development into the twenty-first century: (1) the future demand for natural resources deriving from agricultural use, (2) the demand for natural resources in their nonagricultural uses, and (3) the supply of natural resources and the technology to complement or substitute for them. Each is a complex and uncertain subject, and some careful speculation may help to bring them into perspective.

## AGRICULTURAL COMMODITY DEMAND PROSPECTS

There is a widely held consensus, even by economists, that the prospective expansion in domestic demand for food in the twenty-first century by itself poses no major threat to the U.S. resource base. The combined effects of increased U.S. population and economic growth suggest an increase in aggregate demand for food of slightly less than 1 percent annually by the year 2000.

A precipitous increase in petroleum prices could lead to a heavily subsidized program to produce ethanol, adding to demand for resources by expanding production of grains and other agriculturally based foodstuffs. Otherwise, the growth in demand for agricultural commodities for ethanol production will be marginal at least to the year 2000.

But if there is general consensus on prospects for domestic demand, the same cannot be said for export demand. Projections of recent years in annual growth rates in exports range from slightly over 2 percent to more than 6 percent to the year 2000 from the relatively high levels of the 1970s. It does not require an elegant computer model to conclude that long-term growth at the upper range of that projection could induce significant pressure on the U.S. resource base by the turn of the twenty-first century.

However, there are several reasons for believing that the high rates of growth in U.S. exports in the 1970s will not be sustained even to the year 2000. First is the likelihood that the price increases that would attend such growth would dampen foreign demand and encourage production outside the United States. Second, some of the events of the 1970s that triggered rapid expansion of U.S. exports may have been cyclical or transitory rather than long term. Finally, there is cautious but growing optimism that the developing countries from which much of the potential growth in food demand comes will continue to enhance their own agricultural capacity.

In the context of current perspectives a plausible range of future annual growth rates in agricultural exports might be 2 to 3 percent. In combination with relatively slow growth in the domestic market, total demand might grow at an average annual rate of 1.5 to 2 percent into the twenty-first century. However, the foreign demand component of that projection is highly uncertain and could be highly unstable, thus complicating the planning, development, and management of natural resources.

The development of public and private policies to cope with inherently unstable agricultural markets strikes me as one of the most important choices we face. Our decisions could have important implications for natural resources well beyond the year 2000.

## NONAGRICULTURAL DEMAND PROSPECTS

Although agriculture is the largest single user of land (90 percent of all nonfederal land) and water (80 to 85 percent of annual consumption) and both a supplier and a substantial user of energy, demand for natural resources also derives from other uses: urban, industrial, recreational, transportation, and as amenities. The so-called cropland and water crises have arisen, in part, from the expectation of sizable transfer of these resources to nonagricultural uses.

In an increasingly interdependent society of the future, several complex, sometimes countervailing, forces will shape competition for natural resources. On the margin the value of water and generally the value of land in nonagricultural uses exceeds its value in agriculture. Thus, where markets are unfettered and efficient, agriculture will be in a weak competitive position for use of those resources—much as it is now. Somewhat related is the likelihood of continued erosion of the political power of agriculture at the national level and in many states. By the twenty-first century agriculture will find increasing difficulty in obtaining or even maintaining "special interest" policies for water, other resources, or, for that matter, agricultural commodities themselves.

The growth in nonagricultural demand for resources will be highly uneven, given projected demographic and economic growth patterns. Agriculture near urban centers and generally in the southwestern states will experience the greatest—sometimes irresistible—force of competition.

There are reasons, however, to anticipate somewhat less economic pressure on the agricultural land base in the next three decades than in the past 30 years. From 1967 to 1975 the average annual net transfer of land from agriculture to other uses was about 3 million acres, 875,000 of them cropland, actual or potential. Population growth rates are slowing. The dramatic migration from metropolitan to nonmetropolitan areas of the 1970s will slow depending upon the costs of energy and social services. The rate of household formation will probably decline beginning in the late 1980s as a result of the age composition of the population. And housing starts may also be retarded because of high construction costs. Construction rates for new airports, water and highway systems, dams and reservoirs—all significant claimants upon cropland in the past—have already slowed. Of course, there are important regional and local exceptions to such generalizations. In the Sunbelt states, for example, competition for land and water will intensify, posing critical choices among agricultural, urban, industrial, and other enterprises in the use of resources.

After the escalation of petroleum prices in the 1970s, the quest for alternative energy sources led to considerable apprehension that development of the nation's huge coal reserves would diminish the land and water available for agricultural production in some regions. With declining energy prices, a reordering of federal energy policy, and recision or deferral of plans for construction of several major synfuel plants, apprehension has receded, at least temporarily. Although the impact of such development could be significant in one locale and another, the estimated half-million acres of land that might be occupied annually by coal mining to the end of the century—much of which could be restored—scarcely constitutes a threat to even regional agricultural productive capacities. Of more concern are the possibilities of large-scale water transfers for energy development, such as that recently announced by South Dakota, whereby 16 billion gallons of water will be sold annually and moved by pipeline to coal fields in Wyoming for subsequent shipment of coal slurry. Although currently little or no increase in energy prices is foreseen for the decade ahead, the petroleum-based U.S. economy remains vulnerable to supply interruption.

Summing up, competition and demand for resources will continue to increase into the twenty-first century. However, the "demand–pull" thesis which undergirded the more extreme food-resource scarcity scenarios of the 1970s seems overstated in the context of current perspectives. Still, we might expect as much as 20 to 25 million additional acres of agricultural land—8 to 10 million acres in cropland—to be converted to nonagricultural uses by the year 2000. Considering those requirements and the additional cropland which might be needed to accommodate domestic and export demand for agricultural commodities, a plausible "guesstimate" is that total additional demand for *cropland* might be 35 to 50 million acres by the year 2000, a sizable increase over the current "cropland reserve" estimated to be about 127 million acres.

## THE SUPPLY OF NATURAL RESOURCES AND TECHNOLOGY

Is the supply of U.S. natural resources adequate to meet even the needs I have sketched in? Will the higher-valued uses for land, water, and energy so reduce their availability in agriculture as to lead to much higher real prices for both the natural resources and food? The answer is that we can't be sure, even disregarding the uncertainty of the size of future demand for food and resources. Markets, particularly markets for natural resources, are imperfect; they are constrained by various public policies, such as land zoning and federal water policies. Society may choose in the future, as it has in the past, to give agriculture preferential treatment in the price levels for and access to such resources as water and to regulate the use of resources to attain non-market, noncommercial objectives.

The use of resources is determined by human choice and powerfully influenced by social and economic criteria. Scarce resources are socially valuable resources. As a resource becomes scarcer and more socially valuable, users conserve that resource by substituting other resources and by adopting resource-saving technologies and management practices. Such substitutions dramatically affected the performance of U.S. agriculture in the past century.

The amount of cropland in actual production in the United States in the mid- and late-1960s was virtually the same as in 1910—about 325 million acres. Yet agricultural output in the 1960s was 2.5 times greater than in 1910. That tremendous increase in output and productivity was achieved by altering the resource mix. Relatively cheap energy and water were substituted for land, technologies that enhanced the productivity of natural resources were introduced—improved seeds, machine designs, animal species, and so on—and management was systemically improved by education and better information. Major adjustments were made in production patterns within and among regions of the country. Between 1950 and 1980 use of capital in agriculture nearly doubled, labor inputs were reduced by 70 percent, use of land increased only slightly, and productivity increased 50 percent. Clearly, natural resources have not been a limiting factor to U.S. agricultural production. The substitution among natural resources and between them and other factors of production have been enormously expanded, particularly in the past half-century. Thus, even should less cropland and less water be physically available to agriculture in the twenty-first century, as now seems likely, the question of adequacy of those resources must be placed in the context of the economic availability of other resources, technology, and management.

Fragmentary evidence suggests that to the year 2000 the current stock of technology, coupled with resource adjustments by producers within and among regions, is adequate to sustain growth in aggregate agricultural output at nearly the projected rate of growth in demand—1.5 to 2 percent—without major increases in real food prices.

The more critical natural resource variables for agriculture in the same period, particularly in the West, appear to be the availability and price of water and energy, rather than land. In the absence of subsidized, large interbasin water transfers, water will be increasingly costly in the southern Great Plains, and substantial amounts of land may have to be converted from irrigated to dry-land farming systems. Transfer of water from agriculture to meet demands in growing urban centers in the West and Southwest, whether through the expanding markets for water rights or through other institutional arrangements, could induce major adjustments in agriculture in those areas. Quantification of Indian and federal claims to water of the Colorado River and other water sources in the West pose potentially unsettling issues for agriculture. And it seems evident that in the next decade or two there will be no sizable federal investment in large-scale water development projects. Public policy for water in the West is moving from that of developing additional supplies to that of managing the increasingly more valuable current supplies. On the other hand, there appear to be opportunities for expansion of irrigated agriculture in other parts of the country through private investment.

Adjustments that reduce agriculture's productive capacity in one locale may be offset in other locales. During the next several decades a series of marginal agricultural adjustments to higher-priced water are likely—such as more efficient water application, reduced rates of application, shifts from lower- to higher-valued crops, and shifts in resource use and production patterns within and among regions of the country. The potential to conserve resources from such adjustments is substantial. For example, it is estimated that current water application efficiencies of about 50 percent could be increased to 85 percent by changing application techniques—a 70 percent gain. And in the West as a whole, the quantities of water needed to meet projected urban and other nonagricultural uses to the year 2000 are small relative to the total quantities now used in agriculture. Nevertheless, the water issue will be the source of many difficult, controversial choices. One of the major challenges is the development of more effective institutions to reduce distortions caused by policies predicated upon the premise of an abundant, low-priced natural resource.

Agricultural adjustments to the higher energy prices of the 1970s have already been substantial—conservation in use of energy-based products through such technologies as minimum till and integrated pest management. Pierre Crosson of Resources for the Future predicts that by the year 2010 as much as 50 to 60 percent of the nation's cropland might be farmed by means of conservation tillage. Although agriculture is vulnerable to any major interruption of energy supplies, further moderate, gradual increases in energy prices could apparently be accommodated without major impact on the nation's food supply.

Likewise, the so-called cropland crisis of the 1970s, seen in the light of resource substitution, seems less foreboding than it was popularly depicted at the time. The national cropland base was estimated in 1977 to be about 540 million acres, consisting of 413 million acres of cropland and 127 million acres of pastureland, rangeland, forestland, and other land of "high" or "medium" potential for growing crops. But technology and future economic incentives could make additional acres capable of producing commercial crops. Relatively little is known about the costs of converting potential cropland to a productive state or how much acreage would be brought into production at various price levels for agricultural commodities.

The annual net conversion of 875,000 acres of cropland between 1967 and 1975 has been discussed in dramatic terms, but it constituted only slightly more than one-tenth of 1 percent of the 540-million-acre cropland base. Nevertheless, the future use of our cropland warrants careful attention. Preoccupation with a single national level statistic can be misleading because all land is not created equal! Soil characteristics differ, and, taken in combination with climate and management variables, a seemingly unpromising soil may have unique characteristics for production of high-value crops. Thus land in one area is not necessarily a perfect substitute for land in another area, in either a physical or an economic sense. The conversion of cropland that is occurring unevenly throughout the country is in response to complex economic and social forces, and actions to preserve the cropland or plans to regulate its rational, economic use at the local level are appropriate. The issues and choices of land-use planning to serve the multiple demands for its use are likely to be increasingly important at local levels in the twenty-first century.

With respect to the stock of technology currently available or likely to come "on-stream" in the next decade or two to permit substitution for natural resources, I believe we can again be cautiously optimistic. At a recent interdisciplinary symposium concerning long-range needs for and conservation of natural resources, the consensus of scientists suggested that yields for major crops most probably could be increased 40 to 50 percent by the year 2000. Impressive gains in livestock productivity were cited as possible within the next decade or two. And many suggest that with additional investment in basic research major breakthroughs to enhance both crop and livestock yields are possible by the year 2000 or before.

What is technologically possible is not necessarily economically feasible. Economists and "hard scientists" have collaborated far less than desirable to project the "technological–economic path" by which the results of research might be brought to fruition and adopted. Scarcity of food and natural resources in the form of higher real prices would stimulate both the adoption of available technology and investment in research. But the lead time required to produce new technology is substantial, and discontinuities between the supply of and demand for technology over time are possible. In the past, society has chosen to make substantial investments in agricultural research, even at times when current technology was contributing to the creation of current economic surpluses in agriculture, on the premise that current investment in research was a form of social insurance against long-term shortages. One of the choices we face in looking toward the twenty-first century is how much to invest in agriculturally related research during the next decade or two for a payoff that may not occur until well into the twenty-first century.

## AGRICULTURE AND QUALITY OF THE NATURAL ENVIRONMENT

The relation between agriculture and the quality of the natural environment poses another set of issues of growing importance and controversy. One issue turns on the impact of agricultural production systems on the quality of natural resources. There are those who contend that the current "high-tech" agricultural production system is a major source of environmental degradation in the United States and that it threatens the safety and quality of our food supply. Others suggest that the system is dependent upon chemicals, particularly petroleum-based chemicals, making it highly vulnerable to petroleum supply interruptions and higher energy costs. And still others contend that the system is not sustainable in the long run as a result of its self-defeating tendency to impair the quality of natural resources upon which it depends. An opposing view holds that technology and improved management regimes are available or can be developed to ameliorate, if not eliminate, the worst of the environmental abuses attributed to a "high-tech" production system. These spokespeople contend that no alternative practicable system is available unless we are prepared to pay much higher prices for food.

Three major problems complicate resolution of these issues. The first is the lack of complete scientific evidence concerning the basic relationships involved in the controversy, for example, the fate of chemicals after they leave the farmer's field. The second is the difficulty in evaluating the social costs of the environmental effects of sedimentation, salinity, and so on, which derive from agricultural production. The third is the inadequacy of the institutional mechanisms required to bring these costs to bear on agriculture even if they could be accurately evaluated.

In a recent comprehensive report on the subject of the resource and environmental effects of U.S. agriculture, Pierre Crosson and Sterling Brubaker of Resources for the Future speculate on such effects to the year 2010. They conclude that the major threat among the troublesome environmental problems associated with agriculture is that of soil erosion through its effects on water quality and potential productivity losses on agricultural cropland.

Air pollution, which derives largely from sources external to agriculture, is of growing concern not only because of its immediate effects on agricultural production in urban areas such as southern California but also because of its potential longer-run effects in the form of "acid rain" and "greenhouse effects." Much additional scientific research is needed before reliable assessments of the impacts of such phenomena can be drawn. However, in the twenty-first century such issues could readily become the source of increasing social concern and require difficult public choices on a global basis.

The issues surrounding agriculture and the quality of the natural environment are neither transitory nor ephemeral. Nor are solutions simple or absolute. It is impossible to reduce the environmental risks of a "high-tech" agriculture to zero: trade-offs between food production and quality of the environment are inevitable. By the twenty-first century the choices will be more complex, more difficult, and more important to both agriculture and the remainder of society.

A long view of world and U.S. agriculture over the past century, or even the past three decades, justifies cautious optimism about the capacity of the sector to adjust to what is clearly an uncertain and potentially highly unstable economic and political environment. However, if the population grows into the twenty-first century at rates now projected, formidable new challenges and adjustment problems will come into being, particularly in the developing countries of the world. Peter Drucker has reminded us that we live in an age of discontinuity where unpredictable events in an increasingly interdependent world can be highly destabilizing. It may be that our best strategy is to hope for the best but be prepared for something less!

I see no immutable imperatives to the year 2000 to suggest an approaching crisis in U.S. agriculture or in the availability of natural resources for future development of agriculture. However, neither do I accept the proposition of the technological determinists that a steady stream of technology automatically will continue to cause the real prices of food and natural resource services to decline as in the past. What ensues in the twenty-first century requires purposeful actions of many types, including investment not only in research and development but also in human capital, as well as improved institutions to facilitate adjustments in agriculture and in agriculture's use of natural resources. Those adjustments will derive from several sources—economic, social, technological—and from a political climate in which it will be increasingly difficult for agriculture to maintain "single interest" access to increasingly valuable natural resources.

Complex, critical choices will have to be made, individually and collectively. One concerns the development of institutions to encourage more efficient use and socially desirable allocation of water. Policies and institutions to guide rational, more orderly, and farsighted use of land will pose other choices on the basis not of an impending national crisis but of long-term needs to serve multiple uses. Not all persons, groups, regions, or sectors will benefit simultaneously or proportionally from such choices and adjustments, although we should seek methods of adjustment deemed socially equitable.

Some of the most difficult and critical choices we may face in the twenty-first century turn on the relationship of agriculture to the natural environment, including those that Emory Castle, president of Resources for the Future, terms open access resources, which lie outside the operation of commercial markets. We are not well prepared to address scientifically or quantitatively the trade-off terms between environmental quality and provision of other goods and services. However, even in the absence of adequate scientific, economic, and social data, difficult choices may be required. The development of more coherent, holistic public policies involving agriculture, natural resources, and the environment will require much greater attention in the future than the issue has received in the past.

And there are critical choices to be made with respect to investments, public and private, in research to maintain or broaden our options in the use and conservation of natural resources and the environment.

I have chosen to focus my speculation on the early part of the twenty-first century, leaving to others the possibilities of food production in space, by hydroponics, from synthetic materials, or without natural resources. Even so, you must agree, as Mark Twain once observed, that the fascinating thing about forecasting the future is the possibility of such wholesale conjectures from such a trifling investment of fact.

# 2

# PRIORITY ISSUES IN LAND, SOIL, AND FORESTRY

RICHARD BARROWS

SANDRA S. BATIE

W. DAVID KLEMPERER

Underlying agriculture's current and future problems are old questions concerning natural resources: to what extent are private landowners to be limited in the use of their lands for the public good and how should public lands best be used for the good of all? These issues are involved in the competition for land between farming and housing, in the controversy over dumping of hazardous wastes, in logging on both public and private lands, in the sharing of the water supply, and in many other areas.

Soil erosion, a problem that is still not well understood, will require much more prudent targeting in the coming century if current conservation strategies are to be effective.

Proper management of woodlands—private, industrial, and public—demands comprehensive policies to balance recreational and commercial uses and to allow the market in forest products the freest possible play. The price of timber will probably continue to rise, and perhaps that fact alone will bring more careful management to our forests.

Although persistent, problems concerning natural resources are not insurmountable. Assuming that monitoring and coordination among levels of government and concerned sectors will continue, no disasters are foreseen.

# FUTURE ISSUES IN THE USE OF LAND

## RICHARD BARROWS

Land-use conflict in the United States has been omnipresent. The same basic conflict between private and public interests in privately held land has been repeated in different guises through many generations and in many regions of the country. An example involving the forests of my own region—northern Wisconsin and Minnesota—illustrates the point. In 1867 a special Wisconsin legislative commission declared "One of the most serious evils this state has to contend with is the purchase of large tracts of land by persons who reside in some other states or who, if residing here, still have no permanent and living interest in the land. . . . Their interest should be made to yield to the men who are to become permanent occupants of the land, those who would improve and adorn the land rather than injure it."

Private interests, pursued on private lands, conflicted with the public interest in land use. In response to these "serious evils," the Wisconsin state legislature passed a law in 1868 offering tax incentives and grants to landowners who would plant trees on a part of their land.

About 30 years later some Wisconsin leaders were beginning to realize that agriculture was not well-suited to northern cut-over areas, which were already populated by a large number of immigrants seeking to establish working farms. The Forestry Commission in 1898 concluded, "It will be the part of wisdom, therefore, for the state to adopt a policy which will encourage the use of such lands for the purpose of raising timber crop, rather than agricultural crops proper." Again, it was argued that the private interest in privately held land conflicted with the public interest in land use.

About 40 years later, this time in the Minnesota part of the region, the conflict between private and public interests in private lands was squarely addressed by the Minnesota State Planning Board of 1938: "No large private owner in Minnesota has ever attempted to treat his forest as a crop and to maintain the land in productive condition. . . . The handling of the forestlands of the state, therefore, is becoming more and more of a public responsibility."

The conflict continues today in the region in different forms, and it can be seen in the history of most other regions of the country. At issue is the right of the public to protect itself against "serious evils" that might result from exclusive pursuit of private interests.

The conflict between public and private rights in private land is almost guaranteed by our political and economic system and will almost certainly form the basis for land-use issues and land policy far into the future. Rural lands in agricultural and forestry use are principally involved, but many of the same basic issues also apply to urban land use. Conflicts also arise over the management and ownership of the public lands, but that is another subject.

Conflict over land use can be found in any society in which land, or some products or services of the land, are scarce. In almost any society many prefer a different pattern of land use and may attempt to advance their interests by whatever legitimate means are at their disposal. In the centrally planned economies of Eastern Europe conflicts over land use are resolved largely within the public bureaucracy charged with economic planning and physical development. In the market economies of Western Europe and North America conflicts over land use often lead to a protracted struggle between those who are able to advance their interests through the market and those who have little power in the market and must resort to the political process to block the actions of the owners of the land or capital. The bases for land-use conflicts, however, are similar.

In Switzerland a current conflict concerns the use of some scarce valley land in Rosenthurm, a budding tourist area, for a military training base. In Sweden debate centers on power plant siting, hazardous waste disposal, and development of agricultural land for recreation or urban use. In Taiwan the issue is how to limit expansion of vigorous new industry onto prime rice lands that the government believes must be preserved to protect national security. In India one current debate centers on whether (and how) to reforest highly erosive lands now used for grazing and crops. In Botswana a current issue is whether certain communally owned grazing lands ought to become individually owned in the hope of attaining improved range management.

In the United States land-use conflict is almost guaranteed by the nature of our political and economic systems. Those who created the intellectual foundation for the republic—Jefferson, Adams, Monroe, Madison, and others—clearly believed that a free democratic society could exist only in a system of widespread private property ownership. The right to freely use and exchange property was the ultimate guarantor of personal political liberty. If government were to control land, individuals would be dependent on government for their economic survival and their political liberty would be threatened.

The liberal economic philosophy of free markets and competition also implies minimal public interference in private land-use decisions. In a competitive market prices will allocate productive resources among competing uses. Land will be allocated to its highest-valued use through market exchange in which buyers with higher-value uses in mind outbid those who would use the land in a lower-valued use. The price of land is determined by its profitability not only in the current year but for many years into the future. Individuals and corporations have strong economic incentives to accurately estimate the present and future earning capacity of the land. Thus, in the theory of the market economy, the land resources of the society pass into the ownership of those able to use them most productively.

In a democratic society and a market economy there is little rationale for public involvement in private land-use decisions. In fact, public action to modify land use is viewed as a threat to political liberty and a source of economic inefficiency.

These philosophies of political and economic liberty are tightly woven into the fabric of American government and society. Although few people would subscribe to these theories in their pure form, the belief in the system of private property runs very deep, and the general mistrust of government action to control private land-use decisions is certainly more than a transitory post-Watergate phenomenon.

Yet these beliefs also lead inevitably to land-use conflicts and the demand for public action to control private land-use decisions. The *ideal* of the rugged individualism of the independent free-market entrepreneur clashes with the *reality* of our increasing economic and environmental interdependence. An individual's land-use decisions will often have effects on the natural environment beyond the boundaries of his or her property. These environmental "side-effects" are not often captured in market prices and will not be included in the calculus of private benefits and costs that guide individual land-use decisions. Under these conditions the virtues of the perfectly competitive market are diminished.

For example, a farmer may correctly calculate that terraces to control soil erosion are not profitable for him to adopt when switching from forage to row crops on a ridge top parcel. Yet the resulting erosion may hinder operations on valley farms, especially if the erosion carries chemicals onto fields below. The benefits of terraces to all farmers may outweigh their cost, but to the ridge-top landowner only his own benefits and costs are relevant. The market fails to provide for an efficient level of conservation effort.

Interdependence is a fact of life, and increasingly sophisticated technology, increased population density, and rising per capita incomes intensify this interdependence. Clear-cutting aspen was not much of a land conflict until rising per capita income and high-speed transportation systems transformed some forests into recreational areas with economies dependent in part on "beautiful scenery" for their survival. Staunch defenders of free markets and private property find themselves supporting public regulations controlling private action on private lands. This is nothing new in the United States: public regulation of private land use was common even at the time the ideas of political and economic liberty were being woven into our system of government.

This inherent conflict between our belief in private property and economic liberty and the reality of life in a modern world almost guarantees a long succession of conflicts over land use at the levels of the neighborhood, local government, state, and nation.

Several current conflicts involving rural land use can be traced to recent changes in technology, income, and preferences of lifestyle that increase our interdependence. For example, the conversion of agricultural land to nonagricultural use is an issue in many parts of the United States. Technological change in the transportation system placed vast amounts of rural land within commuting distance of cities. Increased per capita income created for many families the option of living in a rural area and working in a city. A few years later America's preference for surburban or exurban living was heightened by the environmental movement, which increased awareness of the quality of the natural and physical environment. The combination of changes resulted in a rapid and far-flung expansion of housing for urban workers in previously rural areas. Several types of land-use conflicts resulted.

Particularly on the East Coast, state governments were urged to restrict development on agricultural lands in order to preserve open space near cities. Since any landowner's decision to develop his or her property affects others in the area by destroying part of the rural landscape, interdependence is the cause of conflict. In other areas, particularly in the Middle West, farmers moved to prevent scattered urban development in order to protect their farms from problems of nuisance, trespass, and higher taxes. Interdependence is again the key—one farmer's decision to convert his land will adversely affect his neighbors.

Other changes fuel these land-use conflicts. The romanticization of the "rural lifestyle" in the 1970s led to demands to preserve rural land or "rural culture." The "world food crisis" of 1972–1974 led to demands to protect the production potential of U.S. agriculture. Again, interdependence is the key—the decision of a landowner to convert his land is perceived by others as destroying a lifestyle option or limiting the ability of the United States to respond to future food needs.

Similarly, soil conservation, hazardous waste disposal, wetland protection, shoreland development, and coastal zone protection led to land-use conflicts. In each, the principles of private property and minimal governmental interference conflict with the reality of interdependence.

Many of the issues we face today will probably be with us two decades into the future, just as they were with us two decades ago. For example, population decentralization will probably guarantee that conflict between agricultural and urban use of rural land will be an important issue in the future, even though the rate of new household formation will decline. The problems of hazardous waste disposal may well persist as an issue for many years. Problems of soil conservation and shoreland development are not likely to quickly disappear.

Other land-use issues that may be important in the future can only be recognized in the barest outline form today. The relationship between rights in land and rights in water will probably assume more prominence in the future, and some redefinitions of rights will probably emerge as agricultural, industrial, environmental, and urban demands for water increase. The issues of groundwater use and drawdown, now very visible in the Ogallala region, may spread to other regions as groundwater irrigation increases. At the same time the relationship between land use and groundwater quality will probably become more obvious, especially with regard to the use of very powerful agricultural chemicals and the disposal of hazardous wastes. Conflict over mineral rights may well intensify in the future, especially if our access to minerals from politically unstable producing areas becomes more insecure.

It is difficult to develop policy to resolve resource conflicts when the consequences of not acting are not known and even the probability of various outcomes is in doubt. In some cases uncertainty arises because of insufficient understanding of the resource (e.g., groundwater), or because the benefits and costs of a U.S. policy depend on future decisions taken by other nations (e.g., minerals or agricultural land). In other cases there is no easy means of assessing the risks. For example, what is an acceptable level of risk of cancer from increased groundwater pollution due to underground storage of hazardous wastes? As a society we have little experience in incorporating an explicit assessment of risk or uncertainty into our public decision-making process.

Future land-use issues are not likely to be confined to state boundaries—take hazardous waste disposal and acid rain. Responsibility is likely to gravitate to the federal government. Much of the future debate about land-use policy will concern the most appropriate relationship among federal, state, and local governments.

In the past, the federal government has addressed environmental or land resource issues largely through its regulatory power. This is likely to change because many of the issues are so complex at the local level that direct federal regulatory action is almost impossible. For example, the federal government simply cannot plan the uses of agricultural or forestlands, no matter how compelling the national interest in these land uses. More likely, future land policy will involve a combination of state and local policies under a very general set of federal standards or even voluntary federal guidelines. Parts of the Coastal Zone Management program may serve as a rough prototype for cooperative programs involving federal, state, and local governments. The Reagan Administration's proposal for grants to states wishing to pursue innovative soil conservation programs may also be an example of future directions in federal/state land policy relations.

The antiregulatory bias that is currently fashionable represents a basic dissatisfaction with the results of almost exclusive reliance on police power to attain natural resource policy goals in the last decade. Regulation will continue to play an important role in land-use policy, but regulation is likely to be more often combined with tax incentives or grants to alter the economic incentives faced by those who are the targets of the regulation. Such combinations of powers are likely to make regulation both more effective and more palatable.

Policies adopted in the past 5 years to resolve land-use conflict between agriculture and urbanization have often combined several powers of government. A 1977 Wisconsin state law provides tax incentives to farmland owners if local governments adopt strong regulations to guide nonfarm development in agricultural areas. A few local governments on the East and West Coasts have adopted development rights purchase programs, which in some cases combine the regulatory and spending powers. Agricultural-district laws adopted in several states involve combinations of the powers of taxation, spending, and eminent domain.

The challenge of future land-use policy is to develop innovative combinations of federal, state, and local policies. But no matter how clever the policy's intellectual foundations or technical applications are, the most important and most basic issue will continue to be the appropriate balance of public and private rights in private property. A balance must be struck between our philosophical belief in private property and free markets and the reality of interdependence in an increasingly complex world.

# PRIORITY ISSUES
# IN SOIL CONSERVATION

## SANDRA S. BATIE

The United States has had a 50-year-long experience with programs for soil conservation, but the programs have only recently been assessed as to their effectiveness in reducing soil erosion losses and improving soil productivity. This is, in part, because reliable data on the dimensions of the soil erosion problem were not available until 1977 when the Soil Conservation Service (SCS) undertook the National Resource Inventory (NRI). Also, it was not until then that the U.S. Department of Agriculture was required to appraise the soil, water, and related resources of the nonfederal land of the nation and to annually evaluate program performance in achieving conservation objectives. These requirements were part of the Soil and Water Resources Conservation Act of 1977 (RCA).

Analysis of the NRI data as part of the RCA process has provided the conclusion that significant erosion occurs on only a small proportion of the nation's total cropland acreage. Table 1 shows the amount of erosion that exceeded 5 tons per acre per year on land cultivated in 1977. (A soil loss of 5 tons per acre per year translates into an approximate net loss of 1 inch of soil every 30 years.) Combined annual rates of erosion did not exceed 5 tons per year on over 60 percent of the cropland. In contrast, almost 70 percent of the erosion over 5 tons per acre per year was concentrated on 8.6 percent of the cropland.

Severe erosion stemming from corn, soybean, and wheat production is limited to specific regions. Critical erosion areas for cropland devoted to soybeans and/ or corn are southern Iowa and northern Missouri, the Mississippi Delta, the Eastern Piedmont, and the Texas high plains. Serious erosion problems also

Table 1. Combined Sheet, Rill, and Wind Erosion on Land Used for Row and Small Grains,[a] 1977

| Tons of Soil Eroded per Acre per Year | Acres (×1000) | Tons of Excess Erosion[b] (×1000) | Percent of Acres | Percent of Excess Erosion |
|---|---|---|---|---|
| 0–4.9 | 203,247 | — | 60.2 | — |
| 5–9.9 | 67,152 | 133,487 | 19.9 | 8.5 |
| 10–14.9 | 24,976 | 180,444 | 7.4 | 11.4 |
| 15–19.9 | 13,162 | 162,112 | 3.9 | 10.3 |
| 20–24.9 | 7,557 | 131,322 | 2.2 | 8.3 |
| 25–29.9 | 5,610 | 125,269 | 1.7 | 8.0 |
| 30–39.9 | 6,348 | 187,754 | 1.9 | 11.9 |
| 40–49.9 | 3,507 | 138,188 | 1.0 | 8.8 |
| 50–74.9 | 3,307 | 180,473 | 1.0 | 11.5 |
| 75–99.9 | 1,214 | 98,311 | 0.4 | 6.2 |
| 100–149.9 | 643 | 73,428 | 0.2 | 4.7 |
| 150–199.9 | 388 | 64,972 | 0.1 | 4.1 |
| 200 and over | 420 | 99,125 | 0.1 | 6.3 |
| Total | 337,531 | 1,574,885 | 100.0 | 100.0 |

[a] Includes summer fallow.

[b] Number of tons exceeding 5 tons per acre annually.

*Source:* Computed from National Resource Inventory data, USDA, SCS, 1978.

occur when wheat is produced on the Palouse, Nez Perce, and Columbia Plateau areas of west central Idaho, eastern Washington, and north central Oregon.

Despite this concentration of erosion problems, soil conservation programs are widely dispersed. For example, the Agricultural Conservation Program (ACP), which provides federal funds for sharing the costs of establishing soil conservation practices with farmers, distributes funds in essentially the same proportions as it did when the program was first initiated in 1936, despite many changes in agricultural practices, crops, and crop locations. As a result, less than 19 percent of the cost-sharing to date has been used for the highly eroding lands. More than half has been devoted to lands with erosion rates of less than 5 tons per acre per year.

There are several reasons for the wide dispersal of benefits. The U.S. Department of Agriculture's Soil Conservation Service (SCS) and the Agricultural Stabilization and Conservation Service (ASCS), which implement most of the federal conservation programs, need the popular support of farmers and thus have considerable incentive to spread program benefits widely. The eligibility criteria of conservation programs are therefore not strictly related to the magnitude of an applicant farm's erosion problem. The ASCS county committees have accepted applications on the basis of first come, first served. If a farmer meets the eligibility requirements, there is a probability of receiving some assistance regardless of his fields' erosion rates because the program selection criteria are not linked to that. Furthermore, because farmers participate voluntarily, the SCS and ASCS must work with farms and farmers as they find them, no matter what the farmer's limitations in capital and managerial ability.

Even if program managers had attempted to focus assistance on farms with the greatest erosion problems, the lack of data on the nature and extent of soil erosion would have limited them. The SCS and ASCS have also never had the authority to take into consideration other factors, such as tax treatment, leasing arrangements, or loan policies, that may have encouraged soil depleting practices.

Despite the regional concentration of erosion the magnitude of the problem is considerable. RCA analysts estimate that national soil erosion losses from croplands total 2 billion tons per year. This is of concern because soil erosion can affect both the productivity of our nation's croplands and the quality of our water and air.

Soil erosion reduces the capacity of the soil to hold water and removes plant nutrients. The severity of these effects differs widely by region, soil type, farming practice, type of crop, and annual growing season. Also, farmers have been able to compensate for lost soil fertility from erosion through the use of fertilizers. They have also changed tillage practices, used different types of machinery, and adopted new varieties of plants to offset the yield reductions caused by soil erosion.

Nevertheless, experts conclude that the loss of even 1 inch of topsoil may be significant where topsoil depth is already shallow and plant roots can reach less fertile subsoils. Where topsoil is quite thick, soil erosion can occur for years or even centuries without affecting productivity.

Soil erosion is a main contributor to water pollution. Water runoff from croplands carries sediment, nutrients, herbicides, and pesticides—all water pollutants when found in excess. Sediment can reduce the lifetime of many inland lakes and reservoirs and increase the cost of maintaining navigable channels and rivers. Excess nutrients can lead to the decline of water quality of rivers and lakes, thereby raising the cost of purifying public water supplies, as well as being detrimental to wildlife.

Technical solutions to the problem of soil erosion exist; these include terracing the steep lands, retaining crop residues, conservation (reduced) tillage of croplands, contour plowing, and catching water runoff in basins. Studies of conservation tillage show erosion reductions ranging between 50 and 90 percent.

But the high costs of other conservation practices are barriers to their adoption. So are the investment incentives of rising land prices, which fail to heavily penalize lack of stewardship. The personal preferences of farmers, the physical characteristics of the land, and the incentives posed by federal commodity programs clearly play roles too. Finances, however, are a main obstacle to implementing soil conservation practices.

There are exceptions, however. In many areas of the nation conservation tillage, contour plowing, and plant residue retention, which leaves the harvested field covered, are used effectively, in part because they do not cost the farmer and may even increase profits. The economics of soil conservation, however, is such that in most cases it does not pay a farmer to conserve.

Studies estimate that additional cropland required by the year 2000 will range from 26 to 113 million acres. Many of the additional cropland acres probably will be on more erosive lands, and thus we can expect more soil erosion problems in the future unless the adoption of soil conservation strategies increases significantly.

The size of future harvests will depend not just on soil erosion but also on many other interacting factors, including the cost and availability of soil substitutes (e.g., fertilizers and new technologies), the cost of soil rebuilding, and the management of soils, as well as physical attributes of the plants and soils themselves.

While there is still much uncertainty surrounding the importance of soil erosion, this uncertainty makes the case for public concern and public action. The gamble on the importance of soil in future production technologies is a gamble with the next generation's inheritance. Furthermore, many of today's soil erosion problems are also today's water and air pollution problems.

The policy dilemma is to both protect air and water quality and preserve the option to use more soil in the future, while not being overly conservative. Soil conservation is not costless any more than soil erosion is. To spend too much on erosion control or to spend dollars inefficiently means that society will forego other valued services such as schools, roads, or aid to the destitute.

The RCA data and subsequent analyses suggest public strategies for achieving significant soil conservation in a cost-effective manner. These include the targeting of conservation efforts to those regions experiencing the most severe erosion problems; encouraging or requiring farmers to adopt low-cost conservation practices, such as conservation tillage, residue retention, and contour plowing; and employing some cross-compliance strategies. Many of these strategies have the advantage of preventing erosion rather than restoring damages after the fact.

The public has long shared with farmers the cost of implementing conservation practices and is likely to continue to do so. Some form of cost-sharing is probably essential, particularly for investments that will improve water quality but do little to improve soil productivity. A refinement of the current cost-sharing strategy is targeting. This involves a selection of certain regions, or areas within regions, to receive most of the cost-sharing investments. Targeting dollars to areas of greater erosion potential should not mean, however, that every highly eroding area is selected. Some areas are so severely eroded that thousands of dollars per acre could be spent with little improvement in the land's productivity.

Establishing criteria for selecting areas is a problem. If maintaining soil productivity is the main objective, then it may make sense to target areas with fertile but shallow top soils, regardless of erosion rates. If water quality is the primary goal of conservation, then the targeting might be focused on areas with high erosion rates, regardless of topsoil depth.

If targeting is to be an effective policy, important questions remain. One is to pinpoint who will decide where the funds are to go. Presumably, targeting could be conducted at nonfederal levels—in the spirit of the Reagan Administration's "New Federalism"—with federal block grants or matching grants to local and state entities. But there is firm opposition in Congress against any form of nonfederally administered cost-sharing that would rival the ASCS program. This opposition suggests that matching grants will not be generously funded in the future.

If, however, targeting is conducted from the national level, it will be seen as a threat by many of the 2950 soil conservation districts (SCDs) in the nation which contain croplands that are not eroding severely. SCS personnel and their farmer clientele in those districts are fearful that a national or state targeting policy threatens their current funding levels, and they argue that they should not be penalized for having had better soil stewards or for having less erosive lands within their boundaries. Even the modest proposal of Secretary of Agriculture Block to target 25 percent of the conservation budget by 1986 is criticized.

But without reform little improvement in conservation will be possible, particularly given limited budgets.

Another problem concerns T-values, or soil-loss tolerances. T-values are defined as the maximum annual soil losses that can be sustained without adversely affecting the productivity of the land. The USDA has assigned T-values that range from 1 to 5 tons per acre, depending upon the properties of the soil. However, the validity of these numbers is doubtful. T-values have been set too high on many soils to assure the long-term productivity of the land; in other cases soil-loss tolerances have been set too low. Furthermore, T-values do not reflect the impact of technology on crop yields or the costs and benefits of soil maintenance.

However, changing T-values will cause difficult political problems. SCS staff could lose their credibility with farmers who will have worked hard to care for their land within T-values only to see it reevaluated with a stroke of a pen. Of course, criteria other than T-values may be used but unless well specified they would not guarantee a better distribution of targeted funds. It is important to develop legally, economically, and agronomically defensible standards for soil conservation because the choice of targeting criteria is crucial to the ultimate impact of conservation programs.

Cross-compliance strategies are incentive programs. Under these programs, the farmer receives extra benefits from other agricultural programs for implementing soil conservation practices or loses benefits for not implementing them. In what has been termed the "green ticket" approach farmers might receive higher price support payments for their crops if they had participated in a soil conservation program. In a "red ticket" approach a farmer who did not participate would lose specific federal program benefits. According to agricultural economist L. W. Libby, cross-compliance strategies "inject a bitter coherence into federal programs for agriculture" by changing "the situation where one program rewards the farmer for nonconservation behavior . . . while another bribes him to conserve."

For cross-compliance strategies to reduce soil erosion effectively, the participants would have to be farmers whose lands have serious erosion problems. And the most likely program candidates for cross-compliance are the price support programs. But commodity programs are not evenly spread across the nation, so cross-compliance linked to them would have considerably more impact in some states than others. Texas has much of its acreage in corn and cotton, both of which have strong commodity programs, and the state of Washington has many wheat farmers who elect to participate in commodity programs. On the other hand, west Tennessee, like many other regions in the upper Mississippi Valley, has a substantial amount of land in soybeans for which few commodity programs exist. There are many questions which must be asked before a cross-compliance program can be implemented:

- With what other programs should the farmer be asked to cross-comply?
- If cross-compliance strategies are widely implemented, what will be the impact on farmers by type of farm and geographical location?
- If cross-compliance strategies are voluntary, what will happen to participation in other commodity and agricultural programs?

Conservation tillage, another popular and potentially cost-effective strategy, is profitable in many regions and effective in reducing erosion rates. It allows land to be harvested but involves less tillage of the soil, thus conserving soil. However, there are many problems as well. Conservation tillage usually requires increased use of herbicides and insecticides, which not only require skilled farm managers but also may reduce water quality and involve problems of pesticide-resistant insects and weeds. More research is needed on ways to remove obstacles to the adoption of conservation tillage (e.g., weed control and socioeconomic factors) and reduce any attendant environmental costs.

There are, of course, other conservation strategies, such as the funding of research designed to reduce the need for land through improving yields per acre. Experimentation with various strategies in the nation's 2950 soil conservation districts is also important, especially because conservation funding is probably not going to be increased. It will be important to redirect current programs to achieve more soil retention or improve water quality per conservation dollar spent.

Because soil erosion problems are as diverse as the nation's topography and people, there is reason to be suspicious of any centralized, inflexible policy. We are apt to achieve greater soil conservation by encouraging a variety of policies rather than by adopting a single national solution.

While soil conservation and water quality protection are widely held public objectives, so too is the protection of private property rights. Exactly how far the right of private ownership extends to protect a right to let one's land erode is still not definitively resolved. However, the regulatory approach has passed at least one constitutional test (in Iowa), and its acceptability would probably improve if farmers were guaranteed financial assistance in covering the cost of complying with a regulated conservation standard. Acceptability would also be greater if regulations were directed only at farmers with the greatest erosion rates. Since conservation tillage, residue retention, and contour plowing practices can be economical investments, farmers could be required to adopt such low-cost techniques with little negative effect on income and with considerable reduction in soil erosion losses.

There are many political and financial constraints to the adoption of any new conservation program: vested interest in old programs, limited budgets, limited personnel, traditional views of property rights, and conflicts with other policy objectives. Policy changes, whether at state or federal levels, are therefore likely to be incremental. That does not mean that they will be trivial, however. And while the public climate is in flux, it is not the climate of a few years ago. Recent polls reflect broad public awareness of soil erosion and a willingness to support conservation efforts.

Many, and perhaps most, farmers perceive themselves to be stewards of the land, but there is still a large discrepancy between attitudes and behavior. The challenge is to translate the strong desire of society to avoid scarcity and to maintain a quality environment through laws and regulations that will motivate the farmer to conserve our nation's soil when and where it is appropriate.

# ISSUES IN U.S. FORESTLAND USE

## W. DAVID KLEMPERER

Several studies of forestry in the United States show pessimistic resource projections of the type in Figure 1. The vertical axis might be labeled cubic feet of wood, tons of forage, or visitor days of various types of forest recreation. The intended message is clear: some form of shortage looms ahead.

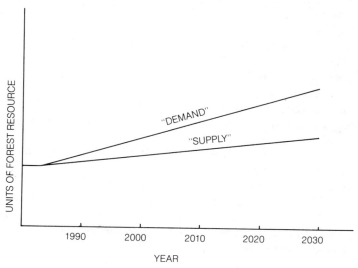

Figure 1.  The "gap analysis" of forest resource supplies and demands.

## ARE WE FACING A WOOD FAMINE?

From an economic standpoint, the "gap" is meaningless, since it is based on projections of current relative commodity prices and existing management trends. As with any priced commodity, a higher projected price per cubic foot of wood will raise the supply line and lower the demand line. In a market economy there always exists a set of prices at which demand exactly equals supply for a renewable and salable commodity. Thus past forecasts of wood famine in the United States have never materialized, and economic theory suggests they never will as long as private property rights continue to be enforced and wood is sold in a competitive market.

The near-complete destruction of forests in countries such as Haiti and Nepal is often cited as an example of what could happen in the United States. But we are not headed in that direction. In those cases timber property rights were not enforced as they are in the United States, well-functioning timber markets did not exist, and forest owners had no incentive to provide timber.

The figures speak for themselves: As of 1976 we harvested annually only about 66 percent of the yearly U.S. commercial forest growth of 21.7 billion cubic feet. Projections for the year 2020 estimate a similar annual growth, with the cut only slightly below growth. The inventory of total commercial standing timber is estimated to increase from 711 billion cubic feet to 887 billion cubic feet over the same period.

The quality of timber and the mix of species have been changing to smaller-sized trees and a larger proportion of hardwoods. But technology and consumers have adapted easily. For example, red alder on the West Coast, once a "weed tree," is now considered excellent furniture stock. The formerly useless aspen now produces paper and flake-board. Former cull trees—trees of poor quality—and mill waste now are used to make particle board, papers, and large laminated beams stronger than solid timbers. Use of veneers extends the quality hardwood resources. Yellow poplar might be used for construction except for the lack of lumber grading and marketing systems. We can look forward to continued technological advances that will stretch the availability of existing timber resources.

As of 1979 U.S. wood exports were 2.1 billion cubic feet, and imports (primarily from Canada) were 3.7 billion cubic feet. Projections suggest that our net import levels will rise from 1.6 billion cubic feet in 1979 to 2.6 billion cubic feet by 2030. These net imports amount to about 11 percent of total consumption.

In each of the three categories of forest ownership—industrial, private, and public—a different situation is found.

## INDUSTRIAL FORESTS

On industrial forestlands the market is effectively stimulating wood output. In the early 1900s when wood prices were extremely low, "cut-out-and-get-out" forestry was common and economically rational. However, with decreasing inventories of old growth and rising timber prices investment has increased dramatically in areas such as reforestation, precommercial thinning, fertilization, damage control, genetically improved seedlings, site preparation, and timberland purchases. Forest industry acreage rose from about 60 million acres in 1952 to nearly 69 million acres by 1977 (or 14 percent of total commercial forestland). The figure projected for 2020 is 73 million acres. Annual timber growth on industry lands increased from roughly 2.6 billion cubic feet in 1952 to 4.1 billion cubic feet in 1976, with growth projected at 4.5 billion cubic feet by 2020.

## NONINDUSTRIAL PRIVATE FORESTS

Nonindustrial private forests (NIPFs) comprise the largest sector of U.S. commercial forestland, about 58 percent of the total. Although timber inventories and growth on these lands have been increasing, annual NIPFs' harvests per acre of total NIPF ownership stood at about 21 cubic feet per acre in 1976, compared to 57 cubic feet per acre for the forest industry as a whole. Although foresters are often concerned about "low productivity" on NIPF lands, it has been pointed out that these acres tend to be concentrated in areas where markets and land quality are poor. In more profitable timber-production zones, where industry lands are concentrated, NIPF productivity does not lag far behind that of industry. Thus the problem is less one of ownership than location, a situation unlikely to change in the next century.

Several studies have found that many NIPF owners are not interested in timber management investment because other investments are more attractive or because the nontimber outputs of their forests are far more important to them than wood production. This explains why NIPF owners have not been particularly responsive to government programs urging them to practice more intensive forestry. Their management decisions are probably socially advantageous, and only if wood products become relatively more valuable should the situation be any different.

## PUBLIC FORESTS

Public forests, the bulk of which are administered by the U.S. Forest Service, are often under attack for not being managed efficiently. Critics argue that the annual value of harvests plus nontimber outputs when viewed as a percentage of timber inventory value is far lower than the rate of return available in alternative investments. Their suggested solution is accelerated harvests accompanied by more intensive management on better sites and less investment on poorer sites. Opponents of this view fear the environmental damage from accelerated harvesting.

Many of the arguments about economic inefficiency in our National Forests stem from the tendency of the U.S. Forest Service to follow an "even-flow" harvest policy with the goal of maximizing long-run, sustainable timber yield within certain budget constraints. The result is low rates of return on capital value. In addition, the Forest Service's desire to provide timber sometimes leads to harvesting costs exceeding receipts. These deficit timber sales have occurred on certain high-logging-cost areas in the Rockies and have also damaged prime recreation areas.

With current trends toward more concern about government spending, I believe we can expect a future focus on greater efficiency in public forest management.

## THE QUESTION OF TIMBER PRICES

Since the timber market seems to respond effectively to price signals, the major issue is not timber famine but timber price. Dramatic increases in wood prices have already occurred, and in 1982 the U.S. Forest Service expressed concern that unless timber output is increased substantially softwood stumpage prices could more than double in real terms over the next 40 years. On the other hand, one can argue that unless timber prices increase substantially many timber-growing investments will yield inadequate rates of return. Higher wood prices are exactly what is needed to stimulate more forestry investment, especially on nonindustrial private forests. Not only do NIPF owners respond to higher prices with increased management intensity, but the forest industry responds with assistance and long-term leases of NIPF lands for timber production.

An effort to dampen wood price increases through government programs that encourage private timber investment by regulations, direct subsidies, or tax subsidies can have unwanted results. Indeed we would have more timber, but that gain would be less than the loss in other social benefits. If more dollars are forced into forestry, less will be invested elsewhere. One can surmise that the investment elsewhere was earning a higher rate of return than the new forest investment; otherwise, the reallocation would have been made willingly in the market. Thus such government-induced investment can actually reduce total returns from society's capital. Also, the subsidy would decrease future wood prices and reduce forestry rates of return still further.

Would not lower wood prices benefit the consumer? Certainly, but the same holds for any commodity. Government-induced higher output would reduce the prices of oil, steel, concrete, shoes—why stop at timber? If the goal of a timber subsidy is to help the poor obtain lower-cost housing, fiscal theory suggests it would be more efficient to assist them directly through payments or reduced taxes. Stimulating wood output can misallocate investment and also give unintended aid to groups such as the wealthy and the wood processing industry.

There would be certain advantages to higher wood prices in the next century. Higher timber prices stimulate more forest management investment and higher returns to landowners, smaller and more energy-efficient homes, more efficient use of timber, neater logging jobs, more wood-fiber recycling, and fewer waste disposal problems. So higher wood prices are not entirely evil: In fact, they are essential if we want more timber from a market economy.

Government programs to stimulate more reforestation will provide improvements in scenery, water quality, and soil conservation in only certain cases. Given moderate care in logging, most cutover areas in the United States become naturally revegetated with hardwoods, brush, and other ground cover that provide the same benefits. This is not to speak out against reforestation, but only to illustrate that where the investment may be financially questionable a lack of tree planting on selected areas is not necessarily a tragedy for future generations.

## CHANGING LAND USE PATTERNS AND REGULATIONS

One often hears concern over a long-term decline in U.S. forest acreage. However, there is not necessarily cause for alarm, nor is there even a fixed trend. Figure 2 shows a decline in forest acreage and an increase in cropland between 1800 and 1920. This was a natural trend since farm crops brought more social values per acre than wood, which was in overabundance. Commercial forests declined from 509 million acres to 483 million acres between 1962 and 1977, but they had increased by a greater amount in the 1950s.

Future changes in forest acreage are unlikely to be as dramatic as they were around the turn of the century, thanks to more stable trends in population and land development. One projection estimates U.S. commercial forest area at 446 million acres for the year 2030, or about 7 percent below today's area. Land use will no doubt continue to shift back and forth between agriculture and forestry as well as other uses in the free market. Society is well served by the change, given no significant unpriced negative side effects.

The question of unpriced side effects may become more controversial in future decades. Whether we change from forest to nonforest use or simply harvest trees, third parties are often negatively affected. In some cases this has led to government-imposed harvest regulations to reduce logging-caused

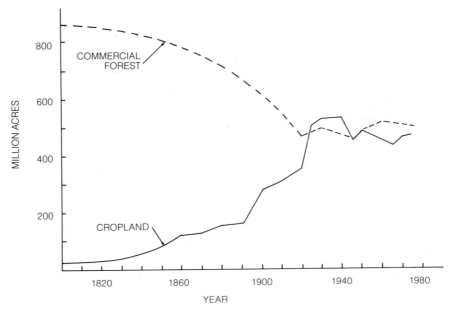

Figure 2.    United States area of cropland and forest. (Adapted from M. Clawson: copyright 1979 by AAAS.)

damage to values such as scenery, fishing, and water quality—most notably on the West Coast. We are likely to see more of such regulations nationwide.

It is rare that we can have a change in land-use activity with absolutely no losses. In theory, we could permit such losses, such as scenic damage from timber harvest, as long as gainers could overcompensate losers. But this is not as simple as it sounds. For example, if clear-cutting is allowed and citizens lose a beloved forest vista (case I), gainers and losers are clearly identified. On the other hand, if regulations prevent or restrict planned harvests (case II), forest viewers gain, while forest owners lose.

The problem is that the measure of gains and losses will depend on the regulatory framework: Case I stresses private property rights while case II stresses amenity rights to outputs such as scenic beauty and water quality—a view toward which the nation is leaning as we look to the future. When we deal with unpriced outputs, the case II framework is likely to lead to less environmental damage than case I. Because of the controversies involved, we are likely to see increasingly more attention focused on these legal and economic questions in designing forestland use regulations where unpriced values are involved.

## MULTIPLE USE IN PUBLIC FORESTS

Under the requirements of the Renewable Resources Planning Act, the U.S. Forest Service currently estimates and compares present values of future benefit flows from alternative National Forest management plans. Economic theory suggests that an optimal management plan is one that maximizes net present value at some appropriate interest rate. Selecting the appropriate interest rate and predicting and measuring nontimber values are difficult problems. A much more fundamental problem arises when several widely differing management scenarios for the National Forest system have very similar present values but sharply differing distributions of benefits, such as timber or recreation, among user groups in different regions and across time. For example, plans with similar present values may provide more recreation or timber in the East and less in the West, or vice versa. Thus cost/benefit analysis will not solve all the distributional problems in public forest management. These are left to the political system and will probably be of increasing concern in the next century.

The question of who pays and who receives is bound to become more controversial. Consider the federal condemnation of and payment for two private tracts of prime old-growth timber for the Redwoods National Park. The first tract cost the taxpayers over $186,000,000, and the second tract, whose purchase is still under litigation, could well cost more than the $320,000,000 already on deposit. In such cases the general public pays and a relatively small group receives the benefits.

The coming decades will see an increasing conflict between interests of preservation and development as the values of both increase, thus bringing more pressure for recreationists to pay a greater share of their bill. By moving more of the nontimber benefits into the marketplace, we will also find more private owners managing their forests to sell outputs such as hunting, hiking, and camping.

Timber famine is really not an issue. Our major wood supply question is that of timber prices: In the future we can expect higher prices, which will stimulate greater forestry investment and more efficient use of existing wood resources by all ownerships. Government intervention to dampen these price increases could reduce society's well-being.

The market will continue fostering changes in forest management and land use. When the benefits of such changes exceed the costs, the best role for government may be to soften the negative side effects rather than to prevent these changes.

Controversy over private forestlands will continue to center on what type of intervention will maximize social welfare and which form of taxation will be most equitable. In public forest management, controversies will center on balancing timber and nontimber benefits and on determining politically the distribution of these benefits. For all forestlands more of the currently unpriced benefits will probably enter into the market pricing system.

With more emphasis on improving efficiency in forest management, we very likely will be able to generate more timber and nontimber benefits from a smaller forestland base. We will be able to have our cake and eat it, too.

# 3

# ENERGY: STRATEGIES FOR THE FUTURE

JOHN N. WALKER

WAYNE H. SMITH

As in many other parts of the economy, American agriculture's dependence on petroleum is a continuing problem with no clear solution in sight. However, this should not conceal the relative energy efficiency of the farming community. The on-farm production of food uses only a tiny amount of the total energy consumed nationally, whereas the rest of the food sector—preparation, transportation, marketing—requires substantial amounts.

Some energy savings can be made on the farm of the future, particularly if electricity replaces petroleum. Innovations in irrigation, fertilization, and tillage will all contribute to energy efficiency, but much more needs to be saved off the farm. Meat production will always be less energy-efficient than the growing of plants, but soon animals will be raised less on grain and more on grass and legumes, yielding savings in energy.

Agriculture will expand its role as a producer of energy, not only with wood, but also with new crops cultivated in water as well as on land. Biomass, a promising alternative energy source will become increasingly important. Biomass production must be distinguished from the growing of crops for food, feed, and fiber and not be considered merely a way to use their residues. Biomass crops can be developed that will not compete for land needed for conventional agriculture.

# ENERGY USE IN THE FOOD SECTOR

## JOHN N. WALKER

By minimizing the risks due to weather, pests, and disease, humans have increasingly expanded their control of the production of essential food supplies. This process has necessitated increasing amounts of energy for agriculture. At first, human energy was predominant. Later, animal energy was substituted for human power, and eventually mechanical power replaced the use of animals. Concomitantly, more and more manufactured sources of energy, such as chemical fertilizer and pest control materials, were needed, and they required energy for their manufacture. The more intensive the agricultural production system, the greater the energy required per unit of production. The trend toward a more intensive agriculture with higher energy needs can be expected to continue well into the twenty-first century.

As the energy put into agricultural production has increased, so too has the energy for the off-farm components of the food sector. They include food processing, transportation, preservation, and preparation. Today the total food sector, including on-farm production and off-farm activities, accounts for 16.5 percent of total energy used in the United States. When forestry and fiber production are included, the percentage of the total rises to 22 percent. A breakdown of the energy use within the food sector is presented in Figure 1. Less than 18 percent of the total energy use in the food system goes toward on-the-farm production activities. The two biggest users are food processing and home food preparation, which consume well over one-half of the energy used in the food sector. When the preparation of food away from the home is included, these three activities account for almost three-fourths of the energy. If future reductions in energy use are necessary, a large percentage will have to come from these components.

A number of authors have attempted to evaluate the efficiency of agriculture by comparing the energy value of the products of agriculture to the fossil energy put into their production on the farm. Ratios for a few selected commodities are shown in Table 1. Although different studies give highly variable results, in general, on-farm enterprises are relatively energy efficient, providing more energy output than required for production. However, for some products, such

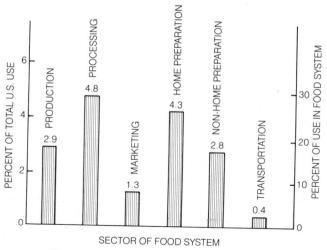

Figure 1.　Energy use in the food system.

as lettuce and meat, the ratios are unfavorable. When the total food chain is considered, the energy efficiency ratios are considerably worse. The only part of the food system which produces energy is the on-farm component. Every other component—processing, marketing, preparation of the product, and transportation—requires energy, with no improvement in the energy value of the product. But energy-efficiency calculations are questionable when applied to

Table 1.　Energy Efficiency for Selected
Agricultural Crops

| Crop | Ratio $\dfrac{\text{kcal Output}}{\text{kcal Input}}$ |
|---|---|
| Soybeans (Ill.) | 4.5 |
| Corn (U.S.) | 3.5 |
| Wheat (N.D.) | 2.7 |
| Peanuts (Ga.) | 1.3 |
| Rice (Ark.) | 1.1 |
| Apples (East) | 0.9 |
| Tomatoes (Calif.) | 0.6 |
| Lettuce (Calif.) | 0.2 |
| Broilers | 0.8 |
| Beef (range) | 0.5 |
| Beef (confined) | 0.07 |

Table 2.   Use of Different Fuels in the Food System (Percent)

| | Liquid Petroleum | LP and Natural Gas | Electricity | Coal |
|---|---|---|---|---|
| Production | 58 | 18 | 24 | |
| Processing | 11 | 42 | 40 | 7 |
| Marketing | 8 | 9 | 83 | |
| Home preparation | | 6 | 94 | |
| Away-from-home preparation | | —100— | | |
| Transportation | 100 | | | |

agriculture. The business of agriculture is the production of food and fiber, not energy. If energy-efficiency calculations are to be utilized, then clearly the off-farm components that affect them most adversely and dramatically need to receive primary consideration. Despite this fact, studies of energy use and conservation in the food system have been directed mainly at the production component.

Although the production of food requires relatively little energy, agriculture is critically dependent upon petroleum energy, which has recently been in short supply. The types of fuel required in the total food system are shown in Table 2. Although the other components with the exception of transportation are not as dependent as production on petroleum fuels, the overall dependency is great. About 40 percent of the energy used by the food system is liquid petroleum fuel and LP (liquified petroleum) and natural gas. This does not include the petroleum used in electric power generation. In the agricultural production component 76 percent of the energy use is in these two forms.

The energy use by on-farm activity is presented in Table 3. Mobility is a

Table 3.   Energy Use in Farm Production

| On-Farm Activity | Percent |
|---|---|
| Fertilizer | 31 |
| Irrigation | 13 |
| Farm vehicles | 13 |
| Preharvest field operations | 12 |
| Harvest operations | 10 |
| Grain and feed handling and drying | 8 |
| Livestock care | 6 |
| Pesticides | 5 |
| Other | 2 |

key factor in several categories, particularly farm vehicles, preharvest field operations, and harvest operations (35 percent of total on-farm use). In the future electric vehicles may have a place in some farm chores, but a portable fuel will continue to be needed for these three activities. In the other areas alternative energy forms can be used. Electricity can be used for irrigation pumping, coal can be used for fertilizer manufacture, and biomass and solar energy can meet all or some of the energy needed for crop drying and livestock housing. Changes in this area will be made when the alternative energy forms become sufficiently low in cost to justify conversion.

## ENERGY USE AND CONSERVATION IN FOOD PRODUCTION

Increasingly, intense agricultural production has used energy in three main ways: to reduce labor, to increase productivity per unit of land, and to reduce the risk of crop failure and maintain quality. Essentially, the farmer used more and more energy for these purposes for economic reasons. However, other factors also played a part: the elimination of drudgery, complying with institutional constraints, and various technical and biological necessities.

### Reducing Labor

From 1949 to 1970, when energy input increased 45 percent, corn yields increased 138 percent. During this period labor decreased 61 percent. In terms of productivity per unit of labor the improvement was 500 percent. From 1950 to 1979 the number of farm workers declined from about 10 to 4 million, while tractor horsepower on farms from 1951 to 1978 increased from 101 to 238 million. The cost of labor during this period increased 383 percent, while the price of gasoline increased only 150 percent and fertilizer only 84 percent. Clearly, in relative terms, the value of both energy and energy-intensive products grew much less than labor and hence were increasingly bargains. Although the cost of energy, particularly petroleum products, can be expected to rise in the future, the rise should not be greater than 30 to 40 percent since the production of synfuels from oilshale will become economically viable. Energy can be expected, therefore, to remain a bargain in comparison to other inputs through the twenty-first century.

In the draft of the revised report on Energy Use and Production in Agriculture by the Council for Agricultural Science and Technology (CAST), a farm worker in 1980 is estimated to cost $36.70 for a 10-hour day. Electricity, costing only 5 cents, can provide the equivalent amount of mechanical work. Comparatively, gasoline operating an engine with a 20 percent efficiency would only cost 10 cents. The economic incentive for substituting energy for labor is clear, and it will continue well into the twenty-first century. The real beneficiary of this reduction in production cost has been the consumer, since the returns to the farmer are traditionally only marginally above the cost of production.

As an example of the impact of labor costs on production costs, a comparison can be made of various methods of weed control on corn. Mechanical cultivation increases corn yields over no weed control by 50 percent; if herbicides are used for control, yield is increased 67 percent; and if hand cultivation is utilized, the increase is 70 percent. Profitability is another matter, however. In 1976 prices the net return to the producer increases 46 percent if mechanical cultivation is employed and about 60 percent if herbicides are used; but due to the high cost of labor, the hand cultivation method results in a 50 percent reduction in the net return to the producer.

The most energy-efficient system is hand cultivation. But the use of herbicides provides almost the same energy efficiency. The energy in the increased corn yield exceeds the energy input in herbicide weed control about 100 times, which is an energy-efficient practice by any standard.

## Increasing Productivity

To meet the increased food needs of the United States and the world increased productivity of the land area suitable for agricultural production must be achieved. The problem is compounded because urbanization and the construction of highways are removing much of our most productive land from agricultural use. From 1950 to 1979 there was about a 10 percent decrease in the land in farms.

There are several possible avenues for improving the productivity of agricultural land. One example was given above in the discussion of weed control in corn. Probably the most significant single practice affecting productivity is fertilizer usage, which is the largest single energy input into agricultural production. The response of corn to nitrogen fertilization (Table 4) shows that the maximum yield is obtained with the highest level of fertilization, and interestingly, the energy efficiency for the highest level of fertilization is only slightly less than the highest level of efficiency. The highest financial returns are received from the highest level of fertilization.

Advances in genetics, agricultural production practices, and increased use of plant proteins by humans in place of animal protein will also contribute to improving agricultural productivity. However, fertilizer use can be expected to continue to rise, particularly in those countries and on those lands where high levels of fertilization are not currently used. This intensive agriculture

Table 4. Response of Corn to Nitrogen Fertilizer

| Nitrogen (1 lb) | Yield/Acre (bushel) | Energy in Nitrogen (kcal × 10³) | Energy in Corn (kcal × 10³) | Net Energy (kcal × 10³) |
|---|---|---|---|---|
| 0 | 35.0 | 0 | 3,563 | 3,563 |
| 70 | 65.3 | 588 | 6,647 | 6,059 |
| 140 | 94.2 | 1,176 | 9,586 | 8,410 |
| 210 | 108.4 | 1,764 | 11,034 | 9,270 |
| 280 | 113.7 | 2,352 | 11,574 | 9,222 |

*Source:* Doering et al. (1977).

will continue to depend upon substantial inputs of energy and energy-derived products.

In the manufacture of nitrogen fertilizer, in which nitrogen is combined with hydrogen, sources of hydrogen other than natural gas can be used. Coal is one viable alternative. Low-priced ammonia now available from Mexico and other oil-rich nations is a source of lower-priced fertilizer and also saves our natural gas. Some reduction in fertilizer usage can be achieved through expanded use of manure; however, use of such products must be carefully weighed against transportation and distribution costs. The use of manures may also increase the problems associated with weeds and other pests. If industrial and municipal wastes are utilized, there may be a problem with heavy metal toxicity. It is estimated that manures can serve as a source of fertilizer at best for only about 18 percent of the cropland.

Another energy-intensive practice that results in improved productivity is irrigation. It is utilized on about 15 percent of the cropland; however, for selected crops the percentage of the acreage irrigated is relatively high. The energy input to irrigation is shown in Table 5. Although these figures indicate that the energy requirement is large, the impact on productivity can be dramatic. In dry areas of the West irrigation clearly is the difference between excellent yields and no yield. However, the cost of irrigation is becoming critical, particularly in the areas of the West where the water table is dropping and the pumping heads are increasing. In the dry areas as much as 2100 liters of fuel may be required per hectare just for irrigation pumping. If diesel fuel is used, the cost would be nearly $550 per hectare ($220 per acre). Cost savings and energy reductions can be achieved by improved design and maintenance of pumps (estimated to have a potential for saving about 50 percent).

Reductions in energy use are also possible by reducing the frequency and amount of water provided by irrigation. Advances in weather forecasting, im-

Table 5.    Energy Required for Sprinkler
and Surface Irrigation

| Energy Type | Energy/Acre/Inch Water (kcal/acre) | |
| | Surface | Sprinkler |
| --- | --- | --- |
| Electric | 16,400 | 37,600 |
| Diesel | 40,300 | 93,900 |
| LP Gas | 64,000 | 146,300 |

proved soil moisture-sensing equipment, and studies of crop responses to water have allowed irrigation to be scheduled more precisely to crop needs. Future advances, particularly in the modeling of crop responses to water, will help further. Additional savings are possible by reducing pressure at which irrigation systems operate through improved system design and efficiency of water use. Trickle irrigation, which places the water directly in the plant root zone without any distribution loss, greatly improves efficiency, but the system cost is much greater than for most of the other systems. Unless sizeable energy reductions are made, with corresponding cost reductions, irrigation usage will decline and so will the agricultural productivity of the land now irrigated. This development would require in turn even more intensive use of the nonirrigated acreages.

Another method of reducing energy inputs into crop agriculture is the expanded use of reduced or no-tillage agriculture. Tillage is one of the more energy-intensive machinery operations. Table 3 indicates that total preharvest field operations account for about 12 percent of the total energy used. In corn it has been estimated to be as much as 10 percent of the total, and corn production consumes 25 percent of all energy used in agriculture. Yet, if all of the corn in the United States was produced by no-tillage methods, the energy saved would only amount to 200 million gallons of fuel, which is less than 0.1 percent of the U.S. total energy used. Unfortunately, not all of the energy can be saved. No-tillage agriculture requires increased herbicide and pesticide application. Also, if comparable yields are to be attained, part of the fertilizer must be side-dressed or increased fertilizer placed at the time of planting. Both of these fertilizing alternatives substantially reduce the energy benefit and may actually result in higher energy use. No-tillage planting will likely increase in the twenty-first century regardless of its energy efficiency, because of its effect on erosion and the time savings in soil preparation and planting.

The energy efficiency of meat production does not compare favorably with plants according to Table 1, but it is rather misleading, since animals, particularly ruminants, convert plant material that cannot be used directly by humans into a food source humans can utilize. More than one-half of the total farmland is in range and pastureland, much of which is not suitable for row-crop production. Therefore the importance of animals in the food chain should not be overlooked. It is true that animals are currently fed large quantities of grain that could be used directly by humans. However, as human competition for this food increases in the twenty-first century, it will be taken out of the animal production system, and animals will be produced on more grasses and legumes. The potential for expanded production in this way is substantial. In the cool-season grass range, the interseeding of legumes, such as alfalfa or clover, into grass can increase the productivity of pasture by a factor of 4, with a concomitant increase in forage quality. Although the shifting of ruminant production to grass will change the meat quality to some extent, consumers have shown they will utilize such meat, particularly at a reduced price. Cattle may also be fed more in the future on crop residues and by-products.

## REDUCING RISK AND MAINTAINING QUALITY

Once a crop is planted a major portion of the energy input has been committed. In corn production the figure is more than 50 percent. If drought, pests, freezing, or other calamities occur, the energy efficiency may become very low or even zero. Once the production cycle has started the producer has no alternative but to continue to provide the money and energy required to secure high yields and to conserve the energy already put into the system.

One of the major activities aimed at removing the risk of crop loss and maintaining quality is grain drying. In nonirrigated corn it is the second largest energy requirement. Considerable thought has been given to using low-temperature drying, as contrasted to the typical high-temperature drying method, with a possible energy savings of 20 to 30 percent. But there are several risks. In the Southeast low-temperature drying may be too slow and therefore increase the risk of spoilage. If harvesting is delayed to allow corn to dry naturally in the field to about 21 percent moisture range so that low-temperature drying can be used, the increase in field grain losses will be on the order of 4 percent. The energy in the grain left in the field would exceed by several times the energy saved by delaying harvesting.

Some energy can be conserved by using more efficient drying systems. The use of a combination drying system that employs high-temperature methods in conjunction with low-temperature drying offers potential for energy savings and will receive wider adoption. Grain can be dried to the 20 percent moisture range by the high-temperature process, and then the more energy-efficient low-temperature process may be employed to accomplish the remainder. Also, the dryeration process, in which heat from the hot, dry grain of a high-temperature method is used to remove the final 1 or 2 percent of the moisture in the grain, shows promise for energy savings. Heat recovery systems for grain dryers have also been designed. All of these systems, however, require added capital investment and some level of additional management, which must be weighed against potential energy savings. Nevertheless, as energy costs increase these systems will be utilized by producers.

Another economically viable system uses a portion of the crop residue as a heat source. Efficient biomass gasifiers that utilize only a portion of corn cobs and stalks to provide all the required heat for drying are available. They will be widely adopted in the future. Solar systems, too, may provide some of the energy for crop drying. And for livestock housing, solar systems have great potential.

The control of pests is conventionally practiced on a regular schedule, and pesticides are used at rates that assure control. Within the past few years integrated pest management (IPM) techniques have been developed that use pest population data collected by field scouts combined with crop and pest life-cycle models to determine the optimal time for applying pest control procedures. These techniques have resulted in the use of substantially less pesticides applied less frequently, with corresponding savings in both money and energy. IPM programs have also used biological control methods, which require little or no energy input. They include natural parasites, pest reproductive control procedures, host plant resistance, and cultural techniques. The use of integrated pest management will expand in the twenty-first century.

If catastrophic crop failures are to be avoided, energy will have to be expanded for protection against frost. Foam, wind machines, and more efficient water sprays can be used for frost protection, but they need to be evaluated for their effectiveness and energy savings as compared to heating. Recent advances in frost prediction and in predicting plant injury have greatly improved the accuracy of determining when frost protection is necessary. Additional advances can be expected. The total amount of energy conserved in all these aspects of growing food is small compared to the demand for energy by the rest of the food system.

## ENERGY USE AND CONSERVATION IN OFF-FARM FOOD ACTIVITIES

Since the energy used off-farm is considerably greater than the on-farm use, the potential savings should be greater. The problem is that the cost of the energy in off-farm activities is very small in comparison to the food value, and much of that cost is related to meeting consumer preference. Foods are processed, packaged, and prepared to cater to the highly mobile lifestyle of U.S. consumers, who have indicated a willingness to pay the added cost.

## Processing

Sixty-five percent more energy is used in the processing of food than in producing it. The amount of energy required to process different foods varies widely. Frozen vegetables require more energy than fresh vegetables, and dehydrated vegetables require even greater energy than frozen. The processing energy involved in canning peas is 2.5 times greater than freezing an equivalent amount of peas; however, the frozen peas require large energy inputs in refrigerated transportation, marketing, and home storage. The total energy cost of frozen peas is actually 33 percent greater than that of canned peas. Conversely, dehydration may be relatively energy-efficient if long storage or transportation requirements are taken into consideration.

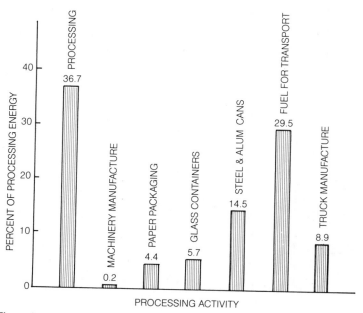

Figure 2.   Percent of energy input into food processing by processing activity.

In a sense, processing is an activity that conserves the energy value of a product, since processing a product usually preserves its quality and reduces its susceptibility to spoilage. This argument, however, cannot be used about a product such as potato chips, which clearly are made to meet consumer preference. A 6-ounce package of chips can cost about the same as a 10-pound sack of fresh potatoes. Increased processing and packaging costs make the difference.

The expenditure of energy in the various aspects of processing is shown in Figure 2. One-fourth of the energy involved is spent on packaging and about another one-fourth is for transportation. More efficient processing systems using reheat and heat recovery systems, more energy-efficient packaging, and more efficient transportation (railroads in place of trucks, for example) can all reduce the processing energy input and will receive wider adoption in the future.

Natural gas, because of its convenience and low cost, is the dominant fuel used in processing (see Table 2). For example, more than half of the energy used in processing sugar beets, canned fruits and vegetables, frozen fruits and

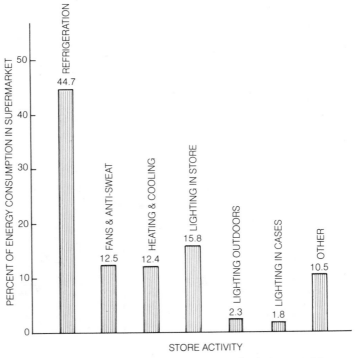

Figure 3.   Percent of energy use in a supermarket by store activity.

vegetables, and poultry and eggs is natural gas. Until energy costs or availability change there will be little incentive for the industry to move to other nonpetroleum-based fuels. And since many processing plants operate for only a few weeks per year, convenience of fuel is a prime consideration.

## Marketing

In marketing, electricity is the dominant form of energy used, the major portion being for operation of refrigeration, as indicated in Figure 3. Any major reductions in energy use in this area will require a major change in both processing and marketing by shifting from refrigerated-type packaging to canned or dried products. Ultra-high-temperature milk processing, which eliminates the need for refrigeration, is one example of such a process currently being adopted. But such processes will not dramatically change the total energy use in marketing in the twenty-first century.

It is of interest to note that supermarkets with a bakery or delicatessen area, which cater to the consumer demand for convenience, use significantly more energy than supermarkets without such areas. The opportunities for energy savings by covering refrigerated display chests, reducing lighting, or utilizing heat recovery from refrigeration for store heating can contribute little in energy conservation. Space heating of stores represents only about 6 percent of the total energy use in marketing.

## Home Preservation and Preparation

Sixty percent of the energy use in home food preservation and preparation, as shown in Figure 4, is for refrigeration. Since frost-free refrigerators utilize 60 percent more energy than the newer, energy-efficient ones, home energy consumption could be considerably reduced by conversion to the newer, less convenient models. The heating and cooking of food consumes about one-third more of the energy. Microwave ovens currently provide a more energy-efficient method of preparing foods, and they are being widely adopted and even becoming common in homes in developing countries.

But energy conservation in home food preservation and preparation will probably receive little attention since it is only 14 percent of home energy use. Space heating and air conditioning accounts for 69 percent and water heating 10 percent. Any major reduction in home energy consumption in the food area, therefore, will depend upon changes in the food processing area, which affect the home energy use.

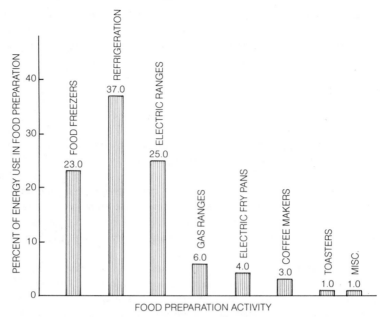

Figure 4.   Percent of energy use in food preparation in a home by preparation
activity.

## Food Preparation Away from Home

The growth of the fast-food industry has been well documented, with one meal in three now being eaten away from home. Today there are more than a half-million food service establishments preparing food for consumers. The energy use in these establishments is substantial, being nearly identical to the energy use in food production. Energy conservation in this area cannot readily be instituted and will occur only as the owner/operator determines there is an economic benefit. Heat recovery can provide some energy savings in selected establishments, but any major change in energy consumption in this area will probably depend upon a major shift in food consumption patterns (i.e., a shift away from dining out).

## LOOKING TO THE TWENTY-FIRST CENTURY

Opportunities for substantial reductions in energy use in the food and fiber system exist. However, since the major portion of the energy use occurs in the off-farm areas, conservation efforts must be directed toward all components of the food system if the maximum potential savings are to be realized. If appropriately planned and instituted, changes can be made without drastically affecting the total food availability or quality of the material produced. Significant reductions in the off-farm segments will be very difficult to institute, however, since the total cost of the off-farm energy inputs is relatively small, and, therefore, business has little incentive to change. On the contrary, the use of energy to reduce labor and other input costs has resulted in the cost of food being relatively low.

In the production sector reductions in energy usage must be carefully instituted. Some practices that will save energy in one area may well cause an increase elsewhere or may significantly reduce yield. If yields are reduced, the efficiency per unit of production will actually be less, even though a reduction in total energy usage is achieved. Reduction in yields, if they occur widely, may in turn result in inadequate food supplies.

Agriculture will make improvements in energy efficiency as we move to the twenty-first century, but they will be modest. Agriculture in the future will look much as it does today. A recent article on EPCOT at Walt Disney World in Florida stated that "most of the crops will be the same, but they'll probably be healthier and more abundant as a result of genetic engineering, cloning, and new techniques of land management. Some visible differences will be new crop species in some environments, many more fields in the deserts, new farm machinery, new sources of on-site farm energy such as from both the sun and biomass conversion. But mostly, agriculture will be fields, orchards, ranches, and grazing livestock." This agriculture will be more energy-intensive than today's, as higher and higher yields are sought from a fixed or declining land base.

Both on- and off-the-farm changes in energy will largely come about because of economic factors. However, institutional pressures will be needed to bring about change, particularly in the off-farm sector. In the final analysis agriculture, whose primary responsibility is to produce food, has been highly efficient and can be expected to continue to be so in the twenty-first century.

# ENERGY FROM BIOMASS: A NEW COMMODITY

## WAYNE H. SMITH

Producing more energy than required for subsistence is a necessity of civilizations and natural ecosystems alike. During the last century worldwide development has been rapid because of abundant energy readily available from fossil sources that require only modest cost to harvest and process.

The Industrial Revolution was possible because of this energy abundance. Similarly, the availability of external energy has allowed agriculture to avoid its own energy production, so that all of its activity could be channeled into food/feed/fiber commodities. Before the availability of external energy, agriculture produced forage and grain for animal power to perform on-farm work and produce food and other farm commodities. For example, in the 1920s about 25 million draft animals a year were used for this purpose in the United States, according to W. E. Tyner and J. C. Bottum of Purdue University. The substitution of chemical and mechanical energy has rapidly increased and now allows a single farmer to produce enough food for about 78 other people—56 domestic, 22 foreign.

However, the declining supply of fossil fuels is causing us again to think about agriculture becoming an energy producer, but with the difference that this time the requirement is mainly for liquid or gaseous fuel feedstocks, not feed for draft animals.

In the near future expansion of biomass as an energy source will occur mostly from the utilization of the residues of forest-product processing or unmerchantable tree species. Use of these wastes and residues and those from agricultural crops is frequently restrained by problems of convertibility, seasonal availability, unpredictability of supply, limited quantities, and harvesting and storage. In many discussions of renewable energy, the terms "biomass" and "wastes/residues" are used synonomously. In this discussion biomass includes all recently formed plant matter including wastes/residues, but especially crops grown for their biomass that will be converted to useful energy. Emphasis is placed on biomass crops because establishing a biofuels industry will require the deliberate production of such crops to ensure a sustained feedstock. Efforts to develop these crops have frequently been aborted by the "food versus fuel" debate.

However, recent analyses by Professor D. O. Hall of the University of London suggest that food/feed worldwide exist in surplus and that the surpluses could further be enhanced by modest behavioral changes regarding diet—eating less meat. In the developed countries today the problem is that farmers do not have crops to grow that are in demand so that the enterprise can be profitable without government intervention. Thus the supply of crops presently grown often exceeds demand. Admittedly, there are hungry people in the world, but this is due to economic, cultural, and distribution problems, not because of supply deficits. Solving these problems that cause hunger to exist must be a high priority.

A frequently identified constraint on biomass production is land availability and associated potential environmental degradation. Both problems often are identified because of the assumption that biomass would be produced on marginal lands. But many acres of land, even prime land, now used for crops that are in surplus could be used for biomass crops if such crops were available and could be grown at a profit. The Payment-In-Kind program (PIK) of the U.S. Department of Agriculture proposes to idle 83 million acres of farmland in 1983. In addition, the USDA Soil Conservation Service has indicated that 44 million acres in the adequate rainfall region of the country could be converted to cropland if needed. This land would come from other uses—forests, pastures, or untended woodlands. Combining these with "set-aside" acres places the total land available at more than 100 million acres. Others have determined that, worldwide, only 41 percent of the land suitable for cultivation is now cultivated. Overlooked in most analyses is the potential for growing biomass on freshwater and/or marine bodies.

Much hope for the future is placed with the modern biotechnologies that promise to markedly improve cultivars and to overcome production constraints. These longer-term strategies include recombinant DNA technology and the related technology of protoplast fusion, which make genetic engineering of plants possible. Cell and tissue cultures are already proving beneficial, but we must await basic research advances before the other technologies significantly affect productivity. In a recent survey by K. M. Menz and C. F. Neumeyer reported in *BioScience* in 1982 scientists expressed great optimism regarding their expectation for five biotechnologies to increase maize production. If these scientific advances occur and are applied, less competition for land should develop and new biomass crops could be accommodated.

Farmers need new crops that are in demand and can be grown at a profit. The world needs new energy sources. Biomass crops could go far in contributing to both of these needs and could do so without adversely affecting essential production of food/feed/fiber commodities. This section describes a process for this development.

## BIOMASS CROP DEVELOPMENT

Growing crops deliberately for their energy commodity value is being explored worldwide. Energy crops would add energy as a farm commodity to the present food/feed/fiber commodities now grown. Developing biomass crops must be as deliberate as developing food/feed/fiber crops and follow similar scientific methods.

Such an effort must recognize the following points in designing and implementing the biomass energy program:

1. Crops and production systems currently available were developed for food, fiber, or feed, and not for energy purposes.

2.  Biomass conversion technologies currently available were often developed for spirits or other industrial applications, for waste treatment or disposal, and not for energy production.

Toward this end potential biomass crops should be systematically evaluated in the various plant resource groups: woody, freshwater aquatics, grasses and related roots and tubers, halophytes (salt-tolerant plants), hydrocarbon producers, and marine plants. To develop biomass crops the first research steps are the following:

1.  Screen species for their local and regional adaptability and for their potential to produce abundant biomass.
2.  Identify within each new biomass crop candidate genetic variability that would be important to its development—noting ranges in growth rate, pest resistance, chemical composition, and sensitivity to environmental stresses.
3.  Develop cropping strategies by determining optimum plant density, planting dates, harvesting period and frequency, and cultural inputs to maximize energy yield per unit area, time, and invested energy.
4.  Assay the plant products for their convertibility to useful energy forms.
5.  Initiate a breeding program to genetically improve the plant species for energy cropping purposes.

In Florida more than 200 varieties, species, and cultivars showing potential for biomass production are being field-tested in the various environments as part of a cooperative program with the Gas Research Institute. The evaluation process recognizes the differences between agricultural crops grown for food/feed/fiber and those that would be grown for energy (Table 1). These potential biomass crops must fill environments that range from temperate to tropical, where soils vary from loams to sands that may be flooded some or all of the time or subject to periodic droughts. Once yield and convertibility have been determined, promising crops must then be evaluated in various cropping strategies compatible with farmer/grower production objectives. In this way, biomass crops and the systems to grow and economically harvest them can be devised to allow the marketplace to select alternatives. Energy crops can be developed specifically with energy production objectives, just as food/feed/fiber crops have been tailored to supply these commodities. Thus a farmer will have the opportunity to select and grow crops that provide the greatest net economic return, while being environmentally acceptable and contributing a net energy return.

The importance of developing crops with characteristics especially suited for energy purposes and culture systems designed for this production objective

Table 1.   Characteristics of Agricultural Crops Desired for Food/Feed/Fiber Compared to Those Possessing Potential as Biomass Crops

| Food/Feed/Fiber Crops | Biomass Crops |
|---|---|
| 1. Small portion (seeds, fruit, leaf, etc.) of total yield used | Entire plant comprises yield |
| 2. Quality (nutrient contents, digestibility, milling characteristics, taste, palatability, appearance) important determinant | Qualitative constraints few (mainly relating to convertibility) |
| 3. Chemicals often required for pest protection | Biological resistance or low susceptibility to insects and diseases |
| 4. Fertilization, especially with nitrogen, usually at high rates | Tolerant to and productive in low-fertility environments |
| 5. Irrigation or drainage to prevent moisture stresses frequently required | Species tolerant of drought or flooded soils |
| 6. Cultivation or herbicides usually required to prevent competition | Aggressive and dominates competitors at early ages because of rapid development of dense canopy |
| 7. Harvesting system designed to protect quality of selected plant parts | Bulk harvested and not sensitive to qualitative factors |
| 8. High quality land required | Species tolerant to marginal lands |
| 9. Crop establishment by costly seeding or transplanting | Low-energy planting with regrowth after harvest from coppice or sprouts |
| 10. Low energy yield ratios: intensive crops ≤ 2; corn–grain 2.8; corn–whole plant 5–9 | High energy yield ratios: low-intensity crops ca. 20 |

cannot be overemphasized. For example, A. G. Alexander, working in Puerto Rico, developed an "energy cane" production system that resulted in the same sugar yield as a conventional cane crop; but the total biomass yield (250 green metric tons per hectare) was considerably greater. In Florida we have found that culture systems designed for biomass and not food/feed/fiber values generally produce a yield increase of twofold or more.

Many advances that can be made in biomass crop development must make use of conventional scientific principles as well as the new biotechnologies which could allow quantum steps forward in reasonable time. Specifically, I am referring to recombinant DNA techniques and tissue/protoplast culture. For example, in energy crop production a pressing question is "How will the nitrogen be supplied?" We know that using nitrogen fertilizer made from natural gas is not economically sound. While nitrogen-fixing companion or rotation crops should be used if possible, biotechnology could provide another option; namely, the insertion of the nitrogen-fixing genes into the biomass crop or the microorganisms that grow around the plant's roots.

Protoplast culture techniques provide a vehicle for introducing desirable genes into a single cell and regenerating a plant from that cell. All cells of the regenerated plant would possess the new genes. Fusion of protoplasts from related or unrelated species also provides a new strategy for breeding and selecting for useful variations—such as increased growth rate, pest resistance, and content of chemicals easily convertible to useful energy forms. Regardless of plant origin, field propagation (especially in high-density energy farms) presents a problem. Emerging tissue culture techniques coupled with "fluid-seeding" technology (planting of seedlings in a gel that prevents dessication) may provide a low-cost way to propagate "genetically true" energy crops. Similarly, by inserting genes that impart tolerance to temperature and high concentrations of salts and organic compounds and by inhibiting by-products in existing fermentation organisms, we could greatly improve the conversion processes for forming either alcohol or methane.

M. Calvin in a 1983 paper in *Science* outlined the process for using genetic engineering techniques, specifically the transfer of the gene controlling the production of valuable hydrocarbons useful for fuel from the high producing *Copaifera* (a tropical tree) to *Euphorbia* (a shrub that is grown in a wide array of environments). He reasoned that the application of these techniques will make it possible to use the green plant, the best solar-capturing device known, to produce needed energy materials.

## BIOMASS CHARACTERISTICS AFFECTING CONVERSIONS

Biomass is composed of a variety of chemical substances; among the simplest and most troublesome is water. The dry matter content of fresh plant matter varies from a few percent (e.g., aquatic plants) to about 50 percent (e.g., wood). Various combinations of organic compounds and mineral ash form the major constituents of dry biomass. The ash content is usually a few percent of the total, except for marine plants and certain tropical land plants that accumulate silica.

The organic constituents vary widely in chemical composition and in their proportion in plants. Fermentations by microorganisms of simple organic compounds, such as sugars and amino acids, in oxygen-free environments that yield energy-rich liquids and gases are the principal biological prophecies for producing energy from biomass. Nonbiological prophecies involve chemical treatments in various temperature and pressure regimes that usually result in gases. Some of these can be converted to liquids, such as alcohol.

Sugarcane and sweet sorghum may possess up to about 10 percent sugars, as the disaccharide (sucrose) or the monosaccharide (glucose). Artichokes contain inulin (polysaccharide made up of fructose sugar), and many plants such as potatoes and cereal grains store energy as starch (a polysaccharide of glucose). Sucrose, starch, and inulin are easily hydrolyzable with amylase enzymes or dilute acids. Hydrolysis is the process in which linked chains of sugars are broken into monosugars by chemically adding water. High concentrations of these compounds favor biological conversions to alcohols or methane. For structural integrity, plants contain varying amounts of cellulose and hemicelluloses. These polysaccharides are difficult to break into their monosaccharide constituents (hexose or pentose sugars) by hydrolysis. This is especially true when the celluloses are embedded in a matrix of lignin, a complex phenolic substance that lends rigidity to the plant.

Chemical form and composition are the key determinants of the way in which biomass is used. Because of the low energy density (BTU per pound) of bulk biomass, long-distance transport is not reasonable. Dewatering (e.g., field drying), extraction of energy-dense compounds (e.g., oils, resins), densifying (e.g., pelletizing, baling), and size reductions that improve uniformity (e.g., chipping, chopping) can improve the transportability and, in some cases, adaptability to conversion systems. Wood, a lignocellulosic biomass form, is currently the most abundant, affordable, and reliably available biomass, but considerable research is necessary to improve its utilization for energy. While drying, reducing the size, or pelletizing improves the wood for certain conversions (gasifiers producing low-BTU gas, or direct combustion), these steps do little for improving biological convertibility to either liquid or gaseous high-BTU fuels which often are those in shortest supply.

Other pretreatments are necessary if lignocelluloses are to be biologically converted to fuel forms. Following size reduction (e.g., chipping), the celluloses have to be separated from lignin (usually by steam/pressure explosions) if enzymatic production of sugar-using cellulose enzymes (usually from the fungus *Trichoderma*) is employed. Otherwise, the breaking of the chains to simple sugars by hydrolysis is accomplished by treatment with aqueous acids that convert the hemicellulose to pentose (xylose) and the cellulose to hexose (glucose) sugars. While both of these procedures are technically possible, they are not presently economically affordable without advances in technology.

Fermentation processes for producing energy forms are well known and have been practiced for centuries. This technology has been advanced mainly for the purposes of producing high-value spirits, high-value industrial chemicals, or for waste disposal. Efficiencies for these purposes are not suitable for energy production. For example, a fundamental problem in alcohol fermentation is the inability of the microorganisms involved to tolerate stress conditions (or conditions that would improve process efficiency), such as increased concentrations of alcohol or elevated temperatures in the fermenting matrix. The identification or development of organisms that could withstand increased concentrations and temperatures would reduce the amount of water to be removed from the alcohol (an energy-demanding step) and make the process more efficient by allowing the use of vacuum distillation at elevated temperatures. These improvements, coupled with increases in biochemical reaction rates and improved engineering processes, would greatly enhance alcohol production from biomass. Alcohol, presently used as an octane enhancer in motor fuel, has been a desirable feedstock for the chemical industry and, in the long term, this may be the major use of biomass-derived alcohols.

Methane gas production from biomass has been recognized for a long time. While some of the organisms capable of forming methane have been identified, their specific roles and biological requirements for efficiently converting biomass to methane are poorly understood. Anaerobic digesters are plagued with operating problems because the factors controlling the biochemistry of methane formation have not been adequately defined. Intensively focused research is needed to provide this information so that appropriate methane generators receiving a variety of feedstocks can be designed, fabricated, and operated efficiently and economically.

## TOWARD THE TWENTY-FIRST CENTURY

Fossil energy supplies are finite and inevitably must run out or become too expensive for uses considered routine today. In the United States abundant energy reserves exist, but only 7 percent occur in the liquid/gaseous forms we can readily use. Alternate energy sources are needed for the energy mix of the future. The present energy surplus, thanks to world recession, tends to discourage the needed research and development necessary to bring alternatives on stream.

Another factor that has affected the development of a strong energy research program in the United States has been the historical decline in energy prices relative to other items in the marketplace. Accurate market signals have been distorted by regulatory programs and cartel actions. The fact is that fossil energy supplies *are* finite, and now is the time to do the research so that attempts to commercialize inadequate technology under "clouds of crisis" can be avoided. To wait will cause us to lose valuable time and resources in making an orderly and effective transition to the future.

Biomass is an alternative source of energy with considerable promise. Much needed research has been aborted because of diversions associated with food versus fuel arguments. While there is a food distribution problem, the truth is that the world produces 10 to 20 percent more food than is required to feed its population. In the United States and Europe the main problem with food is its easy overproduction and general overconsumption. Identifying an appropriate way to deal with these surpluses is becoming increasingly troublesome for policymakers in developed countries. For the farmer the need is clearly for crops that are in demand and that can be grown at a profit without subsidy. Biomass for energy could meet this need.

Diverting land and bringing idle lands and waters into biomass production are becoming increasingly possible, since most projections for increased crop production rely on the new biotechnologies that promise to introduce the grand period of the "science power" phase of agriculture. Hence, land and other resources should not constrain the development of biomass as a renewable energy source for the future.

While wastes, residues, and food crop surpluses may provide resources to initiate a biomass energy program, biomass crop production will be required to maintain a feedstock stream of sufficient magnitude to affect energy supplies. A large number of diversified plant species can be grown and should be evaluated as biomass feedstocks. The biological production potential of these crops under various levels of input and operating conditions must be determined. More appropriate technology for harvesting and materials handling must be developed. Breakthroughs must occur in the technology for converting the biomass into useful energy forms. Conversion technologies meeting the requirements of an energy industry must be developed because the technologies now available were developed for the spirits and industrial chemicals industries. The most promising conversion processes to pursue appear to be anaerobic digestion to produce methane, fermentation to form alcohols, biological hydrogen production, and thermochemical processing to form fuel gases.

Numerous problems must be solved by research if bioenergy production/conversion industries are to emerge. Accomplishing these goals will be neither easy nor rapid, but researchers have available traditional scientific methodologies and the new biotechnologies, microprocessors/computer control devices, and engineering advances such as robotics for attacking these problems. The new knowledge developed must be rapidly incorporated into the curricula of the various professionals entering agriculture and associated biofuel industries. Finally, the extension education system is in place and functioning to transmit this new information to farmers, growers, processors, and consumers as it emerges from research.

The incentive provided by the knowledge that fossil fuels must run out and that a need exists for new commodities to improve the profitability of agriculture should also encourage development of new biomass crops as energy sources for the future.

# 4

# PUBLIC POLICY:
# PRESENT AND FUTURE

HAROLD M. HARRIS, JR.
JAMES C. HITE
B. H. ROBINSON

American public policy in agriculture is effectively only a half-century old. Its basic direction has been to support prices for farm products that the rich land and generally enlightened agricultural system have supplied in such abundance. But policy has created problems as well as solving them. For example, while farmers' incomes have improved overall, the production of crops for which there is not a strong market has been stimulated, further depressing prices. Yet American agriculture has grown and prospered and is the most efficient in the world.

Future problems are likely to be more difficult than were those of the past. Policymakers will have to take into account the fact that fewer and fewer people produce our food in increasingly greater amounts. Environmental problems demand policy answers for public as well as private lands. The price of land itself could become a major policy issue. The traditional goal of supplying adequate food through the family farm system will remain, but costs will rise, and policy will have to determine who will bear the burden. Moreover, international issues will be more critical as the world market becomes of much greater importance to American agriculture.

# MILESTONES OF U.S AGRICULTURAL POLICY

## HAROLD M. HARRIS, JR.

In speculating about farm policies in the coming century there are lessons that can be learned from programs of the past. Whether we have learned them is another matter.

Most government intervention in U.S. agriculture has occurred only in the past 50 years. Policy begins, according to Dr. J. Carroll Bottum of Purdue University, with a "dream about new institutions." But as we trace the history of U.S. agricultural policy it will become clear that only a few such dreams have become reality.

## THE COLONIAL PERIOD TO 1862: LAISSEZ-FAIRE PHILOSOPHY

For 250 years after colonization the prevailing philosophy was that government intervention into economic affairs should be minimized. Nevertheless, policies were established that shaped the development of agriculture and rural America.

Land policies were a key focus of the early government. Semifeudal customs of the mother countries, such as hereditary rights to the eldest son, were rejected in favor of policies promoting the right of every human to hold title to real property. The new nation's treasury was empty. Alexander Hamilton wanted to maximize government income from the disposal of public lands, even by selling large tracts to wealthy individuals. But Thomas Jefferson and others believed that family farmers were the backbone of democracy and called for disposing of the public domain at nominal prices and easy terms.

Thus in 1796 legislation was passed authorizing sales of land in blocks of 640 acres for $2.00 per acre. Subsequent laws in 1800, 1804, and 1820 reduced the acreage required, the price, or both. From 1820 through 1841 settlers could purchase as little as 80 acres from the government for $1.25 per acre. In 1841 the Preemption Act was passed giving squatters on public lands the legal right to purchase the land they occupied. The culmination of programs designed to broaden access of farmers to land was the Homestead Act of 1862. It provided full title to 160 acres to settlers after 5 years of residence. By the Civil War most remaining public lands were in less-than-prime farm areas, and abuses of land disposal programs resulted in large mining and timber corporations gaining control of immense acreages. But our lands policy had helped establish the dispersed family farm system that has persisted to the present day.

Another key element of early agriculturally related policy involved tariffs. Only 2 months after Washington's inauguration the first tariff was imposed, with levies ranging as high as 15 percent for luxury items such as carriages. Farmers as suppliers of raw export commodities recognized their stake in foreign trade. The free-trade view has generally prevailed over the dozens of tariff laws enacted. Tariffs were generally modest and viewed more as revenue-raising tools for the government than trade barriers. Not until other policies pushed U.S. farm prices above world levels in the past 50 years were protectionist policies an ongoing part of the scene.

Surprisingly, even in the earliest days there were programs to intervene in the marketplace. Minimum prices on tobacco were fixed by the Virginia Colonial Assembly in 1631. By 1639 a crop control program was adopted in order to maintain prices. A quota was established, and government officials were authorized to destroy inferior and excess production.

In the early 1800s the predecessors of two of the major actors in the public policy process emerged. Agriculture societies, the forerunners of today's farm organizations, were formed, and in their search for farm improvements they requested help from the government. Initial funding for collecting and distributing seeds, collecting data, and performing studies was granted in 1839 through the Patent Office. Thus public research and service activities by the government became a part of agricultural policy. The U.S. Department of Agriculture was created in 1862, but it continued to function under the Patent Office for another 25 years.

## PUBLIC SUPPORT AND EQUALIZING ECONOMIC POWER, 1862–1933

The Morrill Act of 1862 provided federal assistance to the states for the establishment of universities to educate young people in "practical problems" of agriculture and mechanics. The Hatch Act of 1887 provided federal funds for research at agricultural experiment stations associated with these land-grant colleges. In 1890 the land-grant concept was extended to Black agricultural colleges in the South. The threefold mission of today's land-grant system was completed in 1914 with passage of the Smith–Lever Act, which provided federal moneys for agricultural extension programs.

In the final years of the nineteenth century and the first part of the twentieth, the agricultural community influenced passage of a number of pieces of legislation designed to remedy the imbalance of economic power between farmers, small businesses, and the general public on the one hand and the "trusts" on the other. The Interstate Commerce Act of 1887, which led directly 25 years later to railroad rate regulation, owes its passage to efforts of the National Grange to remedy alleged rate discrimination against agricultural producers. The Sherman Antitrust Act of 1890 was actively supported by farm groups, but farmers soon found that under its provisions their farm cooperatives were perhaps more likely to be attacked than were the giant corporations. The Clayton Act of 1914 strengthened the Sherman Act and also provided farm cooperatives a limited exclusion from antitrust violations. The Capper–Volstead Act of 1922 extended and clarified this exemption for cooperatives.

The Federal Trade Commission Act of 1914 directly resulted in a FTC probe of the meat-packing and grain industries. The investigation revealed monopolistic conditions in the packing industry, and a consent decree was drawn up enjoining the five major firms from a number of practices. Later, the Packers and Stockyards Act of 1921 was passed, providing for direct regulation of the industry by the USDA.

Farmer dissatisfaction with interest rates and credit terms led to passage of the Hollis–Bulkley bill in 1916. It established the federal land bank system and set the pattern for government involvement in the availability of farm credit.

The importance of resource conservation was first recognized in this period. In 1891 Congress authorized the President to set aside from public lands forest reserves to be held by the government. The Reclamation Act of 1902, in contrast, sought to settle the West through irrigation development. By the late 1970s federally funded reclamation projects in 17 Western states provided water for 9 million acres of cropland, contributing to severe groundwater shortages in some areas.

## THE PAST 50 YEARS: DIRECT GOVERNMENT INTERVENTION

The single most significant landmark in the development of American farm policy and programs was the Agricultural Adjustment Act (AAA) of 1933. It focused on farm prices, which had never recovered from the post-World War I recession. Pressure grew to utilize the power of the central government to improve farm prices. Two measures were passed by Congress in the late 1920s but were vetoed by President Coolidge. The 1933 Act featured a number of concepts still in place in current programs. One was benefit payments to farmers who reduced production. Under the Act, payments were funded by taxes levied on processors, but because of this tax the Act was ruled unconstitutional in 1936. The AAA also introduced the concept of parity as the basis of supporting prices—the goal was to reestablish the purchasing power of farm commodities to that of the base period, August 1909 to July 1914.

The Soil Conservation and Domestic Allotment Act of 1936 was enacted as an emergency measure because of the unconstitutionality of AAA. Funding for the program was from general treasury receipts. The stated purpose of the Act was soil conservation, but its true goal was price enhancement. The Soil Bank, established by the Agriculture Act of 1956, was a later attempt to reduce production by removing acreage from production.

The 1933 Agricultural Adjustment Act had also authorized marketing agreements and orders. After the Supreme Court ruled portions unconstitutional in 1936, Congress passed the Agricultural Marketing Agreement Act of 1937 to clarify the legal status of marketing agreements. Two classes of commodities were affected—milk and certain other commodities. Regulations for milk involved classification according to use and minimum price fixing from handlers to producers. For other commodities—mainly fruits, vegetables, and tree nuts—producer prices were indirectly affected by control of quantity and/or quality of shipments.

The Agricultural Adjustment Act of 1938 was a more detailed and comprehensive measure than the stopgap 1936 Act. Like the 1933 and 1936 Acts, it relied on acreage allotments, storage, reserve supplies, nonrecourse loans, and marketing quotas to achieve its objectives.

Major farm legislation has been enacted every 4 years or so during the past 45 years. While emphasis has been changed at times, as during the World War II–Korean War period and the 1973–1974 "food crisis," this legislation has represented a continuation of the goals and means set forth in the 1930s. Table 1 presents a chronology of major farm legislation since 1933. The primary shift in focus over the period has been toward more flexibility in price and income goals and toward payments as opposed to marketplace intervention. But little has really changed.

Two other events were highlights of the period. In 1954 the Agricultural Trade and Development Act was enacted. Known as P.L. 480, it permitted the sale of surplus commodities in foreign countries for local currency and for donations of surplus stocks to needy people both at home and abroad. In 1961 President Kennedy created the Food Stamp Program by executive order. The goal was not only to satisfy the nutritional needs of poor people but also to expand the demand for farm products.

## GOALS, PROBLEMS, AND TOOLS

In the broad sense the tasks any economic system must accomplish are these: allocation of resources, distribution of income, and organization of

Table 1. Chronology of Agricultural Legislation, 1933–1983

| | President | Secretary | Major Farm Legislation | Other Important Events | |
|---|---|---|---|---|---|
| 1933 | Roosevelt (Democrat) | Wallace | Agricultural Adjustment Act, 1933 | Depression | Worldwide protectionism |
| | | | Soil Conservation and Domestic Allotment Act, 1936 | | |
| | | | Agricultural Marketing Agreement Act, 1937 | | |
| | | | Agricultural Adjustment Act, 1938 | | |
| | | Wickard | | World War II | Wartime price and wage controls |
| | | | Stabilization Act, 1942 | | |
| 1943 | Truman (Democrat) | Anderson | | Cold War | Marshall Aid Plan |
| | | | Agricultural Adjustment Act, 1948 | | |
| | | Brannan | | Korean War | Brannan plan rejected |
| 1953 | Eisenhower (Republican) | Benson | Agricultural Act, 1954 | Cold War | Emergence of U.S. and U.S.S.R. as nuclear powers |
| | | | Agricultural Trade Development and Assistance Act, Public Law 480, Food for Peace, 1954 | | |
| | | | | | E.E.C. founded |
| | | | | | Treaty of Rome |
| | | | Agricultural Act, 1958 | | Sputnik launched |
| 1953 | Kennedy (Democrat) | Freeman | Feed Grain Act, 1961 | | Cochrane Bill rejected |
| | | | Food and Agriculture Act, 1962 | | |

Table 1. (Continued)

| | President | Secretary | Major Farm Legislation | | Other Important Events |
|---|---|---|---|---|---|
| 1963 | Johnson (Democrat) | | | Kennedy assassination | Wheat Referendum on Mandatory Controls, 1963 |
| | | | Wheat–Cotton Act of 1964 | Vietnam War | |
| | | | Food and Agricultural Act, 1965 | | World food crisis |
| | | | | | Domestic poverty problems |
| | Nixon (Republican) | Hardin | | | |
| | | Butz | Agriculture Act, 1970 | | Peacetime wage and price controls |
| | | | | | First Russian grain deal |
| 1973 | | | Food and Consumer Protection Act, 1973 | Cold War ends | Energy crisis |
| | | | | | World food crisis |
| | | | Emergency Farm Bill, 1975 (vetoed) | | |
| | Ford (Republican) | | | | |
| | Carter (Democrat) | Bergland | Food and Agriculture Act, 1977 | | Recovery from energy and food crises |
| | Reagan (Republican) | Block | Agriculture and Food Act, 1981 | | Worldwide recession |
| 1983 | | | | | PIK program |

the process of economic growth and development. But here is where the policy dilemma begins, for chances are that a policy devised to affect any one of the three will conflict with one of the others. Suppose farm incomes are too low. An institution that shores up farm prices is devised, but it leads to misallocation of resources by creating the incentive to produce more farm products. In fact, it was misallocation of resources—too many people, too much land, and/or too much capital committed to farm production—that led to the low income in the first place.

The problems of American agriculture with respect to these three tasks of our economic system have been as follows:

1. Since the earliest years there has been a chronic excess of all three of the basic resources—land, labor, and capital—committed to agricultural production. This situation derives in no small measure from the immobility of resources once they are committed to farm production and to the continuous gains in productivity.

2. Farmers and farm workers have persistently fared worse by conventional measures of economic well-being than nonfarmers. Within the farm community itself there are severe income disparities. Low farm income arises from excess resources in the industry combined with an inelastic demand, which in lay terms means simply that large crops sell for considerably less total revenue than small crops. In addition, it arises from relative lack of economic and—since the mid-1800s—political power of farmers.

3. The tremendous growth of U.S. agriculture, which is the envy of the world, is in part due to innate advantages in soils, climate, and so on. But for the most part it stems from our past agricultural policies. Land-grant universities, USDA, government credit, a government-financed infrastructure, our opting for an entrepreneurial structure rather than another system, even price support and acreage controls have all played a part. But our very success has meant continued problems with excess resources. Moreover, since larger farmers are the primary beneficiaries of such programs, success has contributed to income disparity within the sector. Finally, past patterns of growth and resource control raise questions about the adequacy of resources for future growth, development, and societal use. Soil loss, water shortages, and environmental quality problems are also the legacy of our growth.

Surveying past milestones of agricultural policy one can make out the following underlying goals:

• Abundant food for domestic consumers at reasonable cost.

• Returns from resources applied to farming equal to those earned in other sectors.

- Creation and maintenance of a dispersed, family-farm-dominated agriculture.
- Conservation of resources for future generations and preservation of the environment.

The major programs that have been invented to achieve these goals—and we have touched on them all above—are surprisingly few in number:

1. Price supports.
2. Supply control.
3. Marketing orders and agreements.
4. Trade restrictions.
5. Resource conservation.
6. Consumer welfare programs.
7. Export enhancement programs.
8. Research and extension.
9. Insuring availability of land, credit, and other resources.
10. Antitrust laws and limited exemptions for farmer bargaining and marketing associations from these statutes.

How well have these tools solved the agricultural problems we have faced for the past 200 years? Since problems still persist, they could be judged a failure. Yet progress has been made. The human resources moved out of agriculture over the years have been truly phenomenal—from 90 percent of our population in the late 1700s to less than half at the end of the Civil War, to 3 percent today. Income disparity still persists, but the gap between farmers and nonfarmers has narrowed. In terms of efficiency U.S. agriculture ranks second to none.

## LESSONS TO BE LEARNED

Whether one gives our past farm policies a passing grade or a failing one, there are lessons to be learned from history that may help us in coming years. Some may seem obvious, but since we continue to repeat our mistakes, they bear repetition.

*Lesson One:*  New policy initiatives tend to take a long time to enact and implement. For example, the depressed condition of agriculture had existed for over a decade prior to the passage of the Agricultural Adjustment Act. And intensifying the problem, the times seem to change more quickly than they used to, in large part because of technological advances. In 1973 farmers were given a signal directly from the marketplace and indirectly from the 1973 Agriculture and Consumer Protection Act to expand production. They did. The situation rapidly changed to one of worldwide recession and food gluts, yet the program tools were still oriented toward expansion. As usual, resource immobility retarded the ability of the agricultural plant to shrink. Is it possible to have a long-range agricultural policy that is flexible enough to adjust to short-term needs?

*Lesson Two:*   Vested interest groups are responsible for the creation of policy. But agricultural policy, once implemented, creates new vested interest groups that maintain programs even after the need for them has expired. I leave it to others to point the finger.

*Lesson Three:*   Agricultural policies are compromises both within the sector and between agriculture and other interest groups. For example, last year tobacco farmers were forced to shoulder the entire cost of operating the tobacco program, a change that was viewed as the only means of saving it. As farmers become even fewer and more specialized, thus further reducing and fragmenting their role and influence, only then will we be able to forge effective compromises— if we are not stuck with trying to save vestiges of the programs we already have.

*Lesson Four:*   Agricultural policy has focused as much on symptoms as on the underlying problems. The current debate over dairy policy serves to illustrate. The problem is commonly identified as excessive program costs of more than $2 billion a year. The underlying culprit is a support price that is too high—stimulating unneeded production that the government must remove from the marketplace.

*Lesson Five:*   Agricultural policy has been far more effective in increasing output than in reducing it. One needs only to look at the response of production to the World Wars and the badly misjudged world food shortage of 10 years ago.

*Lesson Six:*   Agricultural policy has conflicted with other agricultural goals and with other national policies. In recent years the most glaring example of this conflict has been between agriculture and international relations. Obviously, embargoes do not jibe with the goal of expanding exports. Other examples abound.

Recently, a group of us looked at the net impact of the virtual elimination of the Special Milk Program, which had provided milk through schools and institutions free or at a low price. The purpose was to save taxpayers $96 million. But the reduced consumption had to be purchased by the Commodity Credit Corporation under the milk price support program, stored, and ultimately donated. The action actually resulted in an increase in government costs of $22 million. A lobbyist acquaintance of mine calls this "government isometrics."

*Lesson Seven:*   Given a democratic society and a modified free enterprise system the number of possible policy alternatives we have actually tried has been but a handful. In 200 plus years we have reached shockingly few true policy milestones. Most "new" proposals are simply a rehash of what we have done several times before, as in the case of Payment-In-Kind. As we move toward the twenty-first century surely there is the need to dream up new institutions. The question is, will we be bold enough to try them?

# RESOURCE POLICY AND AMERICAN AGRICULTURE

## JAMES C. HITE

Economists have a miserable record making forecasts, even about events only a few months ahead. Of course they can extrapolate trends, but things never continue as they have been going. Even with hindsight something important will happen that we did not foresee. Scientific discoveries and technological breakthroughs, factors likely to be decisive in shaping the world of the next century and its agriculture, are notoriously difficult to predict.

Yet, certain demographic conditions of the twenty-first century can already be perceived. By the middle of the next century the percentage of the U.S. population that has any working knowledge of production agriculture will be very small. Today, less than 3 percent of the population is on the farm. Most urban Americans—and that means most Americans—do not personally know a farmer. In the next century all but a handful of Americans will be two or three generations removed from the farm. What can they be expected to know about the realities of farming?

Even today there is a stark difference between actual farming operations and the calendar art stereotypes that many urban dwellers carry around. Often the modern working farm is not a very picturesque place and, of course, neither was the working farm of a half-century ago. Farming by its nature rearranges nature, substituting controlled environments for wild, pristine environments. Urban people are sometimes quite genuinely shocked when they see what farmers do to the rural landscape. That makes them susceptible to anyone who raises a public outcry, legitimate or not, regarding farming practices that make use of unfamiliar chemicals or produce nonaesthetic changes in the natural environment. Farmers, as a political minority whose national electoral strength is hardly worth cursory attention, will be living with public policies that are fashioned not in response to their practical needs but the perceptions and needs of urban voters.

Such policies will not necessarily be hostile to farmers. The agrarian mystique is so deeply embedded in American history that it cannot but have influenced values subconsciously held by almost all Americans, urban and rural. Survival of independent, owner-operated farms is likely to continue to be viewed by urban dwellers as important to the public weal. There is a not irrational feeling that an economy that does not have room for an efficient, owner-operated farmer probably does not have room for any independent, owner-operated business and perhaps no room for the individual as a free man or woman.

Moreover, there is every indication that a productive agriculture will continue to be important to the nation's balance of trade and to certain strategic defense and foreign policy objectives. The political problem will be to make urban voters understand that a resource and environmental policy that cripples agricultural productivity will have detrimental effects on the economic well-being of everyone regardless of how he or she earns a living. Solving that problem will require more subtle and sophisticated political skills than farmers and farm organizations have yet demonstrated. Still, it should be possible.

Examined closely, the fundamental natural resource and environmental policy question in this century has been: What is the optimal mix of private and public property rights in natural resources? That same question is likely to remain at the center of policy debate in the twenty-first century. The net effect of U.S. environmental policy is to constrain the private property rights of farmers and other landowners to make use of the environment. Many economists argue that it is possible to devise new private property rights arrangements for the use of the natural environment that will cause a free, competitive market to allocate it to its highest and best use. But their arguments are theoretical, subtle, and complex. To date, economists have made little progress in convincing even those politicians who understand the arguments that the public at large will understand and accept something other than direct regulation as a basis for environmental policy.

As the relationships between the use of land and the quality and quantity of environmental amenities become clearer, there could be a growing demand for policy that restricts more narrowly the private property rights exercised by farmers. Land-use zoning is one example of how such restrictions on private property rights might be instituted. Urban dwellers unfamiliar with farming operations are more likely to see the social benefits of preventing farmers from using land in ways that adversely affect air quality or that increase the sediment load of streams than they are to see and understand the social benefits associated with the farming practices that give rise to such environmental degradation. Not understanding the unavoidable trade-offs, the urban electorate may exert growing political pressure to expand government regulation of farming.

There has been much public discussion in recent years about the United States running out of farmland because of urban encroachment. Few economists who have studied the issue, however, feel that national agricultural productivity will be seriously threatened in the twenty-first century by the loss of farmland to other uses. A more realistic concern is that farmland prices will become so high that it will be impossible to sustain owner-operated farms. Technological developments may reduce substantially the land requirements of agriculture in the next century, but the demand for land is likely to grow, pushing up prices at a rate shadowing the going rate of interest. Already, young people not fortunate enough to inherit enough good land to sustain a profitable farming operation find it almost impossible to service the debt necessary to buy it. Even inheritance will not suffice because the settling of estates often requires that land be divided among several heirs, and after a few generations the land-holdings are too small to be viable farms.

Hence, a new land policy for American agriculture may be required by the middle of the next century or sooner. Both the relationships between land use and environmental quality and the likely increases in land prices suggest major changes in the way farmers have held land. That land policy will be dictated not by the practical needs of farmers but by the perceptions, emotions, and objectives of an urban political constituency. One possibility is outright government ownership of substantial farmland that is made available to farmers under lease arrangements similar to those now used for public grazing and mineral lands. Should such a possibility be realized one can be sure that the leases will carry with them enough regulations to fill several volumes of the *Federal Register.*

The conclusion, therefore, is that farmers of the twenty-first century will have a very difficult time controlling the natural resources that are required for farming. Private property rights to use those resources are likely to be curtailed substantially by public policy. Environmental restrictions and regulations are likely to increase and become more burdensome and complex. The political climate, while not hostile to agriculture, is apt to be insensitive to the needs of production using advanced technology and requiring substantial departures from the old romantic images of farming. Lacking electoral strength to exert political influence through the control of blocs of votes, American farmers will need to become well organized and politically sophisticated if they are to influence natural resources and environmental policies in significant ways.

# FOOD AND AGRICULTURAL POLICY IN THE TWENTY-FIRST CENTURY

## B. H. ROBINSON

In less than two decades the books will close on what history will likely record as the century of revolution. Probably no period in the history of the human race has recorded more change. Not only have societies, governments, and economic systems changed, but the very foundations upon which they were built have changed.

Agriculture has not escaped the upheaval of the twentieth century. It has experienced more change in the past 75 years than during the previous 750 years. While the agricultural revolution is the product of many interrelated forces, it is safe to say that public policy has played a major role in shaping it.

Food production and distribution are vital to society, and governments have not been willing historically to depend solely upon the marketplace to produce the desired results from their food system. Rather, the public has protected and managed its vested interest in agriculture through public policies aimed at achieving desired goals through market intervention. One factor that did not change during the twentieth century revolution was the public's determination to continue its involvement in U.S. agricultural development. During the twenty-first century public policy will continue to influence agricultural production and marketing decisions as well as the structure, performance, and viability of the agricultural sector.

While it has been argued that the United States has never had a long-term agricultural policy, I believe that four long-term policy goals for agriculture can be identified. The first, a goal of any responsible society, was and is to assure a stable supply of reasonably priced food to its citizenry. Second, the policy of land settlement clearly supported the family farm concept. The third major long-term goal has been increased agricultural productivity. The fourth, and a more recent longer-term goal for agriculture, is demand expansion, primarily through international markets. While demand expansion has frequently been a minor part of farm policy, it was generally viewed as a mechanism for helping dispose of surplus. Only recently have agricultural exports been viewed as a major factor in international trade and the balance of payments. The fourth goal, which is related to a myriad of economic and political factors, represents a major shift in agricultural policy.

Almost all of the agricultural policies of the twentieth century have been designed to deal with the adjustment problems in agriculture characterized by excess capacity, low farm returns to capital and land, and relatively low prices and incomes. However, these very policies produced conflicts because the income and price stability they provided encouraged further investment in technology and further aggravated the excess capacity problem.

Over time, economic conditions and weather have had as much impact on agricultural income as public policy. The last decade is a good example. Although there were no fundamental changes in farm policy, both record high and severely depressed incomes were realized by farmers. Why? Because the interplay of things like currency exchange rates, central banking practices, government deficits, international politics, and weather are beyond the scope of U.S. farm policy.

Policies and institutions seem to change more slowly than the problems they are designed to address. Even though the policies themselves do not have a good record of solving problems, tradition, politics, vested interests, and compromise tend to forestall major changes. Thus, the last decade of the twentieth century and the early decades of the twenty-first century will likely contain vestiges of the current set of farm and food policies, imperfect as they are.

When agricultural policy is formulated in the next century many questions will have to be addressed. Will public goals for agriculture change? Are the agricultural problems changing, and if so, in what way? What are the conflicts in policies designed to address different but interrelated problems and goals? Will there be coordination among policies affecting resource use, food and agriculture, international trade, and related areas? Who will control the agricultural and natural resource base? Who will have priority use of nonrenewable resources? What will be the trade-offs among production, conservation, and preservation? What will be the environmental trade-offs? Commercial agricultural policies of the twenty-first century will have to address many more questions and issues than did the policies of the twentieth century. It may not be reasonable to expect miracle solutions, but it is imperative that the issues and questions be addressed.

Six factors will underlie the agricultural and food policy of the future: (1) the economic and biological realities of agriculture and their inherent problems; (2) the national and international economic and political situations; (3) the competition for resources and the trade-offs among users of natural and nonrenewable resources; (4) the changing markets of the future; (5) the flow of events and societal changes that shape public opinion about agriculture; and (6) the perceptions of policymakers about agriculture.

Trends begun in the twentieth century will continue in the twenty-first, but will become more complex. The number of farms and farmers has decreased dramatically, but agriculture continues to be characterized by a large number of competitive firms with little influence over market prices. Broadly defined, the food production, processing, and distribution system is a large and growing industry.

Less than 5 percent of U.S. farms account for 50 percent of total cash receipts and 87 percent of net farm income. Twelve percent of U.S. farms account for almost 70 percent of cash receipts and almost all the net farm income. The two smallest sales classes of farms include 72 percent of the total number, but they account for less than 13 percent of total cash receipts. In 1981 they lost money; the smaller farms derive most of their income from off-farm sources.

Concentration of production on a relatively few large farms is the result of technological change, which has made possible large-scale substitution of capital for labor. This change has also tended to require that farms be large in order to obtain advantages. For example, small farms cannot make enough use of certain types of expensive machinery to justify owning it, but they cannot afford to pay the going wage rate for labor and do without the machinery. To survive, it has been necessary to increase the size of the farming operation in order to take advantage of technological improvements. The result has been a large volume of output per unit and a lower profit margin on each unit of output.

These conditions will continue into the twenty-first century, but their impact on concentration will be reduced. Technological development in the next century, which will likely be more biological than mechanical, will offer less incentive for expansion and capital labor substitution. Additional economies of size are questionable in the coming century. However, changes in tax policy, a collapse of land prices (which are under stress from current conditions), or other nonagricultural forces could make farmland a cheap long-term investment alternative and alter concentration trends.

The agricultural industry of the twenty-first century will be composed of two major classes of farms: (1) commercial farms, which produce about 95 percent of total output but account for less than 10 percent of all farms, and (2) a noncommercial class of farms composed of hobby farms, rural residences, and subsistence farms. The noncommercial class will account for 90 percent of all farms, less than 10 percent of output, derive most of their income from off-farm sources, but control a large natural resource base. Policy will have to address the different problems of each group with different tools and techniques. Conflicts will exist particularly as related to resource control and use. There will be increased specialization and integration of the production and marketing of agricultural products.

One market certainty is uncertainty. Currently, the world market absorbs the production of 1 acre out of every 5. Most countries protect and support their internal production and encourage self-sufficiency. Will the world market absorb production in excess of domestic needs at profitable prices? Will the public continue to support agricultural export expansion, particularly in view of the fact that agricultural exports are the product of the natural resource base? Expanded markets are attractive but create new problems and issues, not the least of which is increased uncertainty for agriculture and food prices.

The urban society of the twenty-first century will have little identity with the agrarian philosophy. The major public concern will be with a reasonably priced food supply. People will not be concerned with where milk "comes from," but only that it is available at a reasonable price when needed.

To foster a viable industry some level of stability in planning horizons and expected resource returns is required. But will the public be willing to guarantee reasonable returns to agricultural resources and capital? Will stability and growth be the product of market forces or government programs?

Agriculture in the twenty-first century will continue to be plagued by the vagaries of weather, a relatively inelastic demand for food and related products, market instability, a capacity to produce in excess of market demand, variable resource returns, high capital requirements, decreasing political clout, high production costs, and income variability. New problems for agriculture will include a dependence upon the world market; competition for natural and nonrenewable resources with other sectors and groups; environmental trade-offs; increased macroeconomic linkages; increased uncertainty; increased dependence on performance, conditions, and policies affecting related sectors; and less public empathy for the problems of agriculture.

The overriding agricultural policy goal of the twenty-first century will be to provide an ample and stable supply of reasonably priced food to the public. This goal will take precedence over all other goals, with the possible exception of environmental and resource policy goals. Trade-offs will occur between agricultural policy and resource and environmental policy. The results of the trade-offs will not be influenced greatly by the agrarian ethic, for the political power of environmentalists and preservationists will be growing relative to that of agriculture, and their pleas will be based on a new ethic with roots in aesthetics, health hazards, open space, conservation, and organic farming. The results for agriculture are uncertain. The public will protect its food supply, but who will bear the costs?

Productivity, too, will continue to be a long-run goal of agricultural policy. Technological development, particularly the output-increasing variety, will continue to attract investment. Public investments in research and education to enhance agricultural productivity will continue in the twenty-first century because of population pressures, weather and other environmentally induced crop failures, and general uncertainty and fears about food supplies and hunger at home and abroad. International political pressures will also support continued productivity investments.

Another long-run goal of agricultural policy will be expanded international trade that will be encouraged and supported in those industries with a comparative advantage. Expanded markets for agricultural exports will aid in two problem areas: excess capacity in agriculture and export earnings and balance-of-trade problems. Policies will likely include credit, food aid, and other programs useful for competition in international markets. Programs that conflict with expansion of international markets—for example, high-price support programs and production curtailment programs—will not be looked on with favor.

The fourth long-term goal relates to resource use, conservation, and environmental degradation.

Among the first three major long-run goals for twenty-first century agricultural policy there is little conflict, except, perhaps, between domestic and international markets. Supply allocation between the two markets and the related price effects could become policy issues, especially during short-supply years. The problem could be eased by managing supply with domestic and international reserves, but efforts to encourage reserve stocks in international market channels have not yet been succesful. Population pressures, food prices, and political stability would favor reserve stocks in the twenty-first century.

The major conflicts will be between resource, environmental, and conservation policies and the other agricultural policy goals. Many factors will be involved in determining trade-offs and compromises.

The most significant of these conflicts will be over policy decisions that determine who bears the cost of attaining the long-run goals of agriculture. Historically, farmers have borne many of the costs through adjustment problems and low resource returns. More recently, they have both reaped the gains and borne the costs of the free market agricultural policy and its inherent fluctuations. Currently, the public is bearing part of the cost of adjustment through the U.S. Treasury. In the absence of an equitable distribution of the costs, either agriculture and food policy will be in jeopardy or the agricultural sector will be required to make unacceptable adjustments.

The most critical features of agricultural and food policy of the twenty-first century include (1) proper identification of problems, both short-run and long-run; (2) established goals for agriculture; (3) identification of potential conflicts in policy goals within agriculture and between agriculture and other sectors; (4) recognition of the interdependence between agriculture and national and international economies; (5) responsiveness to changes in problems, conditions, and influences, and the flexibility to institute programs to deal with transitory problems; (6) an equitable distribution of the costs required to achieve long-run goals for agriculture; and (7) a commitment to agricultural efficiency and productivity, but a responsibility for adjustment.

Serious misjudgments in policy influenced by poor analyses of alternatives or undue influence by vested interests could seriously impede attainment of long-run objectives and create severe internal industry problems—problems that could jeopardize the stability and viability of the agricultural sector as we know it today.

Will the policymakers of the twenty-first century have the vision and courage to address critical agricultural problems and long-run industry objectives and design policies accordingly? Or will the twenty-first century, like the twentieth, be plagued by a lack of vision and courage that have given rise to short-term policies that address symptoms rather than problems?

# II

# AGRICULTURAL RESEARCH
# AND TECHNOLOGY

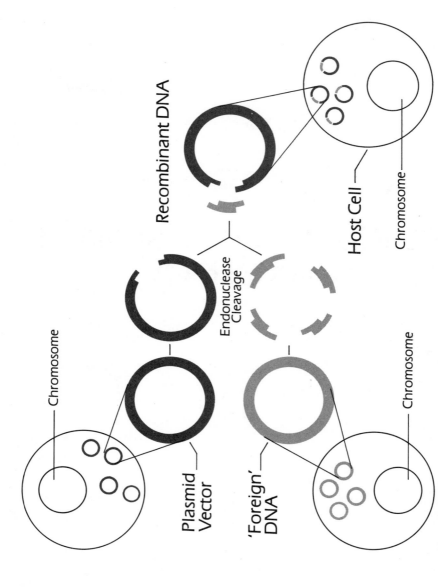

Recombinant DNA. With this technology, plant, animal, and bacterial genes can be dissected, spliced, rearranged, and transplanted from cell to cell and even from organism to organism.

# 5

# THE OUTLOOK FOR AGRICULTURAL RESEARCH AND TECHNOLOGY

RALPH W. F. HARDY

The outlook for agricultural research and technology during the remainder of this century and the first quarter of the next may be outstanding. Science is providing us the basis on which to build new technologies more powerful than any available before. It should be outstanding, for the demands of the future are large. A large part of the less developed world suffers from hunger, while most of the developed world probably suffers from an opulent diet that harms long-term health. In addition, agricultural products are a major resource for certain countries, providing important contributions to their balance of payment position. *Will* the future be outstanding? Only time will provide the answer. My personal judgment suggests modest optimism.

There is increasing recognition of the importance of agricultural research. A compelling case can be made for increasing support for agricultural research, even at a time when some countries have large surpluses of agricultural commodities. The record of technological benefits to animal and plant agriculture has been impressive. The possibilities are outstanding for agricultural technologies based on scientific breakthroughs, such as molecular genetics. Expansion of the scientific base will depend importantly on the ability of the agricultural research system to attract and train scientists with the newer skills. The development and use of new technologies will depend on cooperation among the components of the agricultural research systems: government, university, industry, and private foundations.

International and national recognition of the importance of agricultural research is growing. Several examples will suffice:

- The Congress' Office of Technology Assessment issued a critique of the U.S. agricultural research system and suggested options for improvement.
- A meeting sponsored by the White House Office of Science and Technology Policy (OSTP) and the Rockefeller Foundation produced recommendations on streamlining the agricultural research system in a report titled *Science for Agriculture.*
- Agricultural research was selected as one of the seven areas for which COSEPUP (Committee on Science and Engineering for Public Policy) of the National Academy of Science/National Academy of Engineering/Institute of Medicine produced a briefing report for the OSTP to support the budgeting of additional funds in fiscal 1984.
- The Board on Science and Technology for International Development produced a report on *Priorities in Biotechnology Research for International Development.*
- For the first time, the National Research Council created a Board on Agriculture having equal standing with its other boards and commissions.
- Selected scientific meetings and publications during 1982–1983 also focused on agriculture. The London-based Ciba Foundation, which has held several hundred meetings on health science, convened its first meeting on agriculture, *Better Crops for Better Foods,* and plans to follow this broad meeting with more specific ones.
- CHEMRAWN II (Chemical Research Applied to World Needs) was held in Manila and focused on new frontiers of research in chemistry as they relate to world food supplies. A future-oriented action report will be issued.
- The fifteenth Miami Winter Symposium, an annual meeting on molecular biology research, was devoted exclusively to agriculture for the first time.
- The National Academy of Sciences sponsored a "Convocation on Genetic Engineering of Plants: Agricultural Research Opportunities" in May 1983.

The "doers" of agricultural research are growing more energetic. The traditional ones in the United States are the Agricultural Research Service, the land-grant universities, the state agricultural experiment stations, private foundations, and industries such as those producing agrichemicals, veterinary pharmaceuticals, seed, farm equipment, food and feed products, and forest and forestry products. The new biology is causing several of the traditional institutions to expand their activities. For example, the state of New York has identified Cornell University as a Center of Excellence in Biotechnology for Agriculture.

Several large industries, including my own, have established internal research programs in agricultural biotechnology. In addition, several specialty companies, such as Advanced Genetics Science, Agrigenetics, Calgene, and DNA Plant Tech, have been established to exploit this area. Moreover, private universities such as Harvard, the University of Pennsylvania, Stanford, Rockefeller University, and Washington University have become strong contributors to the knowledge of the molecular biology of agricultural systems. Major foreign research efforts include the Australian Commonwealth Scientific Industrial Research Organization's Plant Industry Division, the Plant Breeding Institute of the Agricultural Research Council in the United Kingdom, and the Max Planck Institute for plants in Germany.

The students of agriculture—tomorrow's leaders in the field—are quite strong in a few important specialties. Admission to schools of veterinary medicine is exceptionally competitive. It is probably more difficult to gain admission to study veterinary medicine today than human medicine. The number of veterinary schools is expanding and can be a strong resource for agricultural research. As young medical doctors have become one of the most important sources of health science researchers, the expanding supply of veterinarians may become an important source of animal science researchers. James Herriot's books and the related TV series may be responsible in some degree for the growing attractiveness of veterinary medicine. Do we need books to popularize other areas of agriculture and agricultural research? Believing that this may be so, the Council on Agricultural Science and Technology has started *Science of Food and Agriculture*, a publication for high school teachers. Such activities may be important in attracting students to agricultural research.

Agricultural researchers have not always made the strongest possible case for the importance of their work and their need for financial support, but the case can be made from many points of view. The primary international reason for agricultural research is to provide the technical capability to produce and deliver the quantity and quality of food needed to meet the demands of the world's people. We are not meeting that goal in 1983. World population, growing at 2 percent per year, will exceed 6 billion by the year 2000, and tastes in food are changing.

Increases in affluence worldwide are expanding the demand for animal food versus plant food. The inefficiency of converting plants to animal protein means that plant production must increase at a greater rate than world population. Some of this increase will come from more agricultural acreage in the southern hemisphere, but most will have to come from raising the yield of existing acreage. Alternate foods or feeds, such as single-cell protein (SCP), will not contribute significantly to food supplies in the short term. Meanwhile, the increased use of crop biomass as a source of energy or demand for chemical feedstocks could place additional demands on the plant production system.

The record of the 1970s is nevertheless encouraging. Total food production was about 25 percent higher in 1981 than 1971. More importantly, per capita food production in 1981 was 2 percent higher than in 1971. Real progress is being made in meeting the world's food needs. This is a credit to our past investment in agricultural research.

Agriculture is subject to an unusually large number of variables. Not only is it subject to economic, technical, social, and political factors such as the availability of capital, raw materials, labor, professionals, and technical inputs plus the vagaries of markets and national and international policy, but it is also subject to other even less predictable and controllable variables such as climate, water, land, erosion, salinity, pests, and genetic materials. The atmospheric $CO_2$ content is expected to double in the next 50 to 100 years. What will this change mean for productivity, climate, temperature, sea level, and so on? Clearly, our food production system is a very fragile one. Research has, and I expect will, provide options to ensure our ability to stabilize our food production system despite these many variables—a most compelling reason for research. And because our environment is affected by agriculture, we need research to provide means of diminishing the negative impact of it.

Agriculture relies on many limited resources. These include water, land, capital, energy, professionals, and genetic materials. Research has and will improve the conservation and the efficiency of use of these limited resources. In some cases, such as energy, research may provide alternatives, just as mechanization provided an alternative to labor. Agriculture on the farm is not a major consumer of fossil energy (3 percent of total national consumption by developed countries), but the agriculture/food category of national economics is a major consumer after the food leaves the farm—about 15 percent.

We need to revitalize human nutritional research. The current outpouring of various books on diets and nutrition documents the thirst of the public for information in this area. The recent National Academy of Sciences' report on the possible relationship between food and cancer and the possibly better-recognized relationship between food and cardiovascular disease raises the longer-term questions to which definitive answers are lacking. It is a pity that such information is not available, but the science has not been done; this sector of research has been even less successful in attracting new talent than all of agriculture.

New crops and new uses for existing agricultural commodities should be discovered and developed. The number of major world agricultural crops is small, and in some cases, like soybeans, their genetic base is also small. Agricultural research is developing triticale, amaranth, and winged bean as new crops. Agricultural surpluses suggest the need for discovering new uses of commodities such as corn and wheat. It will be to a producing country's advantage to develop higher value-in-use products for export in place of exporting low-value commodities. New economic preservation techniques would enhance the marketing of processed agricultural products. What, for example, are the alternatives to refrigeration?

Agriculture is center stage in world interdependence. Consequently, East/West, North/South, and West/West relationships are importantly affected by agriculture. An adequate supply of food is the major need in many developing countries. Countries with the leading agricultural research programs will be viewed by these developing countries as world leaders, with the attendant important diplomatic and economic benefits. The competitive advantage that agricultural research provides the United States should strengthen our position in the world. Investment in agricultural research is as important to our security as investment in military research.

Different sectors of society seek different benefits from agricultural research. The producer seeks higher profits through improved productivity and decreased production costs. The consumer seeks the desired quality and quantity of food at reasonable cost. The citizens of a food surplus country, such as the United States, want to sell the excess food at the highest possible price to those countries that can afford to buy it. In addition, the citizens of a food surplus country should provide food for the hungry who cannot afford to purchase it. The hungry are concerned about the immediate issue of tomorrow's meal, while the citizens of the more developed countries are thinking about the longer-term issues of knowledge, health, the environment, conservation, economics, and lifestyle. Agricultural research is needed to provide facts on which to base answers to these various concerns.

A final reason for predicting a bright future for agricultural research is that it pays. Annual rates of return of 35 to 50 percent on agricultural research investment have been calculated. It seems reasonable to have faith that agricultural research will continue to pay high dividends.

Agricultural research has already produced many important technology inputs in both the plant and animal areas. Crop productivity has grown impressively over the last several decades. Words like the "green revolution" describe the impacts of the new wheats and rices internationally. Corn yield in the United States grew from about 40 bushels per acre in 1950 to 110 in 1981. Overall world grain yield grew about 3 percent per year from 1950 to 1975. Some suggest that the rate of increase may have slowed recently because we have depleted our base of agricultural research knowledge and have failed to adequately replenish it.

Independent of this debate, the impact of technology on agricultural productivity has been immense. These technologies are based on improvements in management, genetics, chemicals, and engineering. They include cultural practices, plant breeding including hybrids, fertilizers, plant-protectant chemicals, mechanization, and irrigation. It is difficult to define the part of the increase in yield provided by each technological improvement. A collective system of inputs is responsible for the increases.

Animal productivity has also been improved strongly by technology. The move from general farms raising a few of all kinds of domestic animals to highly specialized factories such as the egg, broiler, and hog plants and the large beef feedlots is illustrative. The technologies for animal productivity are also based on management, genetics, chemistry, and engineering. They include animal breeding via artificial insemination, the use of feed additives, nutritional supplements, and veterinary pharmaceuticals, as well as new equipment and housing.

I expect the future to be even more productive of improved technology than the past. The established technologies will change in their impact—some will become more important, some will remain at their current level of importance, and some will be replaced by new technologies.

Important disciplines generating future changes and their implementation will include biochemistry, molecular and cell biology, veterinary medicine, soil science, disease and pest science, food science, nutrition, and mathematics. The accomplishments and the potential for science that are expected to lead to modified or new technologies are outstanding.

Techniques to deal with genes at the molecular level are a most important advance. It is now possible in theory to move genes from animals to plants, from plants to bacteria, from bacteria to plants, and so on. There are no barriers to the transfer of genes using the basic technique of recombinant DNA. This capability provides the most powerful tool available for molecular understanding of living systems.

Research at the frontiers of life science cannot be done without the use of these techniques. Understanding generated from these techniques may provide the basis for major new products or processes for agriculture. For example, knowledge of the photosynthetic process may lead to major improvements in the efficiency of this highly inefficient process.

Recombinant DNA techniques may lead to the engineering of cells that produce products or processes. A possible agricultural example is the production of vaccines, such as the one for foot-and-mouth disease, or the economic production of hormones, such as animal growth hormones, to improve the growth or milk production of agricultural animals.

In addition, recombinant DNA techniques may be used to engineer cells that are themselves products, such as seeds or cells. Incorporation of genes for animal growth hormones into mice embryos resulted in mice with up to doubled growth rate (see Figure 1). The possibilities are indeed impressive.

Another important technique is cell and tissue culture—the ability to culture

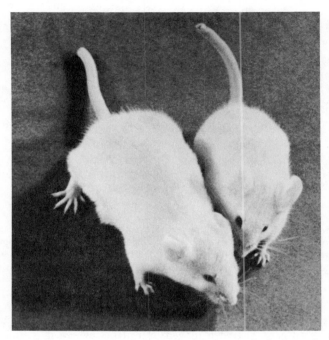

Figure 1.    *In vivo* gene therapy. *In vivo* gene therapy is a method of introducing new or improved genes into a fertilized egg. These genes become part of both the embryo's and the adult animal's gene structure. Introduction of the gene for a growth hormone into these mice caused them to grow to twice normal size. (*Source:*    Dr. Ralph L. Brinster, University of Pennsylvania, School of Veterinary Medicine.)

cells or tissue of plants or animals as one would culture bacterial cells. This technique enables large numbers of cells to be tested for the desired capability. In some species whole plants can be regenerated from the cells. A bonus of this technique is the expression of increased genetic diversity by plants regenerated from cultured cells—a process called somaclonal variation.

Production of monoclonal antibodies is yet another important technique (see Figure 2). In this process a short-lived antibody cell producing a single or monoclonal antibody is fused with an immortal cancer cell to form a hybrid cell which has the longevity of the cancer parent and the monoclonal antibody-producing ability of the antibody parent. Large quantities of a defined antibody can be produced for use as a diagnostic agent and possibly as a therapeutic agent.

What may be the impact of these techniques on agriculture? Somaclonal variation has produced useful sugarcane and horticultural crop variants and is being examined for crops such as potato and wheat. The ability to diagnose

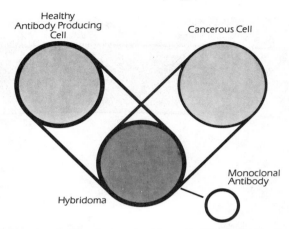

Figure 2.    Monoclonal antibodies. Fusing a healthy antibody-producing cell with a cancerous cell creates a hybridoma which is an infinite outside source of a single type of antibody called a monoclonal antibody. This technique provides new diagnostic testing for plant and animal diseases and may provide a treatment for diseases.

hidden disease is speeding plant breeding. Within recent months the first non-plant (bacterial) gene has been moved into plants, where it is functioning in a model system. Within the next year similar success with an agronomic capability, such as herbicide resistance, probably will be accomplished. Perhaps agrichemical genetically resistant seed couples will be developed. In the longer term, multiple genes such as those providing stable disease resistance will be moved into crop plants. Genes controlling protein quality and yield will be identified and manipulated to improve crop quality and quantity. In time, plants will be given new capabilities, such as the ability to fix their own nitrogen and thereby become independent of nitrogen fertilizer.

A somewhat similar sequence can be suggested for animal agriculture. Vaccines for diseases such as foot-and-mouth disease and the bacterial production of animal growth hormones have been developed and are being tested (see Figure 3). Gene therapy—the addition of genes to embryos, for example—has been demonstrated in model systems with laboratory animals and will undoubtedly be used shortly in domestic animals. *In vitro* fertilization has been achieved recently with cattle, and this technique may eventually provide the ability for almost infinite amplification of desired female genes, as does artificial insemination for desired male genes (see Figure 4).

The techniques of chemistry have also seen revolutionary advances. Computer graphics, the use of the computer rather than the test tube to generate new chemical structures with expected biological effects, is in its infancy. This capabil-

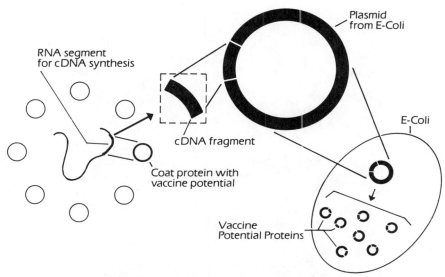

Figure 3.    Vaccines from genetic developments. Using recombinant DNA technology, effective foot-and-mouth disease vaccines are being produced as protectants against this virus-caused illness. After isolating from the virus' shell the gene for a protein which can act as a vaccine, it is cloned in E. Coli bacteria to produce quantities of the protein. This technique produces only the pure immunizing agent. (*Source:*  Agricultural Research Service, United States Department of Agriculture, Press Release, May 18, 1982.)

ity will be coupled with the molecular understanding of biological systems to produce a directed or logical synthesis of agrichemicals, pharmaceuticals, and a new class of products called plant growth regulators. The "cut and try" of the empirical approach which requires the synthesis and testing of a huge number of chemicals will be replaced by the "design and select" of the computer, a bioorganic approach that requires the laboratory synthesis of only a few chemicals to find the product. The techniques of molecular genetics may be combined with the above chemical techniques to produce biomimics—molecules that mimic natural ones without the instability of the natural ones. Recent models have been demonstrated.

What is and will be the impact of chemistry? Feed additives to modify the microorganism content of the rumen are improving efficiency of animal production. Amazingly potent plant protectants, such as the highly active sulfonylureas, are being commercialized. Plant growth regulators are chemicals to increase yield, improve quality, and facilitate the crop production process. Modest successes have been achieved in the horticultural area and in sugar enhancement in cane. One anticipates plant growth regulators that will increase the

Figure 4.   Embryo transplants. Nonsurgical embryo transfer and implantation result in genetically improved farm animals and increased reproductive efficiency. In one experiment, "super ovulation" techniques enabled the genetically superior cow (top left)
to produce eight embryos which were each transplanted to genetically inferior surro-
gates (right) providing eight genetically superior offspring (bottom left), a greatly
increased number. (Courtesy of *Farm Journal*.)

yield of major agronomic crops. For example, most agricultural crops are inefficient since they photorespire—respire under light but not in the dark—a large
part of the carbon fixed by photosynthesis. This photorespiratory reaction is
of no benefit to the crop. This wasteful process is due to evolution leaving
plants with a faulty $CO_2$ fixing enzyme. Plant growth regulators may be found
to correct this fault. The impact would be close to a doubling in the yield of
crops such as soybeans. This represents one of the largest single opportunities
in crop production. Nitrogen fixation to make nitrogen fertilizer is an energy-
consumptive and expensive process. Chemistry may provide new catalysts that
would eliminate the large need for fossil energy and capital and thereby decrease
the economic cost of this $15 billion annual input to agriculture.

Knowledge of plant and animal systems has and will provide the basis for
new technologies. Computers are being used as an information resource to
guide management decisions in a variety of areas, such as integrated pest management. The resource is only as good as the information base, however, and

additional research is needed to provide a better base. Minimum or no tillage is becoming a major cultural practice because of higher energy costs and the need for soil conservation (see Figure 5). Techniques for using water more efficiently will be mandatory. Plants use several hundred molecules of $H_2O$ for each atom of carbon fixed as plant material. The large projected increases in water use for irrigation compound the need for improved efficiency. New crops will be domesticated. Most importantly, plant and animal science will incorporate genetic, chemical, and other inputs into the agricultural system.

In the end, we will use all of these inputs—genetics, chemistry, and plant and animal science—to design genes to make animals and plants which are superior in all characteristics for the use to which agriculture puts them. This goal may be illustrated with the biological fixation of $N_2$. Nature's gene for biological $N_2$ fixation has many characteristics that severely limit its usefulness. It is excessively energy consumptive, slow, wastefully promiscuous because it reacts with substrates other than $N_2$, and requires a special $O_2$ restricted environ-

Figure 5.  Minimum or no tillage. Minimum or no tillage systems simultaneously use little or no plowing in combination with chemical weed control. With this technique, farmers conserve energy, soil, water, labor, and organic matter. However, more careful resource management is required. (*Source:*   Soil Conservation Service, United States Department of Agriculture.)

ment. The ideal enzyme for fixing $N_2$ would require no energy, be specific for $N_2$, be rapid, and be insensitive to air. In the distant future agricultural researchers will design and synthesize the gene to make such an enzyme. The saving would equal the cost for nitrogen fertilizer which may be $50 billion in 2000 A.D. This goal for agricultural research may be viewed as comparable to placing a human on the moon. And in agriculture there are several moons—not just one.

For agricultural research and technology to be outstandingly successful, there are several needs. Among the most significant are the following:

1.    Agricultural research needs the most creative minds and the newest skills. We must attract talent from outside the agricultural research community and we need to provide opportunities to train scientists with the latest skills in the new biology. The Woods Hole and Cold Spring Harbor training activities in the health science areas may exemplify what needs to be done in agriculture. Who will fill this need?

2.    The agricultural research system needs to be streamlined and depoliticized. The best components of the system should be strengthened and the weakest ones eliminated. Care must be taken to recognize the system's strengths and build on them. Production agriculture research judgment useful to growers and industry should be retained. The existence of the system is a strength of agriculture—research to extension—that is not matched by an equivalent system in health care, where research is strong but extension nonexistent.

3.    Different countries have somewhat different agricultural research systems; we need an international comparison to identify the best parts of each. From this study we might design a superior system equal to the challenge and opportunities of the twenty-first century. It is not surprising that a system designed for the late nineteenth and early twentieth centuries needs some modification.

4.    The barriers between the different components of agricultural systems need to be lowered to produce a productively interactive system. Fundamental research should be supported by a type of competitive grant similar to that used in health research but employing a more efficient process than so far used by others. One possibility is oral versus written proposals and reviews. Let agriculture be the leader in improving the competitive grant process.

5.    The quality of scientific leadership in the agricultural system needs to be improved in most cases. People-centered interactions between the components of the research system—private universities, government laboratories, industrial laboratories, and state universities—might facilitate collaboration and the transfer of information.

6.    Support for agricultural research needs to be increased, with an appropriate balance between applied and basic research. The current balance underemphasizes basic research. There is an inadequate base of fundamental information,

especially in the plant sciences. Some areas of research—such as nutrition and health, new uses for agricultural products, and the new biology—need to be greatly increased. More mature, less productive areas should be deemphasized, but established productive areas should be retained.

7. We must communicate the importance of agricultural research, its goals and its accomplishments, to citizens and decision makers in order to attain national and international commitment to this endeavor.

8. We should provide an environment for adequate protection of proprietary products to promote aggressive commercial participation. Genetic solutions may need additional proprietary protection to equal that of chemical solutions.

9. We need to maintain the genetic diversity of plants and animals—a major international resource.

The outlook for research and technology in agriculture can be outstanding. Will we elect to commit to make it outstanding, as we committed to space and placing humans on the moon? The commitment could free the world from hunger and provide a health-promoting diet in addition to providing the excitement and satisfaction that would come from the new and more complete understanding of life processes.

# 6

# NEW DIRECTIONS IN THE PLANT AND ANIMAL SCIENCES

W. RONNIE COFFMAN

J. M. ELLIOT

Real progress has been made in meeting the food needs of the world's population, and the single reason is the agricultural research performed in the past. Research will be just as instrumental in the future—if we have the will to fund it. And we should, because agricultural research can be shown to pay its way. We must encourage the training of the next generation of agricultural scientists, revitalize human nutritional research, and develop new crops and new uses for existing ones. There are, after all, only about 20 major crops that stand between the world and starvation.

We have much to investigate on the frontier of science—in genetics, chemistry, plant and animal sciences, and in biochemistry and molecular biology, to name but a few of the promising disciplines.

We have before us the potential to design genes that will produce animals and plants superior to those we have today. They can be more productive, more disease-resistant, and larger. By researching the processes of digestion in animals, we can improve animals' efficiency in using feed and perhaps substitute new feeds to free cereal grains for human use.

Our need for plant research rests on one startling fact: plants produce more than 90 percent of the human diet. Genetic engineering will be a powerful tool for improving a plant's desirable traits. Most of all, we must better understand the adaptation of plants to their environment—one of the keys to increased agricultural production in the future.

# PLANT RESEARCH AND TECHNOLOGY

## W. RONNIE COFFMAN

Plant products provide more than 90 percent of the human diet, and the remainder comes indirectly from plants in the form of animal products. Fewer than 30 plant species account for most of the production. Of these, only eight cereal crops provide more than half the world's calories. Because eating habits are not easily changed, it is reasonably certain that we will continue to depend on these crops for food for several generations to come. If food is made available adequately where it is needed, the human population will double in the next 40 years. If not, there will be unprecedented human mortality and political turmoil.

On a worldwide basis the breeding of improved crop varieties has been the most important factor in improving food production. The development of hybrid corn is perhaps the most striking example in the United States. In other parts of the world the development of semidwarf varieties of wheat and rice has been far more significant. These achievements have been closely linked to improvements in management technology for water, soils, and pests.

It is not economically or politically feasible to produce food in favored environments and transport it throughout the world or to dramatically alter crop production environments. Thus the future demand for plant technology will be in the area of improved adaptation of crop plants to their present physical, biological, and sociological environments. Technologies that are presently emerging may permit us to attain that goal.

Defined as the use of living organisms or their components in industrial processes, biotechnology is not really new to agriculture. It has been available for centuries in the form of improved plants and animals. Indeed, many of the new "technologies" are actually techniques of conventional plant breeding.

## CELL AND TISSUE CULTURE

Plant tissues were first propagated in cultures in the 1930s, and regeneration of plants from cultured tissues was first achieved in the late 1950s. A series of discoveries in the past 20 years has enabled scientists to study, manipulate, and select plant cells in cultures, as has been done with microbes for decades. We can readily screen for desirable traits that are expressed at the cellular level, but during the process the ability of the selected cells to regenerate plants is often lost. When regeneration is successful the developmental complexity of higher plants may result in the trait not being expressed. Many important traits, particularly those related to plant morphology, will probably not be improved through these techniques until we understand more about their molecular and cellular bases. However, the derivation of important plant products (fragrances, flavorings, medicinals, etc.) directly from cell culture may become economically important in the near future (see Figure 1).

## PROTOPLAST FUSION

A protoplast is a living cell with its cell wall removed. Fusion of protoplasts was first observed in 1910. The discovery in the early 1970s of polyethylene glycol as a fusion initiator allowed fusion to be accomplished at will, even

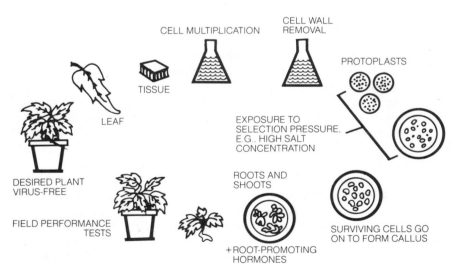

Figure 1.   Plant regeneration. (*Source:*   Office of Technology Assessment.)

between protoplasts from widely unrelated organisms. However, confirmed transfer of genetic material through this process has been limited to related species. The technique has already produced some novel plants, including one derived from the fusion of potato and tomato protoplasts. It is doubtful that useful plants will arise directly from such gross combinations of unrelated genetic material, but it is possible that such plants will contribute to conventional plant-breeding efforts. It also seems likely that this technique will be useful in manipulating traits controlled by genetic material that resides in the cytoplasm, the portion of the protoplast outside the nucleus of the cell.

### GENETIC ENGINEERING

Plant genetic engineering, called recombinant DNA technology or gene splicing, involves moving specific genetic material from one organism to another through a vector, an organism that carries and transmits genetic material from one organism to another. This is now routinely done between microorganisms, and it has recently been achieved with higher plants. However, with higher plants a convenient vector system is still lacking. This technique would appear to offer almost unlimited possibilities for improving plant traits, but the actual potential is still not well defined. The limiting factor will probably be the lack of understanding of the genetic components responsible for plant characteristics.

### MICROBIAL MANIPULATION

Many uses are already made of microorganisms with altered genetic traits. Insulin is now manufactured commercially with a bacterium developed through genetic engineering. The microbial-based insecticides, herbicides, and plant growth regulators that exist may well be improved in the near future through microbial manipulation.

Genetic engineering could soon be important in the manipulation of *Rhizobium* bacteria to improve nitrogen fixation by leguminous plants. A possibility of more distant promise is the modification of free-living bacteria that would fix nitrogen in association with the roots of cereals.

Gibberellin, a microbial-based plant growth regulator, is produced by *Gibberella fujikuroi,* the causal organism of "Bakanae," the oldest recorded disease of rice. Gibberellin is now used to adjust the growth duration and increase the yield of some crops, particularly certain vegetables. New techniques could soon result in the expanded use of growth regulators to improve the production of some of our major food crops.

## EVALUATION OF EMERGING TECHNOLOGIES

K. M. Menz and C. F. Neumeyer, economists at the University of Minnesota, recently evaluated five emerging biotechnologies in a survey of a number of scientists working in those areas (Table 1). The technologies expected to make contributions were plant growth regulators, photosynthetic enhancement, cell or tissue culture, genetic engineering, and biological nitrogen fixation. The survey showed that the first significant commercial applications of these technologies to corn production were expected to occur between 1990 and 1996. Cell or tissue culture was expected to be the earliest contributor to commercial corn production.

The survey respondents also indicated whether they anticipated a positive contribution from each technology by the year 2000 and, if so, the most likely mechanisms by which these contributions would come about (Table 2). Only 44 percent of the respondents expected photosynthetic enhancement to contrib-

Table 1.    Date of First Significant Contribution from Each Biotechnology to Maize Yield and Expected Contribution in the Year 2000

| Biotechnology | Date of First Significant Contribution | Expected Contribution to Yield in 2000 (kg/ha) | Number of Questionnaires Mailed | Number of Questionnaires Received |
|---|---|---|---|---|
| Photosynthetic enhancement | 1995 | 497 | 22 | 18 |
| Cell or tissue culture | 1990 | 195 | 37 | 31 |
| Plant growth regulators | 1994 | 988 | 22 | 19 |
| Genetic engineering | *a* | *a* | 32 | 27 |
| Biological nitrogen fixation | 1996 | 142[b] | 20 | 15 |

[a] This information was not specifically requested in the questionnaire for genetic engineering because the technology is in such an early stage of development in relation to corn production; but see Table 2.

[b] Nitrogen from biological nitrogen fixation was converted to maize yield equivalents as follows: 35.5 kg/ha $\times$ 44/11 = 142 kg/ha maize where nitrogen and maize prices are \$0.44 and \$0.11 per kg, respectively.

*Source:*   K. M. Menz and C. F. Neumeyer (1982).

Table 2. Percentage of Respondents Expecting Contribution to Maize Production from Each Biotechnology and the Most Likely Mechanism for That Contribution

| Biotechnology | Respondents Anticipating a Contribution by 2000 (%) | Most Likely Mechanism for Contribution | Respondents Anticipating Implementation by Listed Mechanism (%) [a] |
|---|---|---|---|
| Photosynthetic enhancement | 44 | Selection for high $CO_2$ exchange within traditional breeding programs | 61 |
| Cell or tissue culture | 70 | Screening for disease stress resistance | 35 [b] |
| Plant growth regulators | 89 | Increasing harvest index | 42 |
| Genetic engineering | 80 | Transfer of single gene trait characteristics | [b] |
| Biological nitrogen fixation | 66 | Nonnodular symbiotic relationship between free-living microbes and corn plant | 66 |

[a] These numbers are the percentages of all respondents who returned the questionnaire. Not all respondents were able to specify the most likely mechanism.
[b] Some respondents listed more than one application, especially for genetic engineering, where no percentage figure was calculated for this reason.

Source: K. M. Menz and C. F. Neumeyer (1982).

ute, but a clear majority of them felt that the other areas of biotechnology would make positive contributions. There was general agreement on probable mechanisms only in the cases of photosynthetic enhancement and biological nitrogen fixation.

## RESEARCH PRIORITIES

An important reason for the lack of agreement is uncertainty about the availability of fundamental information in other research areas. Many of these areas are not receiving adequate attention. For instance, if food production is to be increased on a worldwide basis, plants must be more productive in unfavorable environments. A recent survey of plant research journals showed that only 3.9 percent of the articles were devoted to mechanisms of plant growth in unfavorable environments.

One way to predict the most productive areas for research is to consider the potential productivity of plants under ideal conditions, subtract the current level of production, and then consider the factors that contribute to the gap between these two levels. J. S. Boyer, a U.S. plant physiologist at the University of Illinois, has made such a comparison using data compiled by Sylvan Wittwer and the USDA. The average yields of major crops are equivalent to just over 20 percent of the record yields (Table 3). It is impossible to precisely quantify the factors responsible for this yield gap, but it would seem self-evident that limited availability of water caused by climatic factors and soil characteristics is the major problem. Losses to pests (diseases, insects, and weeds) can be determined to some extent, and there are many other significant factors such as nutritional status of the soil (including mineral toxicities and deficiencies), adverse temperatures, and excess water.

The technology to deal with most of those conditions and to carry us well into the twenty-first century will probably come mostly from conventional research areas. In searching for resources to support new and promising areas of research we must be careful not to undercut our existing capacity to produce technology of more immediate use to farmers.

Increasing our knowledge of the adaptation of plants to their environment is the key to agricultural production in the future. As that knowledge is generated through diligent research, some of the newer biotechnological techniques should facilitate the improvement of plants. Within the foreseeable future these techniques will be applied within the context of conventional plant breeding. According to Norman Borlaug, wheat breeder and recipient of the 1970 Nobel Peace Prize, "New techniques, such as tissue culture and genetic engineering, offer potentially great payoffs and merit research resources in the years ahead. However, we should not neglect the more conventional areas of plant breeding research since they represent the major line of defense today on the food front."

Table 3. Record Yields, Average Yields, and Yield Losses Due to Diseases, Insects, and Unfavorable Physiochemical Environments for Major U.S. Crops[a]

| | | | Average Losses | | | |
| | | | | | Unfavorable Environment[b] | |
| Crop | Record Yield | Average Yield | Diseases | Insects | Weeds | Other |
|---|---|---|---|---|---|---|
| Corn | 19,300 | 4,600 | 750 | 691 | 511 | 12,700 |
| Wheat | 14,500 | 1,880 | 336 | 134 | 256 | 11,900 |
| Soybeans | 7,390 | 1,610 | 269 | 67 | 330 | 5,120 |
| Sorghum | 20,100 | 2,830 | 314 | 314 | 423 | 16,200 |
| Oats | 10,600 | 1,720 | 465 | 107 | 352 | 7,960 |
| Barley | 11,400 | 2,050 | 377 | 108 | 280 | 8,590 |
| Potatoes | 94,100 | 28,300 | 8,000 | 5,900 | 875 | 50,900 |
| Sugar beets | 121,000 | 42,600 | 6,700 | 6,700 | 3,700 | 61,300 |
| Mean percentage of record yield | | 21.6 | 4.1 | 2.6 | 2.6 | 69.1 |

[a] Values are kilograms per hectare.
[b] Calculated as follows: record yield − (average yield + disease loss + insect loss).
*Source:* J. S. Boyer (1982).

# RESEARCH AND TECHNOLOGY IN ANIMAL SCIENCE

## J. M. ELLIOT

The goal of much of the agricultural research that deals with animals is to increase the productivity of livestock through improvement in the efficiency with which forages, by-product feeds, and grains are converted to foods of animal origin. Increased efficiency translates into lower costs to both producer and consumer. The research community has been highly successful in this objective; indeed, almost unbelievable improvements have been achieved.

Consider, as examples, the changes during the last 25 years in milk production of the dairy cow and growth efficiency of the broiler. Since 1955 annual milk production per cow in the United States has more than doubled, allowing a reduction in the number of milk cows from 21 million to about one-half that number. During the same years the amount of feed required by broilers per pound of gain has been reduced from about 3 pounds to less than 2 pounds.

One of the important factors underlying these improvements in efficiency is the highly effective agricultural extension service, the mechanism by which the agricultural community ensures that developments in research and technology are brought to the attention of the producer.

Animal research can be broadly classified into five disciplines—genetics and breeding, nutrition, physiology, health, and management. Each of these disciplines can be subdivided into specializations, each of which has contributed in varying degrees to improving the productivity of animals. All can be considered applied research disciplines, and they are underpinned by more fundamental disciplines such as chemistry, biochemistry, endocrinology, microbiology, immunology, mathematics, and physics. Thus research workers in the animal sciences are in an excellent position to apply the most fundamental developments in science to the solutions of problems of the animal industries. Never has the potential for future developments been as great or as exciting as it is at present.

An examination of current research in each of the disciplines suggests that promising developments are likely to change the direction of animal research during the next few decades. Here are some examples broadly described.

## GENETICS AND BREEDING

The genetic makeup of an animal governs its potential productivity. Whether its potential is ever realized is dependent on the degree to which environmental factors are optimized. This large impact of environment on productivity of livestock makes it very difficult to identify animals of superior genetic merit. Attempts to resolve this difficulty have been the major focus of animal breeding research in the last 30 years, and great progress has been made. The successful development of complex statistical methods applied to millions of records (population genetics), coupled with production testing and artificial insemination programs, has made it possible to accurately identify and maximize the use of animals of superior genetic makeup. The continuing refinement of these techniques, made possible by more powerful computing facilities, has resulted in a rising genetic potential in the species to which it has been applied. A classic example of this is the milk production potential of dairy cows in the Northeast, which in recent years has shown a rising genetic trend averaging about 150 pounds of milk per cow per year.

The gains already realized by this selection technique are permanent, and the potential for future gains has by no means been exhausted, even in dairy cattle. Evidence suggests that its continued application will result in ever-increasing genetic potential in the foreseeable future. In the meantime, however, developments in molecular biology have ushered in a new dimension in genetic research that lies largely in the areas of biochemistry and physiology. Thus after years of rapidly diverging disciplines, geneticists and reproductive physiologists concerned with animal breeding may again be on converging paths.

## PHYSIOLOGY

The discipline of physiology encompasses study of the functions of various organs and systems in animals and thus impinges on many other disciplines, both well-established ones and those just emerging from developments in the basic sciences. The digestive, reproductive, mammary, and endocrine systems in agriculturally important animals are the ones that have been studied most intensively. Great progress has been made in our understanding of the details of their physiology, and applications from each have contributed in many ways to current agricultural practices.

Our knowledge of reproductive physiology has perhaps been exploited more fully than the others at the farm level. Use of frozen semen, the consequence of breakthroughs in research many years ago, has for some time been routine. Today it is possible for an outstanding bull to sire in excess of 50,000 offspring per year. More recently, successful methods of embryo transfer, estrus synchronization, and new techniques for estrus and pregnancy detection have been developed. Progress has been made in freezing embryos, the full success of which will have far-reaching implications in genetic selection of farm animals as well as in research. Special genetic material can be preserved by freezing embryos. Today more than 200 strains of specially bred mice, of potential value as models for farm animals and human problems, are preserved only as frozen embryos.

Much of the research underlying these advances has involved fundamental studies aimed at understanding the basic biology of animal systems. This effort has been greatly enhanced by the widespread use of such techniques as the radioimmunoassay, enabling measurements to be made at the level of parts per trillion. Specific developments with possible practical applications already in the pipeline include the successful limited cloning of embryos. Practical techniques for the sexing of gametes and embryos and for parthenogenesis, or nonsexual replication, of outstanding individuals, which would increase the efficiency of reproduction in gender-related enterprises (e.g., milk production), are almost certain to be a reality before the turn of the century.

Genetic engineering, perhaps the most exciting recent development in biology, is likely to have a two-pronged effect on farm animal research. Its use with bacteria to synthesize specific hormones and other metabolic regulators in abundant quantities will allow animal researchers more rapidly to explore, and possibly to exploit, metabolic regulation of physiological functions such as growth, lactation, and reproduction. For example, there is current interest in the effect of manipulation of hormone levels during short but critical periods of growth and development on mammary development and subsequent milk production potential.

Genetic engineering also raises the possibility of introducing genes into farm animals to influence their production or disease resistance. The effect of successfully introducing the rat gene for growth hormone into the mouse has recently been reported. While this feat, which only a few years ago would have been classed as science fiction, has created great excitement, there are formidable problems to be overcome, including that of gene regulation, before possible applications in the livestock industries can even be intelligently discussed. In the meantime, physiologists are interested increasingly in identifying the physiological basis of genetic differences in animal performance. An impressive array of tools is now available with which to proceed.

## NUTRITION

Research in nutrition is generally focused on determination of the requirements for nutrients and their metabolic functions and on strategies for meeting requirements through diet formulation. Nutrition studies therefore run the gamut from applied studies dealing with the preparation, digestibility, or energy value of feedstuffs to fundamental biochemical studies involving enzyme kinetics.

The requirements for protein, energy (calories), and many of the major minerals and vitamins are now well known, and sophisticated diets can be formulated for today's highly productive livestock. But many gaps in our knowledge remain. Factors important in the regulation of appetite or feed intake and the mode of action of growth promotants, for example, are still poorly understood. Whereas in the past much research with farm animals has primarily involved measurement of production responses to dietary variables, in recent times greater emphasis is being directed toward understanding the mechanisms of digestion and the intricacies of the utilization of nutrients by the body tissues. It seems certain that the major breakthroughs in nutrition research in the future will derive from this approach, which is most productive when nutritionists and physiologists work together. Some promising developments are on the horizon.

By combining an understanding of digestive processes with the technology of "protection" of certain nutrients from fermentation in the first compartment of the ruminant stomach, it should be possible to control or modify the site of digestion of different ingredients in the ruminant's diet. This may have far-reaching nutritional implications since it can greatly affect what is absorbed from the digestive tract. For example, one could enhance glucose absorption, decrease hydrogenation of fats in the rumen, produce polyunsaturated meat or milk, and plan a ration that would provide the tissues with a more nearly ideal array of essential amino acids, thus increasing productivity.

As animal scientists look to the future they recognize that the day is likely to come when it will be uneconomical and perhaps unethical to feed so much grain to livestock. An effort is already underway to find methods to utilize more efficiently forages and by-product feeds and to spare foods that may be directly consumed by humans.

A great deal of excitement surrounds emerging studies suggesting that regulation of the partition of nutrients to different tissues or physiological functions may be one of the ways in which genetic differences are expressed and, thus, may be subject to manipulation. One example is the effect of exogenous growth hormone on milk production. Initial studies indicate that milk production may be increased by 10 to 33 percent without proportionately increasing feed intake, at least on a short-term basis. Several applications are possible. In selecting animals for milk production, blood levels of growth hormone or its secretion rate might be measured. Growth hormone secretion rates might be modified by feeding or management changes. Or exogenous growth hormone might be provided via implants. Interestingly, technology has already assured a future supply of bovine growth hormone for it is now produced industrially by genetically engineered bacteria. More efficient use of nutrients for growth and other productive functions will very likely be made possible by other findings. One of the major goals of nutrition researchers is to develop ways of enhancing the flux of nutrients to muscle tissue development at the expense of adipose, or fat, tissue. The meat animal industries currently produce an enormous amount of animal fat, much of which is discarded.

## ANIMAL HEALTH

The prevention, control, and treatment of disease, both infectious and noninfectious, obviously play key roles in successful animal production systems. The eradication or control of hog cholera, Marek's disease, tuberculosis, and foot-and-mouth disease are but a few examples of successful research applied to practice. Serious problems remain, including bluetongue, scrapie, pseudorabies, respiratory diseases, African swine fever, mastitis, and many others.

As viewed broadly by someone not intimately familiar with current research activities in this discipline, several promising new directions seem to be developing. One of these is a direct consequence of advances in biotechnology. The application of monoclonal antibody and recombinant DNA techniques to more rapid and specific diagnosis of disease and to the development of safe and selective vaccines and immunotherapeutic agents have the potential to revolutionize disease control strategies. A vaccine for foot-and-mouth disease has already been developed. Still in its infancy, this whole area seems likely in the near future to dominate research activities related to infectious diseases. In the noninfectious disease area there is growing recognition that a multidisciplinary approach will be necessary to solve some of the problems. The etiologies of metabolic and so-called production disorders, such as ketosis, parturient paresis, lactic acidosis, reduced fertility, bloat, and digestive upsets, are likely to become much clearer when details of the regulation of normal metabolism are understood. Hence greater teamwork among veterinary pathologists and physiologists, reproductive physiologists, and nutritionists is to be expected in the future.

## MANAGEMENT

Management research encompasses the study of those factors which influence the success of a livestock enterprise—strategies by which animals can be cared for, selected, housed, fed, bred, milked, or marketed. The health and comfort of animals as they are managed for high production is a key concern. Many changes have occurred in livestock enterprises in recent years: to name just a few, the special problems of detecting estrus for artificial breeding in large dairy herds; the health problems of young calves in confinement; the earthy problems of manure disposal in environmentally acceptable fashion as herds and flocks increase in size; least-cost ration formulation; and appropriate record-keeping systems. Intensive systems have been developed whereby lambs, traditionally born in the spring, can now be born at many times of the year, thus making more efficient use of labor and facilities, smoothing cash flow, and avoiding market gluts. Through a simple change in the management of cattle, a method of controlling mammary infection was developed, and it has had an enormous impact on the incidence of mastitis (mammary infection), the productivity of cattle, and quality of the product.

The development that will probably have the greatest impact on management in the coming decades is the automation of many tasks. On many farms computers already formulate rations, allocate feed, summarize records, and provide strategies for maximizing profits. In the research pipeline, however, are several "space age" applications. By combining biology and electronics with computers in their many forms, it is expected that in the near future animals will be reliably and automatically identified, weighed, and fed predetermined amounts of feed at predetermined times. In dairy systems, milk weights will be automatically taken, cows in estrus will be identified for breeding, and sick animals flagged for treatment, possibly even before full clinical problems are obvious.

An examination of research activities at the cutting edge of the several disciplines of animal science suggests that by the beginning of the twenty-first century unprecedented increases in the efficiency of animal production will be possible. It is quite probable that between now and the year 2000 U.S. milk production per cow will have doubled. This will represent a marked increase—perhaps 20 percent—in efficiency of conversion of feed to milk.

An even greater improvement in efficiency in at least some of the meat animal industries can be predicted. This will be accomplished by increasing the number of offspring per female per year, by increasing rates of body weight gain, and by diverting nutrients from synthesis of fat to synthesis of muscle tissue.

Emerging developments in breeding, physiology, nutrition, health, and management, which are raising efficiency at an accelerating rate would have been considered science fiction 10 or 20 years ago. Sustained progress will be impossible, however, without continued public support of endeavors with a record of returning handsomely on investment.

# 7

# FUTURE APPROACHES FOR MEETING NUTRITIONAL NEEDS

ERNEST J. BRISKEY

VIRGIL W. HAYS

ROGER L. MITCHELL

There are stunning disparities worldwide between our ability to produce food and the substandard diets that are commonplace for 70 to 80 percent of the world's population. While 10 percent of the population is currently at the brink of starvation or suffering from severe malnourishment, another 10 percent have all of their food requirements fully met. The 80 percent in the middle would dramatically increase the quantity and quality of their diets if they had the means to do so.

By the year 2000, 85 to 90 percent of our planet's people will live in developing countries, where needs for improved diets will be the greatest. As one authority has stated it, "World hunger is a greater threat to world stability than the arms race."

With minor exceptions the land suitable for agriculture is being fully used, representing only about 27 percent of the earth's surface. Therefore the battle to close the gap between the planet's current dietary standard and what it could be will be fought with the weapons of technology. Because a remarkable research system has been built in the United States, this country can and must play a leadership role in developing these new technologies for deployment around the world. But it is not enough to say that the United States can feed the world. Even if that were possible, it would not be desirable to create an international food welfare system. Rather, the effort should be to speed the arrival of the time when nutritious food is produced close to all who would consume it.

# NUTRITIONAL NEEDS FOR THE TWENTY-FIRST CENTURY

## ERNEST J. BRISKEY

### GENERAL ASPECTS OF NEED

You have only to travel the world to know that hunger abounds. According to the Food and Agriculture Organization statistics, approximately 10 percent of the world's population, some 450 million people, are on the edge of starvation or suffer from severe malnutrition. In such cases nutritional needs are readily discernible. But in reality the rest of the world is not well-fed either. Only 10 percent of the people of the world have all of their nutritional needs being met, and another 10 percent are only modestly to moderately well-nourished. The remaining 70 percent of the world's population would increase its consumption of food sharply with an increase in the standard of living.

Today the issue is more one of income and distribution than food supply; although in regard to the specific foods people want when they have adequate money, our food supply is not that bountiful. But history has shown that we cannot have a welfare world. People must help themselves. They need knowledge, technology, profitability, and organization. Especially, they need agriculture for it is the key to their survival. Agricultural development simply must precede economic development in most parts of the world. As agriculture develops, a nation's economy develops, the standard of living rises, demands increase, and nutritional preferences or requirements change in both quantity and kind with rising affluence.

### NEED ON THE BASIS OF NUMBERS AND PROJECTIONS

The world's population has doubled during the last four decades. One-half of that doubling has occurred in the past 15 years alone. By early in the next century the population will probably be double what it was just 15 years ago. And as we move toward the first quarter of the next century, we will have by a conservative estimate 7 to 8 billion people on earth. Even if world population starts to level off at that point we will still see major shifts in the demographic statistics.

In almost all countries major changes will take place in life expectancy: Elderly people's consumption far exceeds that of children under 5 years of age, who, due to high mortality and malnutrition, contribute little to current food use. Medical information and food supplies, facilitated by computers, telecommunications, and information transfer will add substantially to the longevity of people. Through this next century we might expect a substantial improvement in life expectancy in many of the developing nations—and a gradual and continuing improvement in our own. We will simply have more older people to feed, and they will certainly influence nutritional need. The picture comes somewhat into perspective if we remind ourselves that while it took a million years for our population to develop and grow to 1 billion people, it took only 100 years for the second billion, 30 years for the third billion, and 15 years for the fourth billion.

Rates of increase in population are declining, but before the beginning of the next century we will still have 6 billion people on earth. The distribution of this population, its age, and its relative economic status will be important factors in determining nutritional needs. The less developed world's birth rate is reported to be 33 births per 1000, with a net natural increase of 2.1 percent. Even though these figures are influenced by the effective birth control practices in the urban areas of the People's Republic of China, they indicate that the population of the less developed countries will, on average, double in 33 years. And in some countries, like Kenya, doubling will likely occur in 15 years. It is sobering to note that even today, in an arc from Afghanistan to Japan, including the subcontinent and China, but excluding the Soviet Union, the population is already equivalent to the total world population in 1950.

Fertility rates in the more developed countries, of course, are considerably lower than those in the less developed countries, and have a net natural increase of only 0.5 percent and a projected doubling time of about 116 years. Before much attention is given to this, however, it should be emphasized that 90 percent of the increase will take place in the developing countries. By the year 2000, 85 to 90 percent of our 6 billion people will live in what we today call developing nations.

## NEED ON THE BASIS OF INCOME LEVELS

Population growth is only one aspect of the need equation. There is a much more insidious and important component—food demand. Development is taking place, incomes are increasing, and it is estimated that within 20 years per capita demand alone will increase by an additional 75 percent. As incomes increase, the demand for meat, milk, eggs, fruits, and vegetables increases, while the demand for grain for human consumption decreases.

Many people have predicted increasing diversion of grain from animals directly to human food, but it simply has not happened. When incomes improve, desires for animal products increase. Animal products are becoming increasingly important in meeting world food requirements. All nations are striving for the consumption of increased quantities of meat, milk, and eggs. Increasing quantities of grain exports from the United States are being used to feed ever-expanding livestock populations. In fact, it is projected that by 1985 worldwide use of grain for livestock feed will substantially surpass that for direct human use.

The increase in the use of grain for feed must be considered in light of price and availability. It is true that most of the feed grain is used in middle- to high-income countries, but this is shifting as economics permit. Grain used for feed has continued to increase in the centrally planned economies and in the developing market economies. The presumed competition for grain between humans and livestock, which some like to talk about, has always been a false issue. If less grain were demanded, less grain would be produced—it is almost that simple. Also the world has vast uncultivated land areas where ruminant animals remain the best converters of forage to human food, and require little if any grain. Livestock products are very important for human use. In our country, for example, three-fourths of the protein, one-third of the energy, and most of the calcium and phosphorus in our diets come from animal products.

But cereal grains are vital too. On a worldwide basis cereal grains provide 60 percent of the calories and 50 percent of the protein consumed by humans. When animal products obtained via grain consumption are included, cereals directly and indirectly provide almost 80 percent of our caloric intake.

## DEMAND BASED ON THE NEED FOR VARIETY

Globally, about 20 crops are of ultimate importance. They make the difference between life and starvation. Assuming no immediate or major shift in consumption patterns, they must continue to be taken into consideration in coping with nutritional needs of people in the twenty-first century. These crops are wheat, rice, corn, potatoes, barley, sweet potatoes, cassava, soybeans, oats, sorghum, millet, sugarcane, sugar beets, rye, peanuts, field beans, chick-peas, pigeon peas, bananas, and coconuts (see Figure 1). But in the middle-income and more developed nations of the world the commodity list expands to more than 100. In view of an increasing standard of living in several nations

Figure 1.   These winged beans have huge potential for future development. They are being grown at Walt Disney World Epcot Center: The Land presented by Kraft (© 1983 Walt Disney Productions.)

this growing list of commodities becomes increasingly important as we anticipate fulfilling nutritional needs for the next century.

## THE PRODUCTION DILEMMA

We do not function as an isolated nation. Our exports have increased ninefold during the last 15 years. Food supplies and prices are determined on a worldwide basis. And it is the worldwide availability of raw materials that in the final analysis markedly influences both price and availability here at home. Ultimately, however, more and more food must be produced closer to the people who consume it. Even so, the demand for importation of food by developing nations of the world will be increasing. The United States, Canada, Argentina, and Brazil now account for 10 percent of soybean exports, 90 percent of wheat exports, and more than 75 percent of feed grain exports—more than 60 percent of the grain in international trade is of U.S. origin.

Increases in production in most parts of the world will not keep up with increases in demand on a consistent basis. We are endowed with some of the world's finest agricultural land, water, and environments for production. Most of the rest of the world is not so advantaged. The world has a finite amount of arable land and about 27 percent of the earth's land surface is used for agriculture. Of this, 10 percent is used for raising crops and 17 percent for grazing livestock. Most of the rest is too dry, too cold, too hot, too rocky, or too steep. Thirty-six percent of the land surface is desert, and 15 percent has been deforested.

A recent study projects a food shortfall of 120 to 145 million tons by 1990 in the developing market economies, more than three times that realized in 1975. More than 40 percent of the shortfall was projected for Asia, 25 percent for North Africa and the Middle East, 20 percent for Sub-Saharan Africa, and 10 percent for Latin America. The countries with the lowest per capita income (less than $300 in 1973) were projected to have shortfalls seven times greater in 1990 than they had in 1975. Serious food deficits were projected for a number of countries including India, Bangladesh, and Indonesia in Asia; Egypt, Nigeria, the Sahal Group, and Ethiopia in Africa; and Bolivia and Haiti in Latin America.

Food production has increased by 2 to 3 percent in each of the last three decades. Today, global productivity is at a record high. In fact, one might say that there has never been so much food produced—totally or on a per capita basis. Last year an all-time world record grain crop of 1640 million tons was produced. But these figures do not tell you how many people wanted grain, if they had the income to buy something else, and if other goods were distributed where grain was wanted.

And what is in question, anyway—subsistence or gearing ourselves to provide tasty food that meets nutritional needs? Production of the right kind of food, its distribution and delivery, and the income to buy it pose critical policy issues. Today, food surpluses, malnutrition, and hunger exist side by side in many nations of the world. Surpluses and deficits recur year after year. Average production figures hide year-to-year variations. They ignore the instability of the production levels. Thus droughts, floods, low and high temperatures, crop pest damage, and post-harvest losses bring about alarming year-to-year variations and raise havoc in the absence of reserve food stocks.

These fluctuations in supply also make it virtually impossible to scale up processing on a competitive basis. Basically, what this all means is that food production must be increased three to fourfold if we are to accommodate both a doubling of the population and an increased demand brought about by a higher standard of living. This must be done without adding significantly to our total arable land worldwide (although some U.S. land can be reactivated). In the final analysis, however, 85 to 95 percent of the increase must come from improved technology.

## APPROACHES TO BE TAKEN

In the short term, major emphasis should be placed on implementing and improving upon the best methods we have while continuing our conventional research. Our productivity growth is deteriorating and we must look at some of the reasons. The gap between what food production is now and what it could be constitutes the world's greatest technological reserve. Countries like Japan, Germany, and France show higher percentages of their gross national product going into research. All have higher growth rates than the United States and have had for the last 17 years.

The existing perspective differs from the somewhat longer-term view of 5 to 17 years as we move forward and project toward the twenty-first century. We need human capital if we are to meet our nutritional needs. All potentials are dependent upon people with skills to make institutions function. Our land-grant universities embody this nation's most concentrated and potentially available source of talent to be directed toward helping meet the coming nutritional needs through increased molecular research. The question is, will we have the will to see that they are given adequate resources? Will we note the value of research and not only maintain it but expand it?

Nations must learn that technology and incentives are basic to what can be done in agriculture. In time people should stop thinking of profitability as being synonymous with greed and exploitation. Trained people can help bring about profitability. We will, I hope, have the desire to nurture the strong, involved entrepreneurship of private industry to work together with the land-grant universities to ensure the formation of a bountiful supply of human capital. We need to quantify the extent of the problem, pick our strategies, and build on the experiences of the past. And we must avoid letting our industry's ability to meet nutritional needs fall by the wayside because of insufficient investment in facilities and human capital.

We must expand our own education, research, and extension. Expected rates of return on education are among the highest of any investment that can be made—including the first 8 years of school. There is an urgent need for research to be appreciated and new technologies to be used. What is obviously required is an agricultural research and extension service in most of the developing countries of the world which works particularly in the following areas:

*Animal Nutrition Needs.*    The increased consumption of meat, milk, and eggs on a worldwide basis—conservatively estimated at 25 percent—will alter the nutritional needs of animal agriculture. To meet them we must understand more fully animal physiology and nutrition.

*Plant Nutrition.*    The major thrust will be to conserve the topsoil and develop the biological fixation of nitrogen in nonlegume plants. Cash cropping for export will continue on a desirable basis and under resource control.

*Food Processing.* More foods will be consumed in processed form. Emphasis will be placed on nutrient preservation, sensory quality enhancement, conservation of energy and water, reduction of pollution, and efficiency of operation. By-products will be utilized, and new techniques and packages will be developed.

*Energy Consumption.* Emphasis will continue to be placed on improving overall efficiency and especially reducing energy utilization in production and processing.

*Post-Production Systems.* When food costs inevitably increase, it seems logical that an early gain can be realized by reducing post-harvest losses. Approximately 30 percent of food produced never reaches humans—more so in developing nations.

## LONG-TERM PERSPECTIVES

The long-term problem is one of enhanced production with improved trade and development mechanisms. Agriculture is—and must be—an intermediate to long-term growth industry. To meet the nutritional needs of the twenty-first century, we need new technologies, resource inputs, and economic incentives.

Human talent, trained to develop, embrace, and implement new technologies—including biotechnology—will be the key to our future, for the key is technological change. Our educational system will be challenged more over the next two decades than during any like period of modern times.

Agriculture must also be computerized both to maximize efficiency and to fit in the world market. Computers will make possible the development of robotic agriculture and will otherwise influence the ability of mankind to expand distribution of nutritional supplies at lower costs, especially in the more developed countries of the world.

In 1820 Thomas Jefferson expressed clearly the need of a new, undeveloped country for technical assistance: "In an infant country such as ours," he said, "we must depend for improvement on the science of other countries, long-established, possessing better means, and more advanced than we are." We did, and in modifying the science of other countries to fit our needs, we built a research, extension, and educational system that has made us the envy of the world.

Retaining and enhancing this system for ourselves while seeking it for others should be our goal. We should strongly support Title XII legislation which brings land-grant universities into international development because through Title XII projects we can perhaps make our greatest contribution toward meeting the nutritional needs of the twenty-first century. We cannot just consider nutritional needs of isolated nations when we have worldwide markets. We must help many nations develop their economies if they are going to obtain food. We should also seek development of research centers in this country and abroad linked to our land-grant universities. We cannot consider development without considering trade, and we cannot consider trade without assuring development for export potential. And all need an educational, unbiased input.

In the words of Dr. Clifford Wharton, "world hunger is a greater threat to world stability than the arms race." But we must not work only with the poorest of the poor, for if we do, we stand at the brink of creating an international welfare state. If we give Dr. Wharton's statement the credit it deserves, we will also not allow our technical aid for overseas development to decrease. Technical assistance is geared to helping people help themselves—to increasing incomes. This technical assistance should go to middle-income countries as well, where enormous needs exist and where infrastructures and organizational and societal imperatives are sufficiently in place so as to realize immediate, significant benefits in more adequately meeting nutritional needs.

For all nations, including many segments of our own, we need master plans for development that have feasibility and implementation guidelines. If we are to meet our nutritional needs in the twenty-first century, the urgency of our short-term problems, namely, surpluses, must not be allowed to obscure the importance of considering how to organize an agricultural sector that can survive and flourish for as long as we expect our society to endure.

# IMPLICATIONS FOR ANIMAL AGRICULTURE

## VIRGIL W. HAYS

Animal products—meat, milk, eggs, and products derived from these foods—make very significant contributions to human diets. They are an excellent source of protein of high biological value and provide about 65 percent of the total protein to the diets of humans. These products provide about 80 percent of the calcium (mostly from dairy products), 65 percent of the phosphorus, and substantial quantities of essential vitamins and trace elements. Animal products also add flavor and variety to diet and thereby contribute to psychological well-being.

Significant progress has been made in recent years toward increasing the efficiency of producing animal products; however, we appear to be on the verge of some of the greatest breakthroughs ever.

## GENETIC ENGINEERING

There are new opportunities for breakthroughs in efficiency of animal production by direct and indirect applications of genetic engineering. As a result of the recent accomplishments in molecular genetics, synthetic amino acids, improved vaccines for disease protection, and hormones or other growth regulators are now in the developmental or production stage. This field is so new and the implications are so large that it is impossible to visualize the impact on food production systems by the twenty-first century.

Vaccines for the control of the devastating hoof-and-mouth disease already offer a practical solution to a worldwide problem. This accomplishment should lead to the development of new or improved vaccines for other costly diseases. The presently available microbially synthesized insulin and growth hormone are further examples of the important contribution we can expect through genetic engineering.

The direct application of these new techniques to animals is just beginning to take shape, and the full impact can hardly be predicted. The accomplishments of R. D. Palmiter at the Howard Hughes Medical Institute, Seattle, and his associates are illustrative of what may be accomplished. They microinjected a DNA fragment containing the promoter of the mouse metallothionein-I gene fused to the structural gene of rat growth hormone into the pronuclei of fertilized mouse eggs. Of 21 mice developing from 170 eggs, seven carried the fusion gene. Several of these grew faster and had high levels of the fusion messenger RNA in their liver and growth hormone in their serum. At 74 days of age the mice with the infused genes averaged 54 percent (females) or 38 percent (males) heavier than their respective littermate controls. One of the seven mice transmitted the MGH genes to half of its offspring, indicating the gene was incorporated into one of the chromosomes. Such evidence suggests wide-ranging possibilities for altering the productivity of animals by recombinant DNA techniques.

## AMINO ACIDS

Adequate amounts of high-quality protein is one of the major limitations—perhaps *the* major limitation—in our world food supply. Animals play an important role in providing high-quality protein for human diets; but animals also compete with humans for the limited supply of protein. Availability of certain of the essential amino acids at a reasonable cost would greatly reduce the total protein needs for poultry, pigs, and humans. Developments in genetic engineering suggest that microbial synthesis at a much reduced cost is not only possible but should be realized soon.

Of the ten amino acids essential for growth in nonruminants, including humans, we now have two—lysine and methionine—available at prices that will permit their use in animal diets. The availability of three other essential amino acids—threonine, tryptophan, and valine—could markedly reduce the need for supplemental protein in swine diets. In the United States soybean meal serves as a major protein source for animals, particularly swine and poultry. In swine a finished diet of 150 kg (15 percent) of soybean meal per metric ton of diet is required to supplement corn protein so that the most limiting amino acid, lysine, will be present at a level to meet the pig's need. That level of soybean meal provides an excess of several other amino acids. In fact, the protein from corn alone meets the essential amino acid needs except for lysine, methionine, threonine, tryptophan, and valine. In this example, 150 kg of soybean meal, or approximately 68 kg of soya protein, are required to replace 7.2 kg of essential amino acids (4.2 kg lysine, 2.2 kg methionine, 0.6 kg threonine, 0.1 kg tryptophan, and 0.1 kg valine).

For poultry, for other stages of swine production, or with the use of other combinations of feedstuffs for pigs, the order and magnitude of limitations would differ. From this illustration, however, it is apparent that the availability of amino acids for supplemental purposes would greatly extend our protein supplies. Direct supplements to human foods would further reduce the protein needs. In the very near future microbial synthesis or improvements in chemical synthesis should lead to the availability of these limiting amino acids at costs to permit more extensive use as feed supplements.

The use of amino acid supplements to replace or reduce natural ingredients will necessitate other adjustments and may lead to vitamin or mineral deficiencies which will need to be corrected. Such diet manipulations may lead to the discovery of the need for vitamins or minerals not now recognized as being essential.


## REPRODUCTION CONTROL

Improvements in the reproductive efficiency of food-producing animals is probably the area that offers the greatest opportunities for gains in efficiency of animal production. Failures to show or return to estrus, failures to conceive, prenatal mortality, and early postnatal mortality are all factors that greatly reduce the efficiency of production. It is anticipated that within the relatively near future we will be able to exercise complete control of reproduction through the use of chemicals, primarily exogenous hormones or substances that regulate the secretion of endogenous hormones. In order to reap the maximum benefits we will need to synchronize the onset of estrus, synchronize time of ovulation, synchronize parturition, and predetermine the sex of the offspring.

Some of the improved efficiencies that can result from reproduction control can be illustrated by the efficiencies in the poultry industry. We are not exercising great control of reproduction in chicks; but, simply because of the high rate of lay and the fact that we can hold the ova (eggs) for several days, we can have a large number of chicks at a specific time. By filling a chick starter unit at one time with chicks of the same age we are able to reduce disease by not exposing the young chicks to older birds and by applying appropriate preventive measures to provide the correct nutritional program at all times. We are also able to provide necessary labor and management at the critical periods of the life cycle. However, we are not able to control sex. This is unfortunate since poultry provides an excellent example of the benefits that could be derived from sex control. The male chicks from the layer strains are a total loss to the egg industry and very inefficient producers of meat. At the same time the females of the broiler strains are less desirable than males in rate and efficiency of meat production.

Progress is being made in controlling reproduction in cattle and swine. The synchronization of estrus and to some extent ovulation is being accomplished at present in cattle through the use of prostaglandin $F_{2\alpha}$. Relatively normal conception rates can be accomplished by inseminating at a specified time— 80 hours—after administration of the substance. W. Hansel, professor of physiology at Cornell University, reported excellent results on estrus synchronization and timed breeding in dairy heifers and beef cows by using a progesterone-releasing intravaginal device and an injection of prostaglandin $F_{2\alpha}$. Researchers have developed partially effective synchronization programs in swine. More precision is needed before we can consider that estrus and ovulation synchronization is successfully accomplished; but once accomplished, the timing of parturition should be immediately forthcoming with the application of prostaglandins or other substances.

Precise timing of estrus and ovulation would allow maximum application of artificial insemination and more extensive application of embryo transfer, two techniques that provide our greatest opportunities for rapid genetic progress in both milk and meat production. In our U.S. dairy industry the national average of milk production is approximately 5500 kg of milk per cow per year. A substantial number of the herds on production record programs and making extensive use of performance-tested sires are producing about 9100 kg of milk per cow per year. The difference in production between the above-average and the high-producing herds is not entirely a result of genetic differences, but these figures do illustrate that we have a far greater genetic potential for production than we are now accomplishing.

Effective methods for preselecting the desired sex of the offspring could have a great impact on total production efficiency. Progress is being made in two areas of research that could lead to the solution of this problem. One area is the separation of the X- and Y-chromosome-bearing sperm. A second means of control could develop through chemical or immunochemical techniques which would selectively permit the development of the embryo of the desired sex. Accomplishment by either method would lead to widespread application of sex control in our food-producing animals; however, sperm separation would probably be more effective. It has been predicted that we will accomplish, or at least partially accomplish, the separation of the X- and Y-chromosome-bearing sperm within the next 10 to 20 years.

In controlling sex in dairy production it would be desirable to choose a very high percentage of females and to be very selective as to which females would be the dams of potential sires. Sex control could essentially double the number of females from which to select replacement cows and also assure a high degree of precision in selecting only superior males for performance evaluation.

For beef production under extensive pasture or range conditions, as is done in the United States, Canada, Australia, Argentina, and so on, significant gains in efficiency could be accomplished with greater application of cross-breeding. With sex control, we could opt for females in the first cross-offsprings and these could serve as the dams for three-breed cross-calves. For the three-breed, or terminal cross, we could produce all male offspring because of their superior efficiency in producing beef. The reproduction strategies would differ among species or have different purposes within a species (e.g., milk versus beef); but for each species there can be very significant improvements in efficiency of production from exercising complete control of reproduction, including sex control.

Hansel predicts that by the year 2000 we will have perfected estrus cycle regulation to the point that large groups of animals can be effectively inseminated at a preset time without the need for estrus checking. Furthermore, he predicts that procedures for super ovulation, *in vitro* fertilization, embryo freezing, and nonsurgical transfer of embryos will be developed. These techniques will permit rapid proliferation of superior genetic material in animal species that by nature have a very low rate of reproduction.

## GROWTH PROMOTANTS

A number of chemical substances are now being used to increase the rate and efficiency of production in food-producing animals. These effective agents can be broadly classified as hormone-type substances or antimicrobial agents. The hormones increase rate and efficiency of production by altering the anabolic or catabolic processes, allowing the animal to synthesize tissue at a rate greater than their genetic potential. Antimicrobial agents (antibiotics produced by microbial synthesis, or chemotherapeutics produced by chemical synthesis) increase the rate and efficiency of production by protecting against bacterial diseases or by altering the inhibitory effects of microorganisms and allowing the animals to perform up to their genetic potential.

## Hormone or Hormonelike Substances

A number of estrogenic or progestational substances are being effectively used as implants or as feed additives for the purpose of stimulating rate and efficiency of growth, particularly in beef animals. Summaries of the responses to these agents by Overfield and Hatfield illustrate the substantial improvements these substances bring about in performance of grazing cattle or cattle finished in the feed lot. The improvements in rates of gain range from 9 to 14 percent, and that for feed efficiency range from 8 to 10 percent.

Estrogenic substances alone have been ineffective as a growth stimulant in swine, and androgenic substances have been unsatisfactory because of the undesirable effects on meat quality. However, a combination of androgen and estrogen (methyl-testosterone and diethylstilbestrol), which is approved for use in some countries, has been effective in improving the rate and efficiency of body weight gain while reducing carcass fat in pigs.

Recent developments in producing protein hormones by microbial synthesis should result in adequate quantities of growth hormone at a price low enough to permit its use in animal production. There appears to be great potential for use of a bovine growth hormone to increase growth rate and milk production. Although not adequately tested as yet, the growth hormone should also be effective in improving the efficiency of pig production. A great deal of research is needed to establish optimum dosage, length of response time, and adjustments needed in feeding and management programs before widespread application can be expected. Also, a satisfactory delivery system is needed to provide the optimal level on a continuous or controlled intermittent basis. Developing the delivery system will likely be far more difficult than establishing dosage and response time.

Evidence from short-term trials indicates that milk production may be increased by as much as 10 to 25 percent. L. J. Machlin, a Monsanto researcher, and P. J. Brumby and J. Hancock, New Zealand researchers, continued cows on treatment for 10 to 12 weeks and noted a sustained improvement from growth hormone throughout the test period.

The potential benefits from administering exogenous growth hormone has been recognized for several years, but the advent of microbial synthesis has brought application much closer to realization. The accomplishment of microbial synthesis of insulin and growth hormones opens the door for a new era in regulating growth and lactation in farm animals. Some suggest that direct application of recombinant DNA techniques may make application of exogenous growth hormone obsolete before production and delivery systems are perfected. However, regulation of exogenously produced and administered material may prove to be far more easily accomplished than exercising control of the endogenously produced hormones.

## Antimicrobial Agents

For more than 30 years antimicrobial substances have played a major role in controlling diseases and as growth promotants in poultry, swine, and cattle. A search for more effective drugs, or more effective combinations of existing drugs, is continuing. Such antimicrobial substances improve the efficiency of production by reducing mortality and morbidity in disease situations or by improving the rate and efficiency of growth in the apparent absence of disease. Comprehensive summaries describing their effects and the various factors that influence their effects abound in the literature.

One relatively new application of the ionophore drugs (polyether antibiotics that have been used primarily to control coccidiosis, a disease, in poultry) is as a rumen metabolic regulator in beef cattle. For many years we have searched for ways—drugs and diet manipulations, among others—to alter the rate and pathways of rumen fermentation. The ionophores accomplish this and also improve performance of beef cattle, but at this time one cannot credit the total response from ionophores to their effects on the rumen fermentation since control of coccidiosis in cattle may also be of benefit. However, the improvements in feed conversion on high-energy diets and the improvements in both rate and efficiency of body weight gain on high-fiber diets are of sufficient magnitude to have a great impact on beef production. When added to high-concentrate diets, the ionophores improve the rate of gain by about 3.6 percent and feed conversion by 6.5 percent. Much greater benefits can be expected for cattle on high-roughage diets.

Possibly of even greater significance than the responses observed to date is the model the ionophores provide for altering rumen fermentation. In general, they depress microbial protein synthesis and methane production and increase the amount of propionic acid produced without appreciably altering total volatile fatty-acid production, resulting in improved energy utilization.

The enhanced performance on diets high in fiber can greatly reduce the concentrate feeds required for finishing beef cattle. Although we do not have support data at the present time, it is anticipated that these or related compounds may have a similar beneficial impact on milk production by dairy cows and on meat and wool production by sheep and goats.

As mentioned earlier, the potential for improving milk production to levels of 9000 to 10,000 kg per cow per year is realistic. However, to produce that level of milk we are using diets that contain 55 to 60 percent high-energy concentrates. It has been estimated that milk production on all forage diets may be limited to 4500 to 6000 kg. Reid predicted that forages of 60 to 65 percent digestibility will support 5000 kg of milk per cow per year and that an increase to 70 percent digestibility would allow for production levels as high as 8000 kg per year. The ionophores increase feed efficiency for the growing animal by about 10 to 20 percent, a magnitude that should have a tremendous impact on milk production by cows on all forage diets.

## MODIFYING FEEDSTUFFS

It is well recognized that a very significant portion of the concentrate feeds used in livestock production could be used as human food. Significant gains in total food production can be accomplished by increasing the productivity from those materials that cannot be used directly by humans. A discussion of the potential impact of plant breeding and cropping systems on animal production is beyond the scope of this presentation and will be covered by others. However, I would like to mention two areas of research that will serve to illustrate the tremendous potential for more effective utilization of the feedstuffs now available.

### Treating Crop Residues

In the United States we have more than 200 million metric tons annually of crop residues from corn, wheat, and grain sorghum production, most of which are poorly utilized. The world supply of such materials has been estimated at more than nine times the amount in the United States. Certainly some of this material is used effectively at present as feed, fuel, and fiber, and it is essential that some be returned to the soil. However, there remains tremendous potential for increased animal production from these crop wastes, especially if we can develop practical methods for improving digestibility.

Treating crop residues with sodium hydroxide, for example, will increase the digestibility by as much as 20 to 30 percent, and such treatment will markedly improve utilization of fibrous feeds by ruminants. Treating rice straw or corn cobs with sodium hydroxide results in more than doubling the rate of gain by cattle fed diets high in fibrous feeds (72 percent rice straw or 80 percent corn cobs) and reduces the feed required per unit of gain by 50 percent or more.

Because of its handling qualities, sodium hydroxide treatment is not the final answer to effective utilization of low-quality roughages, but such experimental evidence illustrates that a great deal of food could be produced from materials that are not used effectively at the present time.

### Regulation of Plant Growth

Many of our forage crops progress through a growth and maturation cycle that markedly affects the nutrient availability and acceptability to ruminants. This maturation process is necessary if seed production is a primary goal; but for much of our land area it is desirable to maintain the plants in a vegetable state throughout the grazing or forage-harvesting cycle.

For some time our University of Kentucky researchers have been evaluating the effects of a plant growth regulator, mefluidide, on the productivity of tall fescue (*Festuca arundenacea*) and on the performance of animals grazing the treated pastures. D. G. Ely and other researchers at the University of Kentucky have treated fescue pastures in early spring with mefluidide (280 grams per hectare). The productivity of cattle grazing the treated pastures and the digestibility of the forage are markedly improved. The mefluidide improved digestibility of the regrowth by an average of 6 percent and improved performance of cattle or productivity per hectare by 17 to 18 percent. The differences between treated and nontreated pastures are even more pronounced during the hot summer weather when the quality of fescue is lowest. The total productivity of fescue (dry matter per hectare) does not appear to be significantly affected.

Our success in meeting the world's food needs is highly dependent on a full knowledge of the chemistry of plant and animal growth. Furthermore, success is dependent on using that knowledge to manipulate and control those chemicals and chemical reactions needed to maximize productivity of animals in converting less desirable products into human foods. There remains great potential for increasing animal products for human use with current resources by applying research developments. When such developments are combined with the steady progress brought about by genetic improvement, improved nutrient balance, and more effective disease control, there is reason for optimism about the future contributions of animal products to the world's food supply.

# IMPLICATIONS FOR
# PLANT AGRICULTURE

## ROGER L. MITCHELL

As we look toward the twenty-first century as a focus for long-term planning for the science of agriculture, the following themes very much flavor our thinking.

1. *The Inexorable Growth in World Population.*    While the growth rate has slowed in the past few decades, it nevertheless stands as a primary and continuing challenge. The 4 billion persons now living, the 6 billion expected by 2000, and the 8 billion projected to populate the earth early in the twenty-first century represent an enormous challenge to the food production system.

2. *Resource Conservation.*    Conservation, especially of soil and water, is critical to the food production of the future. In the United States alone it is estimated that the 168 million hectares now in production and the 51 million hectares of potential crop land will all be in use by the year 2000. Soil erosion is considered a major problem on 50 percent of these lands; its control by a range of practices—terracing, crop residue management, conservation tillage—must succeed if we are to retain the resources on which future increased productivity can be maintained or increased.

3. *Post-Harvest Losses of Food.*    These losses deserve major attention and could be the basis for full treatment as one key element in meeting nutritional needs ahead.

4. *Genetic Engineering or Biotechnology.*    New developments are providing a sense of ferment and excitement and seek their place vigorously in the arsenal of approaches to research in plant agriculture. Genetic engineering is a useful point also as a case study for the current debate as to whether sufficient attention is being given to basic research in support of agricultural research.

Any writer of future projections is sobered by seeking to effectively predict the future. When we recognize that traditional agriculture was practiced for centuries and that our science-based, industry-supported agriculture is a phenomena of the past 75 years, we do not make predictions lightly because we are very aware that change can come quickly.

One final caveat is offered in setting the stage for this discussion: Neolithic humans domesticated crops over a 2000- to 3000-year period. During that time more than 3000 different plant species were used as food. In the ensuing centuries we have reduced that number to 29 crop species on which the world's people depend for most of their calories and protein. These 29 basic food crops are supplemented by about 15 major species of vegetables and a like number of fruit crop species that supply most of the vitamins and some of the minerals necessary to the human diet.

Plant agriculture serves as the basis of approximately 93 percent of the human food supply; the remaining 7 percent of the supply—animal products—comes indirectly from plants. Nearly 99 percent is produced on land; just over 1 percent is from the ocean. Thus, as we appropriately seek alternative sources of food and means to grow it, we must also maintain a solid recognition of the evolutionary processes that have brought us to the food production systems we now use and seek to balance efforts at change against these proven experiences.

As we focus on the major activities in plant agriculture which we view as providing the means to meet nutritional needs in the twenty-first century, a recent paper by Gary Heichel, Professor of Crop Physiology at the University of Minnesota, provides an excellent framework for estimating sources of advancement, both from crop management and plant breeding approaches. Heichel offers a five-tiered systems model as shown in Figure 1.

*Level 1.*   The relatively new discipline of molecular biology and other older scientific disciplines.

*Level 2.*   The science of plant breeding.

*Level 3.*   Production, management, and utilization.

*Level 4.*   Transfer of this knowledge to producers and industry (extension).

*Level 5.*   Producer and industry use.

One key point to remember is that the discovery of new knowledge in traditional scientific disciplines often requires several decades for transfer to practical application. Numerous reviews conclude that 50 to 60 percent of the yield increases of our major crops over the past half-century have been due to genetic improvement, with the remainder due to improved cultural practices.

Plant breeders have enhanced crop yields by the increasing shift of biomass from the vegetative to the reproductive organs in various crops (summarized as the harvest index). In several instances, perhaps most dramatically in wheat, this has been done while also reducing plant height to provide standability under higher rates of nitrogen. Further improvement in the harvest index of various species is probable following this technique. Interestingly, the partitioning selection process has had relatively little effect on total biomass, neither increasing nor decreasing it. M. D. Gale and C. S. Law suggest that the high yields of many of the semidwarf cultivars of wheat may arise because of other effects of gibberellic acid (GA), which are not blocked by the endogenous anti-gibberellin. The GA, for example, may disturb apical dominance—the con-

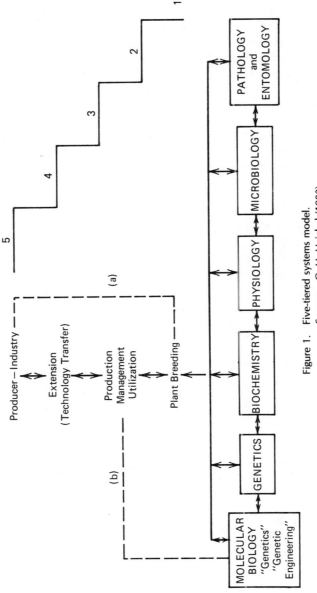

Figure 1. Five-tiered systems model.
*Source:* G. H. Heichel (1982)

trol of the plant apical growing point over the rest of the plant—so as to increase tillering or, at a later stage of crop development, to increase spikelet fertility. Thus plant breeders may be able to grow somewhat taller plants that still stand adequately until harvest, continue to enhance the harvest index, and by these combined steps increase yield per unit of land area.

As a related phenomenon, plant breeders may be able to enhance the nitrogen selectively moved to the harvested portion, thus improving the "nitrogen harvest index." Several related plant growth phenomena (e.g., faster rates of grain growth) can enhance harvestable yield. An understanding of the biology and manipulation of the genetics of these activities can benefit food production.

Recently, 22 scientists estimated the source of crop productivity increases for 2000 and 2030 A.D. Table 1 summarizes their expectations for 2030 in order of their optimistic predictions. (The order for the year 2000 was only moderately different.)

## Plant Breeding

Within the complex of conventional plant breeding practices, the collection of germ plasma and the preservation of genetic diversity are critical needs. Currently, there is considerable attention to this topic. Utilization of

Table 1.   Increase in Crop Productivity by 2030 A.D. Anticipated from Various Crop Technologies (1982 = 100)

|  | Crop Technology | Crop Productivity |
|---|---|---|
| 1. | Plant breeding | 135 |
| 2. | Crop combinations to conserve water | 133 |
| 3. | Genetic engineering | 125 |
| 4. | Growth regulators | 124 |
| 5a. | Nitrogen fixation | 118 |
| 5b. | Multiple cropping | 118 |
| 6. | Photosynthetic efficiency | 116 |
| 7. | Temperature acclimation | 113 |
| 8a. | Forage nutritional quality | 112 |
| 8b. | Crop maturity | 112 |
| 8c. | Transpiration suppressants | 112 |
| 9. | Intercropping | 111 |
| 10a. | Seed coating | 110 |
| 10b. | $CO_2$ climate modification | 110 |
| 11. | Grain nutritional quality | 107 |

*Source:*   G. H. Heichel (1982).

this genetic pool by chromosome engineering is presently a useful procedure for adding disease and insect resistance to crop plants from their wild relatives. Chromosome engineering involves the transfer of chromosomes, arms of chromosomes, and segments of chromosomes from one species to another. Such an approach may have the advantage over specific gene transfer ("genetic engineering") by moving the gene control system at the same time that the gene or complex of genes controlling a certain plant growth response of pest resistance is moved.

## Crop Combinations to Conserve Water

The more efficient use of water in all crop production environments will require continuing attention and serve as one of the primary means of increased food production. An analysis of major U.S. crops shows a large genetic potential for yield that is unrealized because of the need for better adaptation of the plants to the environments in which they are grown (particularly in mild, humid, temperate climates, where much of humanity's food is produced). Record yields are measured from three to seven times average yields.

J. S. Boyer reports estimates that losses to disease and insects depress U.S. yields below the genetic potential by less than 4 percent. The remainder must be attributed to unfavorable physiochemical environments caused by weedy competitors, inappropriate soils, and unfavorable climates. He describes detailed studies in which water potentials below a threshhold of about −11 bars cause inhibition of photosynthesis, transpiration, and nitrogen fixation in soybeans. Newer cultivars maintained higher water potentials by a more dense root system and thus were extracting water more effectively to maintain leaf turgidity. Thus, either by cultivar or species selection, farmers are able to utilize the limiting factor of water more usefully. This trait was selected unknowingly by plant breeders seeking increased yield; it can now be used as a screen in future selections.

## Genetic Engineering

Genetic engineering not only has attracted the attention of many in the scientific community, but has in recent years shown an extraordinary capacity to attract venture capital from the marketplace, perhaps more dramatically than any other single technology in our history. However, there is a need to weigh the likely uses against the proven successes of current approaches.

Genetic engineering will possibly make its most valuable contribution not by the recombination of specific traits into crop plants but from the application of these new techniques to the traditional discipline of plant breeding. To be successful in plant genetic engineering we must begin to develop an understanding of the elements that control gene expression. Thus any consideration of this area calls for an enhanced knowledge of the genome and the basic biology of crop plants, information the new technology can help supply.

Although the transfer of cloned DNA between microorganisms is now routinely carried out, the absence of convenient systems for transferring genes from one organism to another has inhibited similar successes with higher plants. Such developments may be forthcoming, and this arena will be a potential partner in food advances in the twenty-first century.

## Plant Growth Regulators

Plant growth regulators (PGRs) have met some success to date on grapes in California, small grains in Europe, cotton in the southwestern United States, and maize in South Africa. Once the target sites of growth regulators in plants are known, plant breeding selection may be conducted to create cultivars specifically responsive to given growth regulators. PGRs seem especially promising as a means to enhance harvest index and to decrease loss of developing seed. Thus they will enhance the size of the plant part that is to be harvested as well as increase water-use efficiency.

## Nitrogen Fixation

Nitrogen fixation by bacteria in symbiotic or free-living association with plants has been a major contributor to crop yields for centuries. Until the advent of synthetic nitrogen in about 1945, nitrogen fixation was the dominant source of new nitrogen. Increased costs of natural gas—the source of the stock to form $NH_3$ and the energy to carry out the process—have refocused our interests on the biological process. Genetic engineering is viewed as a potential contributor to such possible breakthroughs as developing nitrogen-fixing grasses. However, genetic engineering may make its first contribution by transferring the more efficient enzyme systems of bacteria that fix nitrogen gas ($N_2$) to $NH_3$ from certain strains to others more capable of invading current host plants. For example, improved strains of rhizobia presently available for soybeans will not succeed in infecting the roots in the presence of wild, less nitrogen-fixation efficient strains. Genetic engineering may allow transfer of these better enzyme systems to the wild strains.

Other, more modest steps can contribute effectively in this area; for example, increasing the root proliferation of alfalfa, thereby increasing the nodulation sites and, in turn, increasing the number of nodules and doubling the amount of nitrogen fixed.

### Multiple Cropping

Multiple cropping offers a more intensive use of a given land surface and suggests an actively growing crop present throughout the months of the year when the soil is not frozen and vulnerable to erosion. Multiple cropping can also be designed to have a given crop at the point in its reproductive cycle when maximum sunlight, day–night temperature differential, or other factors may favor increased yields. Multiple cropping of wheat and soybeans has shown productivity increases of 30 percent on land south of 40° latitude. (See Figure 2.)

### Photosynthesis

The ultimate value of plants is their ability to convert solar energy into stored chemical reserves. But experience to date indicates that rates of leaf photosynthesis have little relationship to yield. What seems to be more important thus far is the way photosynthetic products are moved to the harvested part of the plant. Until a more complete understanding of the complexities of photosynthetic and plant growth interactions are gained, however, genetic

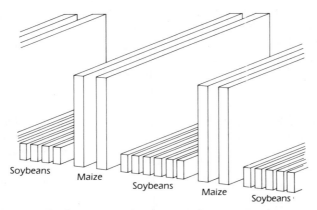

Figure 2.  Intercropping is a system whereby more than one crop is grown on a land area simultaneously. Farmers in China and Taiwan frequently alternate rows of short soybean plants with tall maize plants to produce more crop yield per land area. (*Source:* "Agricultural Production: Research and Development Strategies for the 1980's," Report for Conference, Bonn, October 8–12, 1979.)

engineering of photosynthetic processes may not be successful. So the primary need is continued study in the basic processes.

The other technologies mentioned in Table 1—temperature acclimation, forage nutritional quality, crop maturity, transpiration suppressants, intercropping, seed coating, $CO_2$ climate modification, and grain nutritional quality—will each play a role and justify some attention.

One additional area I wish to mention relates to nutritional value and the research efforts to improve amino acid balance, increase overall protein content, and enhance various other nutritional components. Experience suggests that it is most appropriate to seek maximum potential yields in a reasonably diverse set of crop plants rather than to focus on an ideal composition for any one cultivar or species and thereby to count on a varied consumption of plant and/or animal products to be the means for dietary balance.

Crop yield per unit area of land surface in widely diverse environments holds a continuing key to humanity's need for an adequate food supply. Various reports suggest that additional land resources in the world are limited, so the most effective use of land currently in use and expansion onto some less desirable land are important. Cultivars that produce better yield in the presence of toxic elements such as iron, aluminum, and manganese or in saline soils need to be developed. It is necessary that we understand the physiology of root reactions in these toxic conditions and breed to overcome them. Presently, major efforts to develop triticale and wheat utilize segments of rye chromosomes to enhance tolerance to more acid soils, in effect, soil higher in manganese and aluminum. Recent successes will allow 18 to 20 million hectares of soil in Poland and Brazil to be transformed from very low yields to upwards of 2 tons per hectare. Such increases make excellent contributions to the world food supply.

Mineral nutrition of the plant will continue to require increased understanding. We have alluded earlier to the increasing costs of nitrogen fertilizer and the resultant need to enhance biological nitrogen fixation. Nitrogen fertilizer is the most energy-expensive element plants use. It is six times greater in cost than phosphorus and sixteen times greater than potassium. This leads to strong interest in alternate sources of fixing nitrogen for plant use. One source is by direct release of hydrogen through the action of sunlight or water and then ammonia synthesis. The second possibility is in using the temperature differential between ocean deeps and the surface to gain energy to release hydrogen from water and then to fix ammonia.

Phosphorus supplies, while increasing in cost, are viewed as adequate for many decades. P. J. Stangel suggests that plants may be more effective in phosphorus uptake than previously thought: This is an area that calls for renewed study. In addition, mycorrhizae may be selected or added to the soil to improve the plants' capabilities to absorb phosphorus.

One estimate suggests that there is an 1800-year supply of potash; thus the primary issue is not supply but one of efficient use to minimize cost.

Our success in meeting the nutritional needs of the twenty-first century will come most likely through a steady development of current technologies, with creative use of new approaches, such as genetic engineering as a supplement. The complexity of the plant–soil–climate system does not favor easy one-step advances. Significant advances by more efficient use of current elements—we only get 30 to 70 percent efficiency from the nitrogen fertilizer in use now—will complement approaches of enhanced technologies.

1. We must conserve our soil and water resources to assure that these key natural elements are available to support productivity.
2. Plant breeding not only has supplied 50 to 60 percent of our yield advances to date, but it has also been the basis for allowing improved management practices.
3. Mineral nutrients are expected to be available in adequate supply. The major question will be the cost of nitrogen. Improved biological fixation by a range of approaches—for example, more nodules per plant and more efficient strains of rhizobia—is possible.
4. Crop management to assure season-long use of sunlight will assist in maximizing yield per unit area.

And beyond all of these considerations is the overriding recognition that food distribution is influenced very strongly by an individual's capacity to pay. Thus meeting nutritional needs will also depend on improving the economic status of all parts of the world population.

# 8
# CROPS FROM THE DESERT, SEA, AND SPACE

DEMETRIOS M. YERMANOS
MICHAEL NEUSHUL
ROBERT D. MacELROY

Farming in the future will offer much more than just tilling the soil. The sea, desert, and space hold great promise for increasing the world's agricultural production.

Today the world leaders in aquaculture are Japan and China. The Japanese, in particular, are concentrating in this area because for centuries large portions of their supply of protein have come from the sea. However, the fact that fisheries around the world cannot produce unlimited amounts of fish has led many nations to increase regulation of the fishing rights in their territorial waters. Presently, the Japanese are emphasizing the cultivation of "tame" fish in their own waters. Together with China, they will be in the best position in the future to sell technologies of aquaculture to such countries as the United States, which, with its plentiful, arable land and relatively cheap sources of energy, is just beginning to perceive the promise of tilling the sea.

One of the new frontiers of agriculture lies in the production of renewable resources. An exciting plant, jojoba, has the prospect of becoming an important

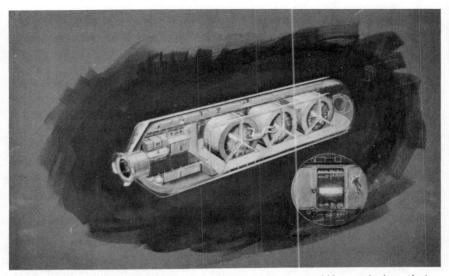

Plant growth module. This module containing plant growth units would be attached to a device, such as a space station, and would serve as a source of food and oxygen and as a sink for carbon dioxide. The plants would be grown in centrifuges to supply some gravity in the otherwise weightless environment. The inset indicates the light source at the center of the centrifuge. (Courtesy of Boeing Aerospace Company.)

source of lubricants and a possible replacement for the oils produced by sperm whales. Durable, renewable, cheap, and easy to raise, this plant could become an important crop for the semiarid regions in which it grows and yields lubricants, which must otherwise come from an endangered species or from our decreasing supply of fossil fuels.

With a space program that is still the best in the world, the United States can transfer back to earth knowledge learned from research outside the earth's atmosphere. It is not that we foresee the time when food raised in space will directly benefit humans below; rather, the agriculture we practice on spacecrafts to make them congenial environments for human beings will teach us more about agriculture on our planet. To give a spacecraft a life-support system suitable for long periods of time we will have to recycle many of the things taken aloft. Crops will be planted; they will use $CO_2$, give off oxygen, and the harvest will provide food for the spaceship's crew. In finding out how to regulate and control such a life-support system we may reach a new level of precision by which to govern some kinds of agriculture back on earth.

# JOJOBA: A SOURCE OF A SUPERIOR, NONPETROCHEMICAL LUBRICANT

## DEMETRIOS M. YERMANOS

The major concern of agricultural research has always been the development of technology for more and better food production. In recent years, however, we have witnessed the appearance of a new frontier in agriculture, the development of energy-related, renewable resources. Several plant species have attracted attention as potential sources of energy, lubricants, and rubber. One of these species, which has already entered the commercial world, is jojoba (pronounced "hohoba").

An extensive effort is now underway in several countries to domesticate jojoba as a potential crop for semiarid regions. In the United States alone over 30,000 acres will have been planted with jojoba by the end of 1983. Jojoba has long been known as a unique plant of the Sonoran desert and a potential industrial crop. Its scientific name, *Simmondsia chinensis,* is somewhat misleading in terms of its area of distribution: Natural populations of jojoba exist only inside the quadrangle formed by drawing lines from Riverside, California, to Globe, Arizona, to Guaymas, Sonora, Mexico, to Cabo San Lucas, B.C., Mexico. Jojoba has not been found growing naturally in any other part of the world. The area mentioned comprises about 100,000 square miles between latitudes 23° and 34° North and includes numerous disjunct populations, some having few plants and others having several thousand. These populations occupy elevations from sea level to 4500 feet and occur on coarse, sandy, or gravelly soils with good water penetration and a pH range of 5 to 8. Jojoba is a dioecious,* perennial, bushy shrub with a natural life span of at least 100 years and possibly over 200 years. In desert areas with about 5 inches of precipitation annually, jojoba grows 3 to 4 feet in height. Where rainfall reaches 10 to 15 inches, jojoba grows to 15 feet in height. Most jojoba plants have a spherical shape; in windswept coastal areas, a low prostrate growth habit prevails much like ground cover.

Jojoba has recently attracted worldwide attention for several reasons: (1) Its seed contains a liquid wax which, in addition to several miscellaneous uses, can serve as a replacement for sperm whale oil, a product derived from an endangered species. (2) It is an extremely drought-resistant crop. (3) It can grow in areas of marginal soil fertility, high atmospheric temperatures, high

---

* Plants are either male or female.

soil salinity, and low humidity. (4) It has low fertilizer and energy requirements. (5) It can be grown and processed with commercially available equipment. (6) It appears to have potential in several semiarid areas around the world with high population density and limited employment opportunities where little else can be grown.

Jojoba plants are either male or female. Flower buds appear on new branches that develop from apical buds on older branches or from axillary buds. Since most of the new growth occurs when temperatures are mild, the greatest number of jojoba flowers appear in the spring and fall. At that time the male flowers enlarge, and the anther sacs turn yellow and burst open, releasing large amounts of pollen. At about the same time the female flowers start to enlarge; when their ovaries get to be about $\frac{3}{16}$ inch in diameter, their three styles are fully extended $\frac{3}{16}$ to $\frac{5}{16}$ inch in length. As long as the styles look moist, turgid, and green, they are receptive. Their receptivity declines as they start to look dry, wilted, and brown at the tips.

Pollen grains seen under the microscope have the appearance of deflated spheres. In this state they may be carried by wind currents to considerable distances: native plants of jojoba, bearing heavily, have been found more than 300 feet away from the nearest male plant. As soon as pollen grains land on receptive styles of female plants, they absorb moisture instantly, become turgid, spherical, and start to develop the germ tube which initiates the process of fertilization. In hot and dry climates the fertilized ovary of jojoba continues to grow in size until May, and the seed is ready for harvesting in late June. In cooler climates maturity may not be reached until the end of July, moving the harvest into August or even September.

As long as the jojoba fruit grows in size, its color is gray-green. As it approaches maturity, it turns yellow, then light brown, and finally dark brown. At that time the ovary walls are desiccated and deeply wrinkled; they split along the three carpel sutures, releasing the seed to fall to the ground. As the fruit color changes from green to brown, the moisture content decreases from about 70 to 9 percent. Each jojoba fruit consists of three carpels, and therefore it has the potential of producing one, two, or three seeds. If the ovule in only one of the carpels is fertilized, a single oblong, spherical seed develops per fruit. If two or three ovules are fertilized, seeds have the appearance of one-half or one-third of a sphere section, respectively.

Although various insects visit jojoba plants, it is believed that pollination is done entirely by the wind. At maturity, the seed weighs 0.2 to 1.5 grams, has a diameter of $\frac{1}{8}$ to $\frac{1}{2}$ inch, and contains 44 to 58 percent liquid wax. Jojoba wax can be extracted from the seed with equipment and techniques commonly used for other oil seeds. The wax is not a "fat" or a triglyceride but a "liquid wax," or wax ester, of high molecular weight. A wax molecule consists of one molecule of a long-chain alcohol esterified with one molecule of a long-chain fatty acid. Jojoba wax mostly consists of two such esters, one with 40 carbon atoms (30 percent) and the other with 42 carbon atoms (50 percent).

Jojoba wax has several advantages over sperm whale oil: It possesses no fish odor and, indeed, has a mild, pleasant, "nutty" odor; the crude oil contains no fats and little besides the liquid wax; it requires no refining for use in most industrial processes; it has a very high viscosity index and very high flash and fire points, characteristics that are important to industry; it takes up larger amounts of sulfur than sperm whale oil; it does not darken to the same extent as other oils on sulfurization; and, even if highly sulfurized, jojoba oil remains liquid in contrast to highly sulfurized sperm oil, which requires the addition of mineral oils to do so. One of the most important properties of jojoba is that it is undamaged by repeated heating to high temperatures.

The meal of jojoba contains about 30 percent protein; it also contains "simmondsin," a monoglucoside not found in any other plant species which acts as an appetite depressant when the meal is fed to animals. Simmondsin may find significant uses in the future in the pharmaceutical industry. At this time, however, it constitutes a problem because it makes the meal unsuitable for feed. Research is under way to develop methods of removing or neutralizing simmondsin.

Jojoba has an extensive and deep root system, it is nonpoisonous, it does not accumulate large quantities of flammable dry material under the canopy, it is perennial and evergreen, and it requires little care. In addition to its potential as a crop, these attributes make jojoba a valuable soil conservation and landscape plant for highway shoulder plantings, rest stops, city parks, and greenbelts around cities suffering from sand or dust pollution.

The prospects of developing jojoba into a profitable, energy-related, renewable resource appear to be excellent. Current expectations are that the small quantities of wax that will be available initially from cultivated plantations will be absorbed by the more lucrative markets, such as cosmetics, waxes, and possibly pharmaceuticals. The real challenge for jojoba will be to penetrate the vast market of lubricants. This may not be too difficult, however, because with the rapid disappearance of fossil fuels, cheap sources of lubricants will also be lost. Lubricants will be needed in the near future as badly as fuels. None of the new sources of energy now contemplated—solar, atomic, geothermic, or synthetic fuels— has lubricants as by-products.

Looking into the near-term future, it appears certain that if the plant science technology already developed for jojoba is applied, seed yields of 3000 to 4000 pounds per acre producing 1500 to 2000 pounds of oil per acre for 7- to 8-year-old plantations are within reach now. Information is currently coming in from the 1982 harvest of 3-year-old commercial plantations in California and Arizona. In all cases seed yield has been higher than that obtained under Riverside conditions from plants of comparable age. Overall yields range between

30 and 80 pounds per acre. In one exceptional case a 40-acre field produced in excess of 200 pounds of seed per acre. With continued genetic improvement and with vegetative reproduction of superior materials expected to start in a year or two, these yields will soon be easily surpassed. When this happens, jojoba will not only have established itself as a new, energy-related, profitable commodity for the semiarid regions, but, even more importantly, it will have become a symbol of how new, ecologically sound, renewable resources may be developed to meet the energy needs for future generations.

# NEW CROPS FROM THE SEA

## MICHAEL NEUSHUL

It is generally agreed that world population and economic growth will bring us to the threshold of resource limitation and that there will soon be a transition from food abundance to food scarcity in world trade. By the middle of the next century food requirements in the developing countries could rise as much as fivefold over those required at present. It seems logical to predict that new marine crops will be grown to help meet the food needs of the twenty-first century.

Marine farming is a recent product of our own generation and it should continue to develop very rapidly given the technological advantages we enjoy and our present knowledge of the physiology and ecology of marine plants and animals. The development of marine farms will be speeded by financial support provided by farsighted governments and private industry. As examples, we need only remind ourselves of how the Rockefeller and Ford Foundations stimulated the "Green Revolution" in land-based grain cultivation. In the Orient, support from the Chinese Institute of Oceanology and the Ocean Engineering Association of Japan is leading to the development of novel marine farming techniques and the domestication and genetic improvement of new marine crop plants.

To illustrate how we might plant and harvest new crops from the sea, let us look first at historical trends that are likely to continue into the future and then observe how farming in the sea, like farming on land, will involve the modification of natural ecosystems. Specifically, I will explain how an open-sea ecosystem could be modified to produce anchovies and sardines and how a seafloor ecosystem could be farmed to produce plant biomass and fish.

Many of the spectacular advances made in land-based agriculture in the twentieth century grew out of research in the first agricultural experiment stations established in the 1840s at Rothamsted, England; Edinburgh, Scotland; and Mockern, Germany. By the last quarter of the nineteenth century a network of experiment stations had been set up in the United States as part of the land-grant college system. This system provided funding for state agricultural universities which have become centers for research in agriculture. The U.S. government in cooperation with the states now sponsors agricultural research in some 500 experiment stations throughout the country.

Based on the successful pattern of the land-grant college system, the Sea Grant Program was established in the United States in 1966 to develop and conserve the nation's marine resources. Scientists now working in newly designated Sea Grant Universities are conducting basic scientific research needed to effectively farm the sea.

By the 1950s, 70 percent of all cultivated land was planted with improved varieties that had not been in existence 20 years before. Eighty percent of the wheat, 86 percent of the corn, 92 percent of the oats, 95 percent of the cotton, and 98 percent of the soybeans that were grown in the United States in 1955 were new genetic varieties. Plant breeding proved to be an inexpensive way to increase crop yield and to produce disease-resistant plants that were able to grow well under less than optimum conditions. This has made it possible to grow crops in areas that were once thought to be unsuitable for agriculture. For example, drought-resistant wheat can now be grown in desert areas of Australia and North America.

Given the sophistication of plant manipulation on land, the further expansion of farming in the vast seas of the world now seems possible. This process has already started in China where genetically improved strains of a brown seaweed (*Laminaria*) are now farmed along much of that country's long coastline, where the plant does not grow naturally. We can expect that genetic studies will produce other new crops to be farmed in parts of the sea not now used.

At one time the resources of the sea appeared limitless. However, today more and more nations are attempting to harvest fewer and fewer fish. The sea, unlike the land, is almost impossible to police or fence, so that once an area is overfished it can take a long time to recover. The decline of the preferred fish stocks of the world has led to the development of a limited entry, or quota system, in some of the fishing grounds, which theoretically allows for the regeneration of fish stocks. However, entire fisheries have collapsed, for example, that of the Californian sardine or the Peruvian anchoveta. Hopefully, future management of natural fisheries will protect them so that ultimately resource enhancement through cultivation will be possible.

The introduction of a 200-mile fishing zone by many countries was an attempt to control access to fishing grounds. This action has spurred efforts to farm the sea, particularly in Japan, a country where a major portion of the protein in the nation's diet comes from the sea. In 1976 the Japanese initiated a 7-year plan with a $870 million budget to reduce their dependence on fishing in distant waters. The plan called for the creation of benthic farming areas where the sea floor would be cleared of existing vegetation to encourage the settlement and growth of selected plant and animal species. The Japanese also experimented with digging coastal channels to be used as farms, treating the intertidal region with caustic soda to clear away existing vegetation, and sandblasting rock to encourage the growth of desirable species. Many kinds of floating rafts, nets, and long lines were developed for use as substrates for off-the-bottom cultivation of seaweeds and invertebrates. The Japanese have also developed large floating fish pens which can be pulled down below the sea surface for protection during storms. These were first tested in 1975 and are now widely used for the cultivation of tuna and yellowtail in unprotected offshore regions where normal floating pens could not be used.

The Japanese propose to double the annual output of their marine fisheries products to 4.8 million tons. They intend to create 2500 artificial reefs and develop one-fifth of their coastline into fishing zones. They also have an ambitious program underway to develop seaweed farms for the production of biomass energy in deep offshore water. This program, funded in 1981, calls for the expenditure of approximately $42 million to design cultivation, harvesting, and processing systems to produce seaweed biomass that will be converted into biogas by anaerobic bacterial digestion. As of 1970, there were about 130,000 acres of sea surface under cultivation in Japan and about 25,000 acres under cultivation in China. This area has increased over the past decade. There is no doubt that in view of their recent successes the Japanese and Chinese will continue to develop their new marine farming technology.

Recent historical trends clearly indicate that part of the "world fishing ground" will be transformed into sites for the marine farms of the twenty-first century. However, the "forests and grasslands" of the sea are still relatively unmodified ecosystems and the would-be marine farmer is still working mostly with undomesticated plants and animals. If we are to develop productive marine farms, we must become familiar with the biological features of the plants and animals that we wish to domesticate. We must develop methods for seedstock production and genetic selection and learn to understand the marine ecosystems that we propose to modify as farms. Two possible farms could be established in the United States in the future: an open-sea planktonic farm offshore where fish would be grown, and a near-shore, on-the-bottom benthonic farm where seaweeds would be grown.

If such farms were established in southern California, the open-sea farm would be influenced by the California Current and the seasonal upwelling of cold, nutrient-rich deep water. This offshore region is highly productive and has in the past supported a major sardine fishery. It is now a source of anchovy. The near-shore farm would be a modified kelp bed ecosystem. Recent experiments, which we hope will lead to the domestication and farming of the giant kelp, will serve to show how this seafloor ecosystem might be modified for marine farming.

Although one might reasonably assume that the ocean is a well-mixed body of water, it is not. The organisms in marine ecosystems are usually confined to discrete masses of water, either swimming in them or sinking through them. Various types of water masses are present in the sea, ranging from the large wind-driven gyres that impinge on the coasts of continents and become coastal currents, like the California Current or the Gulf Stream, to smaller circulating bodies of water produced in the uppermost water layers by wind blowing along the sea surface. Nutrient-enriched "patches" of water a millimeter or less in size are thought to influence the growth of microscopic plankton. The upwelling of deep water, changes in surface circulation, and mixing all modify these patterns.

The larvae of planktonic organisms and the food that they eat exist in discrete water masses, where their success is probably determined largely by differences in larval mortality during the period immediately after the eggs hatch. At this time they are both vulnerable to predators and dependent on adequate amounts of appropriate food organisms being available. Oceanographer Ruben Lasker has studied anchovy larvae and has concluded that they will survive only if they hatch in a patch of food organisms and away from predators. In order to survive, an anchovy larvae must consume about 230 cells of the microscopic dinoflagellate (*Gymnodinium splendens*) per day. They cannot survive on other microscopic cells (*diatoms*) as their only food source. Thus if discrete water masses form and patches of dinoflagellates grow in them, then anchovy larvae can feed and survive after hatching. This results in a large anchovy population that grows and ultimately provides an abundant catch. However, when winds and storms produce upwelling and mixing that disperse the discrete water masses and the populations of food plants in them, then the anchovy larvae die after the eggs hatch and the subsequent catch will be a poor one.

Infrared thermal imagery from a satellite has recently been used to show that when winds from the north blow off the California coast, large areas of coastal ocean mix with the cold water (less than 14°C) that is brought up to the sea surface north of Point Conception. The satellite images obtained daily were used to direct an oceanographic ship from which collections were made, thus showing that spawning anchovies avoid cold water. Instead, the fish spawn in warmer near-shore water (16°C). Measurements of spawning intensity and water temperature indicate that the fish select a large pool of stable warm water, where discrete water masses form, as a site for spawning—a behavior pattern which has obvious adaptive value.

The satellite provides a synoptic view of a vast area of sea surface and records the conditions of the entire spawning area of the anchovy during the spawning season. With this new synoptic view, a prediction of the location of the heaviest concentrations of anchovy eggs could be made. Knowing the food requirements of the first-hatched larvae one could sample and test for the presence of nutritionally adequate food. The marine farmer interested in ensuring that a large number of larvae survive would probably elect to preseed the stable warmer waters with dinoflagellate cysts in order to be sure that the pastures were ready when the larval fish hatched. The reproductive cells of dinoflagellates or other suitable food plants might be grown in a land-based or near-shore hatchery and then spread from boats or even from airplanes. Thus the marine farmer seeking to cultivate an open-sea ecosystem to produce anchovy would have to make sure that anchovy eggs, dinoflagellates, and appropriate water stability were present at the same time and place. Upwelling areas would have to be detected and avoided.

The question of ultimate crop ownership and control of harvesting activities along with other economic factors would have to be dealt with. These questions are easier to approach in the case of benthonic farming, where the seafloor can be leased from the government and, like crop plants on land, the cultivated organisms can be planted in rows and harvested.

Submarine forests of giant kelp, *Macrocystis*, are circumpolar in the Southern Hemisphere but are found in the Northern Hemisphere only along the Pacific coast of North America. The Pacific coast kelp forests were first surveyed and mapped between 1912 and 1915 by the U.S. Department of Agriculture. The giant kelp plants produce masses of vegetation that float to the water surface forming a canopy. The kelp bed canopies were harvested early in the century as a source of potash and are now harvested to produce algal gums and mucilages used as thickening and emulsifying agents in many products.

With the advent of aviation aerial photographs were taken of the Pacific coast kelp beds, and today the beds can even be seen in satellite photographs. Most of the beds have been studied underwater by scuba-equipped divers who began to explore them in the 1950s. These studies were stimulated by concern that the kelp beds were being overharvested. There is also concern that sewage outfalls, offshore oil development, and nuclear generating stations will adversely affect the California kelp forests.

Natural kelp beds can serve as models for what large-scale, near-shore marine farms of the twenty-first century might be like. Natural kelp beds are very productive and stable. For example, the kelp bed in Goleta Bay, California, covers about 1000 acres. It has ranged in size from 650 to 1160 acres since 1955, and it now has a standing crop of approximately 44,000 metric tons, of which about 5700 tons of material are dislodged during storms and deposited on the beach annually. One-quarter to one-half of the plants are lost and replaced every year. Estimates of yield from this kelp bed suggested that it would produce about 3.7 tons of dry ash-free biomass per acre per year. (Yield is expressed in ash-free units because seaweed contains a large amount of noncombustible ash.) We might reasonably assume that a near-shore marine biomass farm would show similar self-renewing capabilities and, perhaps, a higher biomass yield if appropriately planted, cultivated, and harvested.

Whether kelp can be farmed depends on the answers to several questions. Can kelp plants stand up to repeated, severe harvesting? What amount of biomass do they produce? How do "wild" plants respond to high-density planting and fertilization? What is the plant-to-plant and season-to-season variability in growth? How can these plants be "cultivated" to enhance their health and yield?

In order to answer some of the questions, 722 plants (12 tons) were planted in the summer of 1981 on the seafloor at a depth of 7 meters off Ellwood, California. They were planted at three densities—one plant per 1, 4, and 16 square meters. Half the plants were sprayed from the sea surface 4 to 5 days a week with a fertilizer supplement, while the other half were not. The kelp canopy that grew up and floated near the surface was harvested every 3 months. For 10 percent of the plants growing fronds were marked when they first appeared. They were then measured and ultimately harvested (see Figure 1).

The results of these yield experiments have shown that the giant kelp withstands repeated harvesting and responds to cultivation. The yield from this first kelp farm was 10 to 15 dry ash-free tons per acre per year. This exceeded the yield that was mentioned earlier from a natural kelp bed (3.7 dry ash-free tons per acre per year). Close planting reduced light penetration into the farm and resulted in plants that had the lowest individual production and frond

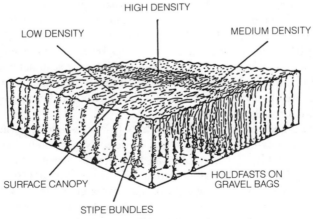

Figure 1.    Kelp yield test planting.

initiation rates and the highest mortality. At the medium-density planting, the plants maintained their frond number, while at the lowest density they had the highest individual production rate (see Figure 2). The test plantings were as stable as natural kelp beds and showed similar patterns of plant loss. The plants on the test farm responded to the application of fertilizer during the low-nutrient, high-irradiance months of late summer. When fertilized, the yield of the medium-density planting doubled and that of the high-density quadrupled.

Fifty-five tons of wet kelp were harvested in the first year of the experiment. Some of the plants consistently produced two to four times the average amount of biomass. This is very encouraging since, if it is possible to select and cultivate such high-yield plants, one could significantly increase farm production. Overall, the production rates of the kelp plants were highest in the spring because of the abundant nutrients and light and lowest in summer because of the high temperatures and low nutrients.

The yield of future kelp biomass farms might be as much as 20 to 30 dry ash-free tons per acre per year if we could devise cultivation strategies that would (1) reduce the rate at which the leaflike blades decay and fall off the floating fronds, (2) reduce the loss of individual plants, (3) allow us to select and breed high-yielding strains, (4) permit harvesting to increase the efficiency of use of light energy, and (5) develop efficient ways to apply fertilizer.

The establishment of commercial marine biomass farms in the future will

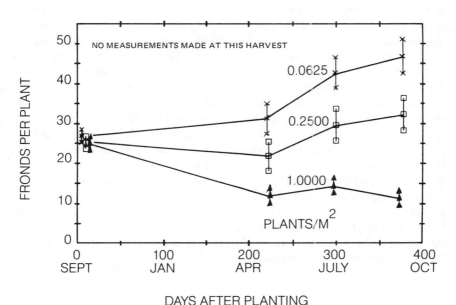

Figure 2.   The effect of planting density on frond production.

undoubtedly lead to conflicts between farmers and fishermen. However, since our test farm was found to be a nursery ground for many fish, it is likely that fishermen would enjoy larger catches if kelp farms were to be planted in areas where kelp does not now grow.

The Food and Agriculture Organization publication, *Agriculture: Toward 2000*, focuses on world population growth and the productivity of agriculture, forestry, and fisheries in attempting to predict what the future holds. The potential of large-scale marine farms is not considered, nor is information on aquaculture in China mentioned. Chinese aquaculture has been discussed by John Ryther who points out that Chinese fishermen harvest 3.1 million tons of marine fish yearly. In contrast, aquaculturalists grow and harvest 17.5 million tons of freshwater fish. If this type of intensive fish cultivation were extended into coastal marine waters, it is likely that marine fish production would be similarly enhanced.

While the technology to farm the sea is being developed in other countries, efforts in the United States are being curtailed. In 1980 Congress passed the National Aquaculture Development Act. Unfortunately, there has been some reluctance to provide the necessary funding, and only now is the Office of Management and Budget considering it. Moreover, the U.S. Department of the Interior, which previously endorsed the development of aquaculture, has now decided to close some 33 hatcheries, and the National Marine Fisheries Service has proposed to terminate some $4 to $5 million of aquaculture research. Furthermore, the National Sea Grant Program was proposed for phase-out and termination in the 1982 budget and was similarly targeted in the 1983 budget.

There are obvious advantages in developing a U.S. marine farming program. The Sea Grant Aquaculture Plan for 1983–1987 proposes to support the legislative mandate of the Aquaculture Act of 1980 to ensure the development of a national aquaculture industry. Such an industry would produce valuable fish products for domestic consumption since more than 60 percent of the fish products now consumed in the United States is imported. This imported food creates a trade deficit of more than $2.5 billion, or about 28 percent of the U.S. trade deficit exclusive of petroleum products. It is likely that increasing demand and further worldwide limitations of wild fishery stocks will result in even more expensive imports of fewer fish. In contrast with the Chinese aquaculture program, fish production via aquaculture in the United States is now only 3 percent of the total fisheries landings. The United States is clearly the world's leading agricultural country. It should also become a world leader in farming the sea in the twenty-first century.

# FARMING IN SPACE

## ROBERT D. MACELROY

It is unlikely that products from crops grown in space will ever contribute significantly to terrestrial needs. However, it is very likely that crops will be grown in space for local needs, most specifically to support space crews during long trips. Preparing for space agriculture will involve very fundamental reexaminations of how plants function on earth, and from such investigations is likely to come information of considerable value to terrestrial agriculture. The most immediate beneficiaries will probably be the greenhouse and controlled-environment agriculture industries, but certain concepts may be adopted by field agriculturists.

Agriculture is practiced under a large number of constraints, but most of these can be gathered into two major categories: demand and the economics of production. Both are strongly affected by energy costs and rely heavily on nonrenewable energy resources for irrigation, transportation, and fertilizers. The intensity of the labor involved in agriculture has decreased while the reliability of production has increased.

Although there will be vast differences between agricultural practices in space and on earth, both are driven by the same economic principles. It will be energy costs that determine whether agriculture is practiced in space. However, both costs and benefits must be calculated in very special ways.

The National Aeronautics and Space Administration has established a research program called the Controlled Ecological Life Support System (CELSS) program to integrate its efforts toward developing a biologically based regenerative life support system. What follows is a summary of current thinking on the possibilities of agriculture in space, mostly generated within the CELSS program, and a description of some of the peculiarities that space agriculture must address. The interesting problems of life support will be described, and space agriculture will be introduced in terms of an integral element rather than a pleasant amenity of a life support system. Concepts that seem likely to contribute to terrestrial agriculture will be identified.

When a crew is launched into space, it must carry the means to sustain itself in good health. The environmental parameters required to maintain health include: (1) enough oxygen for normal breathing, (2) maintenance of carbon dioxide at levels that are not toxic, (3) water and food that is available as needed for drinking and washing, and (4) facilities for proper disposal or storage of waste materials.

While the cabin of a spacecraft is isolated from the vacuum of space, it does leak gases to some extent. The ways in which the environment of the cabin is maintained have not changed appreciably for many years. Oxygen is released from a stored supply to maintain a concentration roughly equivalent to its concentration at sea level. Carbon dioxide is absorbed by lithium hydroxide to form lithium carbonate which is stored. Food is prepared before launch and is often rehydrated before being eaten, with the water being obtained from storage tanks. Waste material is stored until the return to earth.

As the size of the crews and the duration of space flights increase, the cost of resupplying a mission with the materials it needs also increases. The economics of resupply soon draw attention to the possibilities of reuse of at least some of the materials. Figure 1 suggests the "cost" savings in terms of the weight of materials needed to be resupplied to be gained by regenerating water, air, and water plus air during a mission that has a four-person crew with a crew rotation at 90 days and a mission lifetime of up to 14 years. Two methods are usually considered for reusing water: (1) evaporation under reduced pressure (vapor compression distillation), which leaves solids and organic materials behind to be stored; and (2) a critical water reactor. The reactor increases the temperature of the liquid waste material to about 400°C and increases the pressure with oxygen to about 3000 psi. At these temperatures and pressure organic materials are completely oxidized very rapidly. Normally, soluble salts crystallize out of solution, and when the pressure is released, carbon dioxide and nitrogen emerge as gases, leaving highly purified water.

Air regeneration techniques that have been considered include a group of chemical processes that essentially strip off $O_2$ from $CO_2$, leaving the remaining carbon as methane ($CH_4$), which can be vented to space.

## FOOD PRODUCTION IN SPACE

Recent studies have suggested that even long spaceflights relatively close to the earth, such as to space stations, may benefit from food regeneration. The model used in these studies is a vehicle with a lifetime of 15 to 20 years and a four-person crew changed every 90 days. Figure 2 compares three kinds of space missions. In a baseline mission that includes recycled water and air, the weight penalty paid for regeneration of 50 percent of the food would be equal to the weight penalty paid for resupply after a period of about 5.5 to 6 years. Regeneration of 97 percent of the food needed would "break even" after about 7.5 years, and a decision of whether to go with 50 percent or 97 percent food regeneration would be based on a break-even time of about 10.5 years. Such a mission might have scientific, military, and industrial justifications. In order to consider the problems of food production in space, it is worthwhile first to consider the constraints, other than the economic ones, that will affect the methods that might be used.

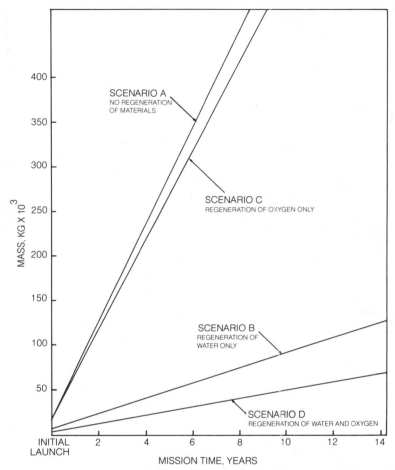

Figure 1.    Mass comparison of physiochemical systems mission: *Leo*—low inclination (4 people). (*Source:*    E. Gustan and T. Vinopal (1982), NASA Contractor Report 166420.)

A food production capability in space will use energy and most likely sunlight
to convert simple carbon compounds into the complex materials that can be
eaten. Theoretically, there are numerous ways to accomplish this. Energy can
be used to drive reactors that chemically synthesize food; or it can be used to
hydrolyze water to hydrogen and oxygen. The hydrogen and some of the oxygen
can serve as an energy source, allowing hydrogen oxidizing bacteria (e.g., some
species of *Pseudomowas* or *Alkaligewes*) to use carbon dioxide from the atmosphere to build cell materials. After processing to remove possible toxins and
nondigestible material, the bacteria could be eaten.

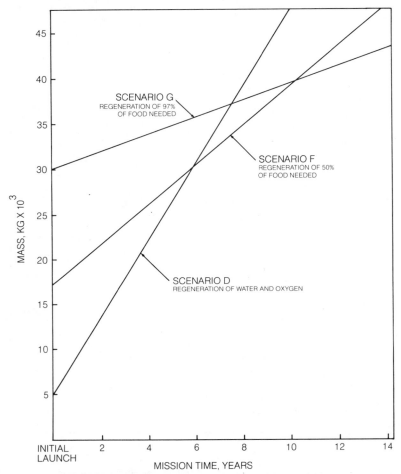

Figure 2.    Estimated breakeven time. Mission: *Leo*—low inclination (4 people). (*Source:*    E. Gustan and T. Vinopal (1982), NASA Contractor Report 166420.)

Alternatively, energy can be used to convert $CO_2$ to methanol on which certain yeasts can be grown, then processed, and eaten. Finally, energy could be used as light by plants. Of all of the possible ways to produce the food materials needed in space, plants seem to be the most appropriate. The use of secondary sources of food, such as animal protein from insects, chickens, rabbits, or fish, introduces severe complications because of the low efficiency of food production by these animals as compared to plants. However, many of the problems associated with such food sources may be eliminated by future developments for spacecraft in more efficiently collecting solar energy and in removing thermal energy.

Chemical synthesis of food is not yet sufficiently reliable to eliminate the possibility of toxic materials. Fungal and bacterial food sources have not been proven nutritionally adequate or to be reliable, primarily because of problems in processing them to eliminate biological toxins that often become apparent only after extensive periods of ingestion. Thus plants, both microscopic algae and higher plants, are of interest in space agriculture.

Unlike the products of terrestrial agriculture, the products of space agriculture *all* have value. These products include not only biomass (food), but also oxygen, transpired water, and $CO_2$-absorbing capabilities, all of them required in the small, isolated structure that a space crew will inhabit. In calling attention to the value of these products the central problem of space agriculture is emphasized: In a space agricultural system it is necessary to recycle all possible chemical elements. Only by recycling can the system remain stable and economically useful.

## THE CLOSED-SYSTEM CONCEPT

Except for the gases that will unavoidably leak into space and the people who will be periodically exchanged, the mass of chemical elements that will cycle in a spacecraft will be present at the start of the flight. Thus it is necessary to consider how to start up the system. For the purposes of the following short scenario the initial system will contain people, food, water, and oxygen for a period of 90 days plus the equipment necessary to maintain the composition of the atmosphere for 90 days. During this 90-day period crops will be planted, $O_2$ will be generated, and $CO_2$ will be absorbed by the growing plants. Bacteria, an inevitable part of the system, will begin to grow around the roots of the growing plants and consume $O_2$ and generate $CO_2$. As the plants grow to maturity, the food they have produced will be harvested, separated from the inedible portions, and stored or eaten.

This scenario immediately raises many questions about the dynamics of movement of materials. How much $O_2$ will the plants produce and how fast? How much $CO_2$ will be required by the growing plants and at what rates? How will $CO_2$ stored in the inedible biomass be released? How much $O_2$ will be consumed by the root bacteria? At what rate will denitrification occur? Can nitrogen be fixed to supply the plants with the quantities that they need? What seed germination scenarios are appropriate? How should excess $O_2$, $CO_2$, and $N_2$ be removed from the atmosphere and stored? Such questions as these must be addressed when planning for the initiation of a closed system. Other questions of an ecological nature must be considered. The systems will resemble a factory process system, and each of the components, including the biological elements, will be part of a system in which all individual activities will affect all other components.

## PROBLEM OF CONTROL

Any life support system, including those with biological components, must be integrated and capable of operating as a single entity. The predominant goal is to maintain the health and safety of the crew: all other considerations are secondary. The crew's requirements are stable supplies of food, water, and oxygen. To establish and maintain stability the system must be controlled.

The biological components, and particularly the plants, will be the major drivers of materials within the system, but because they will be operating under their own genetically determined control systems, they may behave in ways that are inimical to the integrity of the whole system. To prevent undesirable plant reactions more information must be obtained about their physiological responses and how plant metabolism can be controlled. The CELSS program was initiated for this purpose.

Plants respond to their environments in complex ways. Given stable environments they germinate, mature, and senesce in ways that serve the plant and are reasonably predictable. However, when we examine some of their requirements we begin to suspect that plants can be controlled sufficiently to allow them to act as life support system components, with their behavior dictated by the needs of the system. Plants require water, carbon dioxide, oxygen, mineral salts, fixed nitrogen, and light in certain quantities and with certain qualities at certain times of the day. In addition, they require water vapor (humidity) within certain ranges, some air movement, and also specific temperature ranges. These requirements vary significantly with the phases of the plant's life cycle.

Imagine, for example, a closed life support system in which it is desirable to decrease the rate of food production because sufficient stores of food have been accumulated. Such a decrease may be accomplished in several ways: the plants could be harvested or they could be "turned down" by adjusting the temperature or the light level or the nitrogen supply. While any one of these parameters might be manipulated, there is strong evidence that there is "coupling" among them: If the temperature is decreased during a certain growth phase, the result might be the initiation of a different growth phase. Or decreasing the nitrogen supply without altering the concentration of carbon dioxide might trigger a metabolic response that results in prevention of fruiting or even the death of the plant.

Obviously, a large amount of information must be collected about the behavior and responses of plants if biological management is to be a practicable aid in controlling a life support system. It is in these areas, among others, that preparations for agriculture in space may contribute to terrestrial agriculture.

A large number of materials must be recycled in a spacecraft containing a biologically based, regenerative life support system. The recycling will depend upon components that move or convert materials: people, plants, bacteria, and a waste processing system. The most likely waste processing system is a physical and chemical process, such as the critical water reactor mentioned above.

## CONCEPTS APPROPRIATE FOR TERRESTRIAL AGRICULTURE

Recycling materials in space involves regeneration (plants), consumers (crew), and waste processing. Stability in a recycling system will require that some materials be stored part of the time; thus reservoirs are essential. But only by holding the weight and volume of equipment to a minimum will the system be economically feasible. It is therefore important to reduce the amount of materials involved to the minimum safe level. This will mean recycling materials faster than on earth, and reducing the volumes of material waiting to be processed to the smallest practical levels. Natural, biological methods of controlling recycling are not appropriate. A large measure of cybernetic control will be necessary: namely, computers that can sense flow rates of many materials, that can predict the behavior of the system as a whole for periods ranging from minutes to years, and that can modify flows so as to bring the system to some stable state.

The introduction of computer control methods into a mixed biological, chemical, and physical system is fraught with problems. To date nothing so complicated has been attempted, and both the theoretical and practical aspects of control will have to be developed. The differences between control of a bioregenerative life support system and a petroleum plant are illustrative. Chemical processing is fast, the processes are well defined and easily measured, and the responses of the reactions to changes in conditions are reasonably linear. In contrast, a bioregenerative system has very slow response times. Its components, in particular the organisms, are not and never will be well-defined, and their responses to environmental changes are not linear. Nevertheless, significant progress is being made in developing the techniques to accomplish this task.

As the CELSS program has developed, it has become obvious that some aspects will have application to commercial agriculture. Of significance in this regard are studies of nitrogen utilization by crop plants during various phases of their growth, studies of crop density, investigations of the interrelationships between bacteria and plant roots, studies of hydroponic and aeroponic plant culture in controlled environments, and studies of the selection of higher plants as candidates for a bioregenerative system. Indirectly, work on developing system control may have wider significance. It is partially directed at the development of reliable models of plant growth, taking into account such factors as nitrogen availability, temperature, $CO_2$ and $O_2$ concentrations, and humidity. Such models have the potential for being used as predictive models, particularly in closed environmental agriculture. Finally, later products of the program, the control of selected crop plant species through environmental manipulation as well as the determination of maximum crop yields in controlled environments, promise to contribute conceptually to the problem of sustained, high-density agriculture.

It is doubtful that any crops grown in space could become commercially justified. A possible exception is a situation in which the separation of some important biological component derived from a plant can best be done in space. Most plants have adapted to the gravity of Earth and are able to cope with terrestrial ionizing radiation levels. In space, plant growth may have to be practiced within centrifuges capable of supplying some gravitational force (see Figure 3).

But the fundamental scientific work of space agriculture will uncover many plant characteristics that will be useful in practical, commercial agriculture.

Figure 3.    Close-up of plant growth module shows the seedling development centrifuge, plants growing, and cultivation. (Courtesy of Boeing Aerospace Corporation.)

Investigations will be done in environments that are not commercially feasible and which, therefore, have not been previously investigated. Responses to unusual $O_2$ and $CO_2$ concentrations are examples of environments that may prove economically interesting for controlled-environment and greenhouse agriculture. Another area of potential practical interest is likely to be the application of computer models and control to commercial plant production.

# 9

# THE MISSOURI PARTNERSHIP FOR ECONOMIC DEVELOPMENT

MAX LENNON

JAMES B. BOILLOT

JOHN T. MARVEL

The state of Missouri is providing through its *Food for the Twenty-First Century* program a model of coordination and cooperation among government, industry, and its university to advance the state's agricultural prosperity. As Melvin D. George, Vice President for Academic Affairs of the University of Missouri, explained:

> Agriculture may become the major worldwide function of the United States. We may no longer be the automaker to the world or the steelmaker to the world, and we may not be for long the computermaker to the world, either. But for as long as I can foresee we will be the world's best hope for an adequate diet. Although it may sound melodramatic, all of humanity depends on us.

We must all be aware, therefore, of the state of American agriculture today and its prospects for the future. I believe that we may be nearing the end of significantly increasing agricultural efficiency through applied research. A major theme of the Missouri planning effort is that the breakthroughs of tomorrow are likely to be in more basic areas such as biochemistry and genetics, which are outside more traditional agricultural domains and which will involve radical changes in the kinds of problems agriculture tackles and the ways solutions are developed.

In Missouri in a joint effort involving government, the University of Missouri, and industry, we are trying to lay a foundation for this new agriculture for the twenty-first century. For many years agriculture in this country, compared with agriculture elsewhere in the world, has been perhaps our best high technology industry. We must take steps now if we are to continue that high level of achievement, so that, blessed as we are with productive land and favorable climate, we may fulfill our moral obligation to the rest of humanity.

# THE ROLE OF
# THE UNIVERSITY

## MAX LENNON

Competition provides a healthy climate for economic growth, but there are situations when cooperation is a better alternative. America has long needed to bridge the gap between universities and industry. Partnerships can play an important part in planning as well as in efficient, economical production, particularly in the food and agriculture sector. For nearly 100 years the University of Missouri–Columbia has incorporated a triad of teaching, research, and service in its mission. Now, working hand-in-hand with government and the private sector, the University is assuring a flow of innovative leaders as well as technology for food and agriculture. This partnership is striving to preserve and enhance the United States' leadership role in world food production as we enter the twenty-first century.

*Food for the Twenty-First Century* is a cooperative effort within the University by the Colleges of Agriculture, Home Economics, and Veterinary Medicine to develop programs, initiate research, and cultivate leaders for the twenty-first century. The purpose of the joint effort is to analyze the future role of food and agriculture in the State of Missouri and to identify research areas that provide solutions to important future problems and constraints and also to recommend that these areas receive priority in the University of Missouri's long-range commitment to food and agriculture.

Food/agriculture is the state's largest industry, employing one out of four workers and ranking high nationally in production and export trade. Thus, the *Food for the Twenty-First Century* program is important not only to the University in fulfilling its commitments and responsibilities but also to agribusiness and government because of the potential it provides for economic growth. This broad, far-reaching program can help in the following ways:

- Stimulate the state's economy through access to new technology and thereby encourage growth and preserve Missouri's edge in the competitive market.
- Ensure an ample supply of quality food through efficient food production, a wise use of limited natural resources, and a better understanding of nutritional needs.
- Improve the quality of life and enable people to enjoy long productive lives by providing nutritional information.
- Train scientists needed to enrich Missouri's food and agriculture industry through a quality education program founded on strong, well-supported research.
- Encourage continuation and expansion of international cooperative programs and exchanges to provide an opportunity for growth in technology, leadership, and economic development.

*Food for the Twenty-First Century* began when an eight-member planning group was appointed by the deans of the Colleges to gather information and ideas. The group focused on three distinct resources—University faculty, corporate research leaders, and agribusinesspeople. Their discussions focused on agriculture in the year 2000 in order to determine which areas should receive research priority. In addition, farmers and agribusinesspeople statewide provided direction for the program through the College of Agriculture Commodity Planning Process in mid-1982. The planning group based its recommendations on fundamental needs and significant scientific opportunities and used the following criteria for selecting priority research areas:

- Demonstrates economic importance to state.
- Affords significant scientific opportunities with potential for major scientific contributions.
- Builds on present strengths of the University, including interdisciplinary work.
- Provides scientists for the future.
- Falls within the University's goals and responsibilities.

In order to plan for the future the constraints that will limit the food and agricultural industry in the decades to come were identified by the planning group. These constraints include the following considerations:

- New updated procedures based on an improved understanding of basic metabolic regulatory processes in plants and animals must be developed to achieve major advances in productivity and quality.
- Missouri's variable and unpredictable climate as well as the increased regulation of pesticides and concern for a safe environment will challenge the ingenuity of scientists. Techniques must be refined to combat stresses placed on plants and animals, and new methods must be developed for disease and pest management.
- The dependency of current agricultural production methods on limited petroleum resources must be considered realistically, and more energy-efficient plants, animals, and production methods must be developed.
- Continued erosion of topsoil must be reversed by adapting different production methods and crop plants with growth characteristics that help retain this valuable resource.
- The nutrient needs of specialized groups and the effects of nutrition on health and human performance must be better understood to educate the consumer and direct the food manufacturer to meet these needs.
- As the production, processing, and marketing of food becomes increasingly complex the demand for highly trained specialists will also increase. Care must be taken that the demand does not outstrip the supply.
- The variability of climatic and political conditions worldwide often results in an inadequate planning basis for food producers. Policymakers must be sensitive to this dilemma and establish the best possible policies to strengthen the agricultural industry.
- Current economic constraints may lessen public support for research investments, but the public must be educated to the importance of agricultural research so the foundation for the twenty-first century will continue to be built and not erode.
- Solving immediate problems seems to take precedence over long-range planning, but leaders in agriculture must keep the goals and challenges of the twenty-first century in the forefront.

Having considered the constraints, the planning group recommended that the three following major research areas receive emphasis in order to prepare Missouri food and agriculture for the next century.

## METABOLIC REGULATION IN PLANTS AND ANIMALS

The key to future increases in food production is understanding the mechanisms that regulate the metabolism within an organism. These chemical changes in cells and organs provide energy for growth, maintenance, and reproduction. A better understanding of these processes will result in improvements through genetics, adaptation to the environment, and effective use of new scientific technologies. Practical applications in plants and animals would also be numerous.

*Objectives*

1. To develop a more complete understanding of factors affecting metabolic processes in plants and animals with special emphasis on hormonal control.
2. To determine the physiology and regulation of metabolism under abnormal genetic or stressful conditions.
3. To study the interactions involved in resistance of plants and animals to disease, insects, and other pests.
4. To investigate the use of new biological "high tech" methods, such as genetic engineering, to improve growth and reproduction in plants and animals.


## ALTERNATIVE SOURCES OF FOOD AND ANIMAL FEEDS

As preparations are made for the twenty-first century, the basic understanding of conventional food production systems must be broadened to examine and include the nonconventional. Nontraditional methods should be evaluated for their potential for increased efficiency. But before any innovations are adopted safety and nutritional values must be evaluated and consumer acceptability determined.

*Objectives*

1. To explore new production methods and the potential of plant and animal species not now common in Missouri.
2. To find additional methods and approaches in order to more fully use potential food sources, such as wastes and by-products.

## SPECIALIZED HUMAN NUTRITIONAL NEEDS FOR QUALITY OF LIFE

As population increases put additional pressure on food supplies, it will become increasingly important to have human nutritional needs more adequately identified. Improvement in the nutritional well-being of humans is limited by lack of knowledge in three major areas: the nutrient needs for a lifetime of optimal performance, the nutritive value of food, and the factors affecting food acceptance patterns. Research in the future must focus on special groups. In Missouri the increasing "over 65" population is of particular concern. As the academic community plans for decreasing enrollment effected by the postwar baby boom, researchers must anticipate the needs of this group as older adults.

### Objectives

1. To determine the specialized nutrient needs of human groups who face special life situations.
2. To study how different life situations affect what and how much people eat.
3. To determine effects of personal diet choice on health and physical and mental performance.
4. To develop methods to increase adoption of improved nutritional habits.

Recognizing that innovative leaders are needed to implement progressive plans for the twenty-first century, the University of Missouri–Columbia, with the assistance of the Kellogg Foundation, inaugurated a leadership development program in mid-1982. This joint venture is designed to tap agriculture's greatest resource—its young men and women.

As with *Food for the Twenty-First Century*, the leadership development program focuses on the alliance of industry, government, and the academic community. Additionally, the program and its related activities are designed to close the gap between rural and urban interests through interaction of potential leaders and policymakers from the farming and agribusiness communities. The future of farming and its role in the national and international markets will be a common denominator.

The curriculum for each class of 30 young adults is not intended to teach agriculture. Rather, it will stress the part food and agriculture play in our cities, states, and nation as well as their importance to the world. The 2-year program includes seminars and national and international travel experiences. Each session is designed to broaden the participants' outlook and expand their understanding and appreciation of the economic impact of agriculture in the world.

Missouri realizes that strong research and educational programs, enhanced by new linkages with agribusiness and government, are essential to retain its position of prominence in the nation and grasp opportunities in the future. An atmosphere attractive to high technology firms must be cultivated. As competition for "high tech" industry sharpens, the availability of quality education becomes increasingly more important.

Thanks to *Food for the Twenty-First Century* specific linkages between the University and several agribusiness firms are being developed. The benefit to the University will be realized in several ways, including graduate training program enrichment through expanded visiting scholar opportunities, postdoctorate research support, and support for graduate student fellowships and research equipment. Industry will benefit from the expanded pool of highly trained scientists and the improved access to the ideas that emerge from the academic process. As a result, it is likely that ideas resulting from university research will move more quickly into the marketplace and provide more opportunity for economic development.

With a solid commitment to quality education and sound economic development, Missouri and the United States can retain a leadership role in food and agriculture in the twenty-first century.

# GOVERNMENT'S CONTRIBUTION TO MISSOURI'S AGRICULTURAL DEVELOPMENT

## JAMES B. BOILLOT

Any discussion related to incentives for economic growth in the rural areas of Missouri and wise use of the state's land resources must consider the state's complex political structure. Missouri's agricultural base is diverse in character. The northern portion of the state is typically Midwestern; the southeast Mississippi Delta is historically Old South; and the Ozark Mountain area stands for conservative independence. The state has two major metropolitan areas: St. Louis on the eastern border, with its industrial base dominated by aerospace, major agribusiness concerns, and automobile manufacturing; and on the western border, Kansas City, "The World Food Capital," which is a major hub of commerce in the Midwest.

Our Midwest flavor causes some to see Missouri as the "backdoor to the East," while others consider it to be the "gateway to the West." The influence of the South mixes with the conservative independence of the Ozark Mountain heritage and provides a unique challenge to develop a program serving all our people. Our intent to improve the quality of life for all Missourians will require the people of our state to think in terms of "Watch Me" rather than waiting for others to "Show Me." A successful effort will require the coordination and dedication of a university system dedicated to the future, an optimistic private sector willing to assume risk, and a state government possessing a genuine concern for the well-being of its citizens.

Government is the entity that can bring together leaders in our state to provide the direction and philosophy needed to guide the development of long-term plans. Leaders in the private sector and members of the academic community must join together to provide information and alternatives to policymakers, and policymakers must rely upon this information to guide their decisions.

It is generally accepted that government serves as a protector. The most visible functions of government are those of regulator, tax collector, and provider. We believe government should also accept the role of facilitator or motivator. In short, government should "make things happen"! We must address the variables of climate, geographical differences, and rural-versus-urban jealousies if we are to benefit from our state's labor force, natural resources, and ingenuity. Our effort is intended to provide incentives to business and the academic community, to encourage the personal involvement and imagination of our people, and to utilize state government as a coordinating force in bringing together the numerous divergent interests of our state.

The actions I will describe were taken without the benefit of an organized plan of action. Nonetheless, I believe they will contribute to our progress in addressing the concerns of the twenty-first century. Some of these actions result from initiatives that Governor Christopher S. Bond has taken to fulfill his commitment to expanded economic development as a major goal for the state. Others have resulted from the ingenuity and imagination of individual groups or from legislative initiative.

One recent action that utilized the cooperative efforts of government, industry, and the university was our effort to reestablish Missouri's leadership in the wine and grape industry which was destroyed during Prohibition. Because of the climatic and geographical assets which Missouri possesses and the leadership of a Wine Advisory Board, Missouri is once again preparing to establish its position as a wine-producing state. Today, through a coordinated effort among industry, the university, and state government, technical assistance is being provided to producers and restrictive legislation has been removed. Grape production is increasing dramatically, and through strong promotional efforts the quality of Missouri's fine wines is once again being recognized.

One of the more unusual actions we have taken is the development of what is commonly termed the Research Assistance Act. This was a legislative initiative for the purpose of promoting research projects in the university system which will enhance employment opportunities, stimulate economic development, and encourage private investments. The Act allows appropriated moneys for needed research projects to be matched by funding from the private sector. To qualify, research projects must be in the area of agriculture, natural resource management, industrial processes, or information processing, storage, and retrieval. The Act encourages the University of Missouri's Board of Curators to solicit appropriate proposals for research activity which will enhance employment opportunities in the state. The Act also encourages members of the university community to develop projects with appeal in the private sector and urges the private sector to join with the state in funding projects having significant impact upon the development of Missouri.

As we develop plans to guide our state's agricultural growth into the twenty-first century, this Act will play a significant role in encouraging the agricultural community to participate in research activity and provide the necessary data base for our desired economic growth. Although to date funding for this Act has been somewhat limited, we anticipate that the development of a long-term plan will generate needed support for accelerated funding activity.

Another legislative action, the Agriculture and Small Business Development Act, has a significant impact on production agriculture. This Act established a public corporation to promote the development of agriculture and small businesses in rural Missouri through the use of tax-exempt bonds. Funds are provided for capital improvement programs at interest rates lower than are obtainable otherwise. This Act should prove beneficial to young farmers seeking early capitalization for their operations.

In 1975 the Missouri General Assembly enacted legislation that provided differential tax assessment for agricultural lands and thus created a financial "break" to farmers and other Missourians owning agricultural land for production use. For the purpose of general property assessment, land that is actively devoted to agricultural or horticultural use, and has been so devoted for at least five successive years, can be evaluated on the basis of this use. This includes lands devoted to the production for sale of plants, animals, fruits, and other productive agricultural enterprises.

As urban areas expand into traditionally agricultural areas, we have recognized a need to protect those engaged in production agriculture. In 1982 the Missouri General Assembly enacted what is commonly referred to as a "Right to Farm" law, which protects farmers from the threat of suits by others moving into the countryside who may regard a nearby farming operation as a "nuisance." This law provides that any agricultural operation which has been in existence for more than 1 year may not be deemed a nuisance, as long as it is not operated in a negligent or improper manner.

Recognizing the need to provide financial support for rural development, the people of Missouri by popular vote authorized a $600 million bond issue to provide for new construction, repair, and renovation of existing state buildings, economic development, and soil erosion control measures. Although the adoption of this bond issue provides significant benefits in the short term, its impact on economic development and the control of soil erosion will provide long-term benefits for our state. Approximately $90 million of the $600 million bond allocation is earmarked for economic development. The language contained in the bond proposal regarding economic development restricts the use of these moneys to actions that will result in new or increased job opportunities. For soil erosion control, the bond provides $24 million in cost-share programs to enable farmers to construct erosion control structures and establish production methods that will reduce the loss of Missouri's soils.

Recently, government and private industry began a cooperative effort to develop an Agricultural Trade Cooperative that will provide numerous small-to-moderate-sized agricultural and agribusiness interests with an opportunity to join together to develop a mechanism whereby they can gain access to export markets. This cooperative will offer the necessary expertise and technical assistance to allow these firms to utilize opportunities currently benefiting major corporations. In conjunction with the expanded emphasis on export activity by the Missouri Department of Agriculture, the cooperative provides needed incentive for participation in the markets of the world by the agricultural and agribusiness communities of our state. We believe that any action to assist the agricultural community must consider the role which agricultural exports will play in our state's economic future.

There are numerous unanswered questions regarding future actions. Missouri's diverse background and multiplicity of interests necessitate the development of a coordinated plan. The preparation of this plan must involve all sectors and interests and must be built upon the assurance of adequate support for our academic community to provide the necessary data base.

To develop such a plan the Governor appointed a 20-member task force with representatives from agriculture, industry, state government, and the academic community. The task force has been asked to provide an inventory of the assets of rural Missouri, a review of the incentives which are needed to encourage the appropriate use of these assets, and a review of the factors which are limiting progress. We expect that the task force will consider those opportunities that have the greatest potential in the area of production agriculture and review the potential for development of agribusiness enterprises in our rural areas. The Governor has requested that the task force provide him with a list of legislative initiatives which will be needed to provide the necessary incentives and encouragement required for future agricultural growth. Finally, the Governor directed the task force to review current restraints and suggest those areas where revisions are needed in existing laws.

Missouri must provide a variety of incentives to return economic vitality to our diverse rural areas and help them make the transition to the twenty-first century. To assure equality of opportunity for rural Missourians, the state government must pursue those actions with maximum benefit. Although we can count numerous past accomplishments, we must now develop a coordinated plan to assure that future actions are consistent with the appropriate use of natural resources and that incentives are provided which will assist in assuring the opportunity for an improved quality of life and greater personal growth for rural Missourians.

# AN INDUSTRY PERSPECTIVE

## JOHN T. MARVEL

Missouri is at the center of U.S. agriculture in both the geographic and corporate senses. The state is not only at the crossroads of U.S. agricultural transportation routes; it also is the home of the world's largest feed manufacturer, the largest herbicide company, and the largest U.S. farm cooperative. With much of its most productive croplands along major waterways, the state is admirably located for direct sales of its grains and oilseeds for export. While Missouri is not first in the nation in any major crop or livestock category, it is in the top 10 in many.

To put Missouri's agriculture in perspective and to understand the major external forces that influence it, we must examine the national needs of agriculture. National policies and priorities, or the lack of them, have an impact on Missouri, the nation, and, indeed, the world—and we cannot ignore them.

I would like to briefly review the importance of agriculture, not only to Missouri but to the entire nation, and highlight just a few of the contributions research has made to the agricultural system in the past. Then I would like to address the role that biotechnology can play in improving agricultural productivity for the future. Finally, I will outline a number of areas where research has been inadequate. These shortcomings of research are the major limitation to the rapid application of biotechnology advances for improving farm productivity. This will underscore the sense of urgency I feel about increasing our commitment to funding basic plant and animal research.

Agriculture is this nation's largest industry. Its assets approach $1 trillion and its sales total some $140 billion. The 1981 corn crop by itself was worth $30 billion. Agricultural exports for 1981 totaled almost $44 billion, and they are a major factor affecting the balance of payments. This incredible bounty for the United States and the world is produced by less than 2 percent of the population who actually work on its farms. Their productivity enriches this nation far out of proportion to their numbers. The economic impact of agriculture alone makes it essential that we maintain and, in fact, increase the efficiency of our agricultural system to avoid erosion of our competitive edge with foreign crop producers. A side benefit of improved productivity would be that we could provide more food from less acreage. This would allow more land to be rotated out of production, thereby minimizing energy inputs, tillage-related topsoil erosion, and water use. Hence, we could minimize scarce inputs while maximizing conservation of our greatest natural resource—fertile land.

In addition to the compelling economic rationale for improving agriculture there is an even more urgent reason. Huge areas of the world are not as fortunate as we. Nearly one-third of the world subsists on a marginal diet, and literally millions starve to death each year. Population experts see a pessimistic picture of world food needs in the future and our ability to supply them. While the world's population grows at the rate of 210,000 people each day, constantly adding to the pressure on world food supplies, only 8 of the world's more than 170 countries consistently produce more food than they consume. This is an extremely small margin of countries that can provide food for the rest of the world. United Nations' experts tell us that the world's population will double in the next 35 years and that we will need to produce as much food during this period as we have since the dawn of recorded history.

Superimposed on the increased demand for food is the great variability and unpredictability of climate and pest problems, which cause and will continue to cause substantial losses in crop and livestock production. Heat, cold, drought, and floods will continue to exert pressure on farm yields. Increased regulation of pesticides and concern for a safe environment will challenge our ingenuity to rapidly develop new pest management methods. Today many current agricultural production methods depend heavily on petroleum energy sources, for example, tractor fuel, fertilizers, and grain drying. Such energy sources will likely become increasingly scarce and even more expensive in the future. As if this scenario were not bleak enough, topsoil erosion continues at an alarming pace, and the nation's largest aquifer, the Ogallala, is running dry. Clearly, these conditions warrant concern about our abilities to meet our future food needs.

Science and technology have already contributed greatly toward improved productivity in the form of better farm equipment, improved crop varieties, superior pest control technologies, and increased awareness of farm management techniques. Through agricultural chemistry and biology, extremely successful herbicides, fungicides, insecticides, and fertilizers have been developed. Without these technologies, crop losses would be as great as the total gained from 1981 agricultural exports! Consider the effect that would have on the U.S. balance of payments or on world food supplies.

Plant breeders, the first biotechnologists, have significantly improved the yield performance of all the major crop plants. For instance, the development of hybrid corn and breeding for efficient response to chemical fertilizers has increased corn grain yields by 300 percent since 1940. The research necessary to realize these yield gains through agricultural chemistry and plant breeding has been very productive and cost efficient. The public and private investment in agricultural research brings an estimated annual rate of return of 35 to 50 percent, making it an incredible natural resource for this nation and for the world.

However, much needs to be done. Past advances in plant and animal production were made by simple trial and error or by lengthy selection processes. The potential gains to be achieved using old technology are decreasing. New breakthroughs are needed to usher in the next "Green Revolution" required to meet world food demands in the twenty-first century.

Biotechnology will play a major role in catalyzing breakthroughs that will result in significant advancements in food production efficiency. Most of the early public attention paid to biotechnology focused on potential applications to human health—the possibilities for curing diseases that have been beyond the reach of modern medicine. But the potential applications of biotechnology to agriculture are equally exciting. For example, in years to come it may become possible to induce plants to grow in what are now considered hostile or impossible conditions, for example, extremes of heat and cold, salty or brackish water, and desert conditions. It may become possible to improve significantly the protein content of animal- and plant-derived food sources. A greater understanding of how plants and animals interact with the environment may reduce stress-related yield losses and level out the wide swings in food production. We may even see plants producing their own pesticides and providing their own fertilizer through nitrogen fixation. However, biotechnology will affect agriculture most immediately in two areas: crop improvement via genetic engineering and tissue culture and development of microorganisms that improve plant growth and control plant pests. These areas hold the most potential for improving crop productivity by the year 2000.

Microorganisms and their interactions with plants represent a frontier in plant biology. Very little is known about the role nonpathogenic or nonsymbiotic organisms play in crop productivity, although the roots of crop plants invariably contain tens of millions of bacteria per gram of tissue! The aerial parts of plants are also covered with numerous microorganisms in smaller populations. It is intriguing to speculate on whether these organisms can be harnessed by genetic engineering to improve crop productivity by aiding in pest control.

The idea of using microorganisms to control plant pests is not new: In fact, several commercial microbial products already exist—for example, Dipel®,

Thuricide®, and Elcar®. However, the revolution going on in molecular biology today gives us the ability to selectively move many desired traits into virtually any microorganism through genetic engineering.

To better illustrate the potential of this technology, consider the commercial product Dipel®, a bacterium that produces a protein toxic to cotton bollworms. If one could isolate the gene coding for this toxin and transfer it to microorganisms inhabiting the surface of the cotton boll, the bollworm, when it started to eat the boll, would ingest bacteria producing a toxin and presumably be killed. This type of pest control would be very selective, since only insects eating the cotton plant would be controlled. Furthermore, the protein toxin is effective at one-billionth of a gram on insects, but it is not toxic to mammals even at gram levels. The existence of such a billionfold safety factor would be a great improvement over current pesticides. Finally, if the boll-colonizing microorganisms survive all seasons, which is entirely possible, only a single application would be required. Contrast this to the six or more applications required to control the bollworm with normal chemical insecticides. It is easy to see that significant energy savings would result if such products can be developed.

As exciting as this type of product portends to be, the greatest benefit from biotechnology to agriculture will most likely come from unlocking the genetic secrets of plants themselves. Genetic engineering of plants will be the penultimate technique in breeding improved plant varieties. Among higher organisms plants are uniquely suited for genetic engineering because they can be grown as individual cells and subsequently regenerated into whole plants (Figure 1). This remarkable regeneration process is useful even independent of genetic engineering. For instance, millions of cells may be grown in culture in a very small area. Selections can then be made for cells with desirable new properties such as improved protein quality, herbicide resistance, or disease resistance. Selected cells can be regenerated into plants that have the improved properties selected in culture. This technology, termed "tissue culture," has been used to develop plants with disease resistance, improved cold or salt tolerance, and improved protein content.

An example of this technology, undertaken at Monsanto Company, is the development of Roundup®-resistant alfalfa plants. Alfalfa tissue was placed in culture, and the cells were exposed to a lethal dose of the herbicide Roundup®. Cells that survived this exposure were regenerated into whole plants, and these new plants (or variants) were transplanted into the field and treated with Roundup® as a farmer would use it. Several variants were found to have exhibited resistance to the herbicide in the field. In addition, the other agronomic characteristics of the plant—standability, morphology (structure), and disease resistance—appeared to be unchanged by the selection and regeneration process. The evaluation process is by no means complete, and several more years of

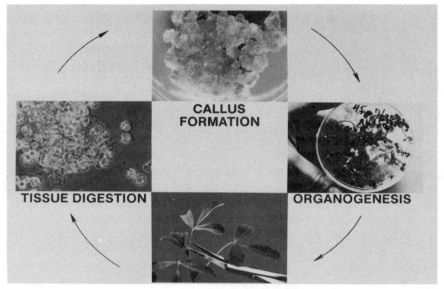

Figure 1.  Protoplast formation and regeneration.

field work would be necessary before this experiment could be declared a complete success. Nevertheless, this illustrates the potential utility of this technology for crop improvement.

However, there still remain a number of problems in the use of tissue culture alone. First and foremost, not all crops can be regenerated from tissue culture. Most notable among those which have not been regenerated are soybeans and cotton, although there appears to be some progress on these crops in the industrial sector. Some regenerable crops, such as potatoes, undergo dramatic changes during culture and may lose many desirable agronomic traits. Most crops cannot be regenerated from single cells, and selections from clumps of cells may produce highly heterogeneous regenerates which will be difficult to use in a breeding program. Furthermore, selection in culture of herbicide resistance, for example, may result in variants with many changes in addition to the ones we are seeking. Finally, genetic transmission through the seed of traits selected in culture does not always occur. Considerable applied and basic research will be required to fully explore the potential of tissue culture for crop improvement.

Genetic engineering offers an extremely selective method for modifying plants in culture. Instead of selecting for desirable traits by screening sheer numbers of cells for mutants (or variants), the gene coding for the desired trait is transferred directly into the cell. The resulting transformed cell now codes directly for the desired trait. For example, genes coding for antibiotic resistance have been inserted into antibiotic-susceptible bacteria, transforming them into antibiotic-resistant organisms. Similarly, if a gene coding for herbicide resistance were available, it could theoretically be inserted into a plant cell. This transformed plant cell could be regenerated into a whole plant which would be putatively resistant to herbicides. The potential advantage of genetic engineering is that a single, designed change can be made in a plant cell as opposed to the unpredictable number of changes which may occur while selecting variants in culture.

Recent research at Monsanto and several other laboratories has finally led to the transformation of plants using the bacterium *Agrobacterium tumifaciens*. DNA coding for resistance to the antibiotic Kanamycin was successfully inserted into tobacco and petunia cells, and expression of the trait was observed. While this is an extremely exciting result, we have a long way to go before we can insert genes into plants and control their expression at the right developmental time and in the right tissue. We need more basic research on plant gene organization and expression to be able to effectively manipulate plant genetics.

To illustrate this point, if the gene coding for lysine were simply inserted into a corn cell and this cell were regenerated into a whole plant, this plant might produce high quantities of lysine in all the tissues (leaves, root, stem, etc.). Since high lysine is desirable only in the kernel, considerable energy could be wasted by the plant in producing lysine in all tissues.

Perhaps the major limitation to genetic engineering of plants is knowing which genes we wish to transfer. There is a woeful lack of knowledge about plant biochemistry, and, hence, we do not understand the molecular mechanics of the traits we wish to transfer. This is compounded by the fact that most traits will be coded for by many genes and therefore will require more sophisticated molecular biology technology to transfer and regulate.

While we are enthusiastic and keenly aware of the enormous long-range potential of biotechnology, we feel that the greatest limitation to its successful exploitation in agriculture is the sheer lack of basic knowledge about plants. Our knowledge of mammalian biochemistry is sophisticated and has allowed immediate translation of biotechnology breakthroughs to products for better health care. The knowledge gap in plant biochemistry is too large to allow rapid exploitation of biotechnology in agriculture. The origins of this knowledge gap are abundantly clear.

A brief consideration of statistics, compiled by the National Science Foundation, shows that since 1972 only 2 percent of federal research funds were spent on agriculture. This is approximately half the moneys spent on the space shuttle during 1981! At a time when grain storage bins are bursting and farmers are paying a high price for their efficiency and productivity, it is tempting to overlook the fact that we must strengthen our agricultural system with increased spending for basic plant research, but this is exactly what we must do. Agricultural research is a lengthy process, and success in achieving major technological breakthroughs is neither guaranteed nor predictable. Waiting until food supplies become scarce will be too late to begin our research since it often takes 20 years of basic research to provide practical applications. Agriculture in the United States cannot stand still and allow foreign agriculture to surpass us, as has occurred in the steel, electronics, and auto industries. To maintain our leadership position we must begin to rectify the neglect of basic plant sciences at once.

What specific steps should be considered? We must increase basic research for plant science to discover new knowledge which will catalyze future practical breakthroughs. We must attract the best available new scientists to agriculture by providing long-term, stable funding. We must strengthen the basic research aspects of the USDA–ARS, and we must aggressively build new bridges between government, industry, and academia.

In 1978 the U.S. Department of Agriculture Competitive Grants Program was established for agricultural research. It is intended to complement and supplement formula funds, and it is designed to focus on basic research issues of high priority to the nation. These grants should be expanded to support a critical mass of scientists in the major scientific areas of high priority to agriculture; specifically, gene structure, organization, and expression in crop plants, plant development biology, plant–pest interactions, and plant–microbial interactions, as well as basic plant biochemistry. It has been estimated that federal support for basic plant research is just under $60 million per year. This amount should be quadrupled over the next 5 to 10 years. These expanded grants should cover graduate fellowships, postdoctoral fellowships, sabbaticals for established scientists, and exchange programs between academia and industry.

Sophisticated spectroscopy techniques need to be applied to plant systems. This has not yet been done because of the simple lack of funds available to bring state-of-the-art instrumentation to plant science laboratories. This should be corrected at once, and consideration should be given to establishing several regional centers of excellence in plant science where such instrumentation would be available.

While basic plant research is the long-term solution to the knowledge gap in plants and, hence, to the exploitation of agricultural biotechnology, academic–industrial collaborations could well be a mechanism for maximizing the application of our current, limited knowledge base. Creative new partnerships between business and the universities might provide highly fertile arenas for hybridization of ideas and may well facilitate practical breakthroughs. Partially aligning the best academic minds in the country on significant practical problems can only expedite our success in biotechnology. The government might even consider providing tax incentives for basic research collaborations between universities and industry.

Academic science could not have grown without large-scale support from government, and the continuation of that support is essential. Although the government occupies a pivotal position in financing basic research, its support will always depend on economic conditions, competing social needs, and political considerations. Industry and the private research foundations can and should act as buffers and protectors of basic research against varying levels of government support.

The successful utilization of biotechnology in improving crop productivity is blocked by high barriers, but given sufficient time and money they can be removed. Surmounting them could revolutionize plant breeding and agricultural output. The key is *research*. While the United States is a leader in basic and applied plant science, agricultural research has been given too little emphasis in the past. American agriculture must receive increased attention from government, academia, and industry *now*. Tomorrow's success in agriculture lies in today's research.

# III

# THE FARMER
# OF TOMORROW

# 10

# THE FARMER AS AN INNOVATIVE SURVIVOR

## ROBERT B. DELANO

American agriculture will remain free and highly productive in the twenty-first century. I expect that in many ways those on the land in the year 2000 will be better farmers and ranchers than is true today.

Tomorrow's farmers must be, and will be, better agronomists and herders, better conservationists, better financial analysts, better communicators, and much better marketers. They will be reluctant to deliver products to any market without full advance knowledge of price. They will insist on being involved in all aspects of marketing, including access to sources of up-to-date price information, orderly marketing through control of storage and transportation, hedging on the futures market, and pricing through negotiation.

The successful farmer of the future will be an exceptional manager of time and energy. He or she will operate from the proven base of family ownership, be politically wise, consumer-conscious, and world-oriented. Working with high technology and backed by more than three centuries of personal freedom on American soil, tomorrow's farmer will continue the historic pattern of invention and innovation in response to the traditional market incentives and signals of the American competitive enterprise system. Although often larger, more highly capitalized and automated, or different in other ways, the farm of the future will mostly reflect past development or extension of trends we see now. This includes the rapid adoption of high technology.

At one time a single farm family could barely support itself in the virgin wilderness of America. Within a century the wilderness had largely been tamed and migration from the farm was well underway. From water mills to horse-drawn hay stackers, farmers invented, developed, and used labor-saving devices. Farmers worked out the principles of the gang plow, the combine, and the silo, and were quick to use the power-take-off units of their early tractors to turn them into portable power plants for feed grinding, wood sawing, and silo filling.

Another century brought a growing partnership between farming and other industries and between farming and science. With these alliances came new crops, better livestock, and every manner of machine to replace muscles and brainpower and to boost farm productivity. As a result, one U.S. farmer now provides for about 80 people, 25 of them living in other countries.

Most nonfarmers still visualize rural homesteads as having a chicken coop for a small laying flock, a few cows for milk and butter for home use and for sale, a pen full of pigs, a large garden for food preservation, and a winter's supply of potatoes and squash in the root cellar. These "general" farmers, involved in producing something of just about everything, have almost disappeared into history. A small flock of chickens are now more trouble than they are worth. Instead, professional poultry people specialize in either broiler or egg production, but seldom in both. To make the enterprise pay, poultry producers must have thousands of chickens existing under near-ideal conditions. This is a matter of personal preference, a decision based on economics and labor. It came at a time when the economics of egg and poultry production required that one either expanded greatly or got out. Similar economic decisions were made in all parts of the agricultural community.

Some people profess alarm at this specialization of American agriculture. I do not. The day may come, however, when America's farming system is too rigidly overspecialized—when the economies of production will argue against production rigidity. If and when this happens, farmers and ranchers will proceed flexibly to reverse their high degree of specialization and return to whatever system the times demand. The freedom of management that allows this flexibility is the beauty of American agriculture and the story of America's success on the farm. The U.S. farmers' adaptive and innovative abilities allowed them to remain in the forefront of the industrial revolution and helped place a firm agricultural base under today's technological revolution.

Currently, U.S. agriculture is thought to be the leader among industries in the adoption of computer technology. Some researchers believe that farmers and ranchers know more about computers and use them more frequently than the "end users" of any other industry. On many farms a new generation of machines determines seed density and planting depth, mixes rations, feeds livestock, computes costs, and projects profit margins.

Rapid adoption of high technology by today's farmers gives a hint of what is to come. It is reasonable to assume that tomorrow "compubots"—computer robots—will do much of the work of agriculture. Memory units will allow farm machinery to "learn" jobs from human operators and later to perform as directed without human guidance. There will be built-in sensing devices for emergency modification of established procedures, with the computers using a selection of "best course" alternatives while signaling for human help. These learning-and-logic machines will log available resources, weather, and financial variables, check them against priorities, sort out appropriate conclusions, and, in their own voices, tell us our best farming options.

We can look forward to the development of an entire new science in monitoring. Monitoring the opinions of citizens, including farmers and ranchers, monitoring the internal temperature of fruit buds and blossoms, monitoring moisture in everything from grain to cloud banks, monitoring worldwide commodity supplies and prices—monitoring and more monitoring.

There will be much ado about reporting. Reporting trends in citizen thinking, including farmers and ranchers, reporting the moving frost that may kill one-fifth of the apple blossoms by lowering the internal temperature of fruit buds and blossoms, reporting the movement of moisture out of the grain kernels and cloud banks, reporting world crop conditions—reporting and more reporting.

This system of monitoring and reporting will limit many of nature's more deadly surprises, such as those that occurred around Mount St. Helens. It will reduce the effects of drought and flood. It is even possible, but not likely, that in the twenty-first century we will have reasonably accurate weather reports. At that point one U.S. farmer might well provide food and fiber for 200 people, half of them living in other countries. Let us take a closer look at that farming future.

In the twenty-first century we can expect the following:

1. More than 90 percent of all farms and ranches will remain family-owned and operated.
2. The current high levels of energy use will be greatly modified.
3. Income from fish production and other types of water farming will begin to compare well with income from farming the land.
4. Megacities will go "up" rather than "out," and farming "green space" will move up to the city.
5. Farmers will welcome and embrace "high tech" agriculture.
6. New technology and new consumer needs will bring a turnaway from "mono-agriculture" and a return to more flexibility in farming.

7. A consistent national farm policy will answer most of the worries about population and its drain on the food supply.

Where, in all of this, will corporation or conglomerate farming fit? The problem of corporate takeover has been greatly exaggerated, mostly by politicians who profess great concern about the imagined demise of the family farm. To date this concern has taken the form of anticorporate farming legislation in about eight states, with the people of one, Nebraska, going so far in the 1982 election as to prohibit by constitutional amendment "outside" corporations from further purchase of farm and ranch land.

Leaders of "save-the-family-farm" movements find it possible—especially in election years—to ignore the fact that the profile of U.S. farm ownership has remained virtually unchanged for at least the past 50 years. About 2 percent of all farms are "corporate" entities. Of these, about 85 percent are family corporations, with 10 or less stockholders. With something like a 4 percent return on investment, true corporate or conglomerate agriculture has been immensely unprofitable. After reviewing the narrow margins most corporate boards of directors become less than enthusiastic about farming as a high-return enterprise.

Nonfamily corporate farming has worked to a limited extent in high-value commodities, such as fruit (pineapple, citrus) and sugar crops, primarily in Hawaii and Florida. Even so, most of the corporate Hawaiian pineapple and sugarcane operations have closed shop and are moving out. It therefore becomes relatively easy to predict that more than 95 percent of all U.S. farms and ranches will continue to be family-owned and operated in the twenty-first century.

"Less oil, less toil, less soil." That perceptive description of a trend toward less energy use in American agriculture, expressed by the president of Deere and Company, will hold true for the future. The savings in oil and toil will come from simplified low-tillage or no-tillage operations which bypass the plow normally used in traditional seedbed preparation. The savings provided by the use of "less soil" will come from more productive varieties of crops and from the expansion of what I loosely call "water" farming, or hydroponics.

There are a number of versions of hydroponics in which crops are grown in nutrient solution rather than soil and of aquaculture which involves fish growing and shellfish culture. Although the hydroponic concept is attractive, it requires expensive, rigidly controlled growing conditions and is labor intensive. Hydroponics will remain impractical long after aquaculture has moved strongly into food production markets.

For the future I see the use of both more—and less—farm machinery: first, a gradual increase in no-tillage operations in which there is minimal equipment use to reduce soil and water erosion and to save energy costs, expanding to eventually involve up to 50 percent of all farming. Second, I also see the introduction of multifunction farm machines capable of performing up to a dozen operations simultaneously. This development may be along the lines of a new "maxicultivator" presently being field-tested by an Italian firm. The machine is capable of completing up to nine functions at the same time. It can aerate the soil while incorporating crop residue, smooth and level, create drainage, plant, fertilize, and apply herbicides and insecticides—all in a continuous movement across a field.

Farmers and ranchers have very real concerns about the security of an unbroken supply of the special fuels, oils, greases, and other petrochemicals needed to farm. In 1980 the International Energy Agency, a watchdog group, predicted the worldwide demand for oil would within 5 years exceed the available daily supply by 3 to 4 million barrels. The group also projected that within a decade the excess of demand over supply would reach 10 million barrels per day.

In his paper "A New Economic Agenda," written in 1979, W. W. Rostow noted that he believed we were "heading into a period when the 1977 energy shortage predictions of the CIA would come true; that is, the demand for OPEC oil would far exceed OPEC production capacity for 1983. . . ." Said Rostow, "The CIA projection may even turn out to be over-optimistic. . . ." As later events proved, even with the Iranian–Iraqi war and the continuing uncertainties of the Mideast, no one has gone begging for crude oil or petroleum products in 1983. What this proves is hard to say, unless it might have something to do with the unpredictability of predictions, including my own.

Nevertheless, I predict that U.S. farmers of the future will not court energy disaster by overreliance on fuels imported from hundreds or thousands of miles distant. Crops of the future will be designated for food, or fuel, with larger-scale, individual farmers and cooperative combinations of farmers capable of growing and processing their own fuel crops. Their emphasis will remain on supply reliability rather than on economy. Meanwhile, nuclear and coal power will prove to be the only practical large-scale energy sources to supply the nation's needs over the next century.

So-called "fish farming" may well be the food production growth area of the future, if for no other reason than because larger food creatures eat more and take longer to mature. To reach their productive stage cattle require several years and great amounts of grain and grass. Conditions vary, but it can be said with some degree of accuracy that it takes about 8 pounds of grain to produce 1 pound of beef.

Somewhat in proportion to smaller animal size, it takes less feed—perhaps 4 pounds and proportionately less time—to produce 1 pound of pork. Within months, fast-growing and short-lived poultry convert grain to protein on about a 2-to-1 basis—2 pounds of feed for 1 pound of chicken. If some way could be found to prevent the waste that is associated with water feeding, fish, especially catfish, could do even better. They might produce meat acceptable to humans at nearly a 1-to-1 feed ratio—1 pound of feed to 1 pound of fish. In addition, genetic manipulation may allow a number of future food animals to reach the same high meat-to-feed ratio now possible with poultry and fish.

It is not hard for me to predict that despite antianimal agriculture activism the consumer of the future will continue to want meat. Even in the twenty-first century most people will prefer to eat a pound of fish or steak to eating a pound of grain. It is possible that future economies of production will tilt consumer preference to more readily available farm-produced fish and away from beef, pork, and chicken.

Futurists at one time held that all meat production would move away from western range and rural feedlots and closer to where the consumers are. The theory was that it is easier to move hay and grain than it is to move cattle or meat. This relocation of animal husbandry has not occurred and quite probably never will. Livestock produce offensive odors and make noise. Most urban people think they are nice to look at, but other than in a few quaint European or African villages few people prefer to live with cattle. I doubt this will change.

The great architect, Frank Lloyd Wright, dreamed of and actually designed mile-high, city-sized buildings that spread the population up instead of out. Each single building was to contain both home and job with shops, manufacturing, entertainment, medical, and recreational facilities for thousands of people. Wright thought one such building, erected every 50 miles, surrounded by green and growing farmlands, could take care of all population needs.

There is a growing uneasiness among city people of the population-dense Northeast about the vulnerability of their food supply. They now must "import" more than 80 percent of their food needs. In response to this concern future metropolitan planners may adopt in some fashion Wright's concept of a single large city building surrounded by a farming greenbelt to protect the local food supply. Modern architects who have examined Wright's mile-high building designs judge them to be sound and their construction feasible. I would expect that some version of these supercity structures will be built sometime in the twenty-first century.

Farm operations, unlike factories, have always been multipurpose production centers. It is feasible that farms of the future will complete the cycle from specialized agriculture to much greater production flexibility. I rather expect present intense specialization within a narrow range of farm crops or livestock to relax somewhat in response to a fantastically complex and efficient telecommunications network. It will be easy for everyone to be fully informed about production and marketing opportunities outside their production specialty. It will include the use of every imaginable type and application of computers, electronic sensors, monitors, and working extensors. There will be robotic feeders and cleaners, automatic tillers and planters, shellers, and pickers. These aids could allow tomorrow's farmer the flexibility to engage in general production if there is the chance to take immediate advantage of shifts in markets and consumer demand.

As general farm organizations, such as the American Farm Bureau Federation, complete enrolling farm and ranch families into their membership, their emphasis will turn to self-identification—to learning more about their members and their needs. A simple listing of members' names and modest operating data will not be sufficient in the information-hungry and action-oriented farming future.

Farm organizations of the future will spend much more time and effort in self-examination. Using modern, computer-based telecommunications, member "identification" will be positive and complete. This will mean much more than an updated account of what kind of farm or ranch, the type and amount of crops or cattle, and the make of tractor. Rather, through "talk-back" telecommunications built in as part of the farm home or office screen, there will be direct knowledge of what the farmer, the farmer's spouse, and other members of the families do and think. Members will have an instantaneous and direct voice in organizational affairs and will pursue their economic and political self-interest.

For some period of time this self-interest will focus on developing and gaining public acceptance for a consistent, national food policy stabilized by removal of the issue from the center of the political arena.

To date "politics," rather than economics or concern for the welfare of farmers and ranchers, has been the driving force in constructing federal farm programs. "Cheap food" policies designed for their appeal to the larger body of consumer-voters have been damaging to agriculture. For example, high price support levels, although politically appealing, have resulted in substantial overproduction for dairy products and many other crops.

These government-stimulated, excess supplies have depressed commodity prices to the consumer's temporary advantage. Over time, however, these apparent savings are doubly paid for in higher consumer taxes for extravagant program costs plus the cost of storing the surplus. Meanwhile, agriculture is weakened, sometimes by raising crops on fragile grasslands in response to government incentives, but always by surplus-depressed prices, weakened markets, and by management uncertainties tied to a politically motivated farm program.

There is growing perception among farmers of the destabilizing effect of politically generated farm programs on markets and market prices. Farmers are aware that, worldwide, governmental intervention in agriculture continues to be the prime factor in creating and sustaining world hunger. The food production formula is not complex. It includes the right to own land, the right to make management decisions including those concerning farm size, the right to move goods, and free access to markets. However, personal incentives and the self-determination necessary for adequate food production is political poison to any centrally planned economy.

Common political roadblocks to world food production include government ownership and control of land, production, prices, credit, transportation, and markets. Removal of these blocks to personal incentive would assure more than adequate food production for virtually all food-deficit countries in the world.

The food production system of the United States—despite a half century of federal farm program mistakes and a far shorter agricultural history than most countries—remains the envy of the world. I am confident that this country will maintain this position far into the future by evolving (with the help of organized farmers) an enlightened, consistent food policy that retreats from political manipulation toward full use of marketplace incentives.

Under these conditions—and with the United States in the lead—I worry little about population growth in relation to the food supply. Viewed positively, both are greatly overrated as legitimate material for worry. There is very little that humankind cannot or will not do once the need is fully understood. That includes limiting population through self-interest and doing whatever is necessary to produce and distribute an abundance of food.

If we are seeking things to worry about for the future, let us concentrate on the debilitating power of political interference. Political interference has shown a remarkable ability to short-circuit just about everything it touches, especially farming and food production—and perhaps in the future, even the weather.

# 11

# NEW ROLES FOR
# THE AMERICAN FARMER

JOHN L. MERRILL
STEVEN T. SONKA
JO ANN VOGEL

Our traditional image of farming is often plagued by inappropriate and sentimental holdovers from the past. Today technology is firmly planted in every farm and ranch, and farmers must understand where that technology best fits into national and international farming and financial systems. The farmer of the future will have to be flexible and develop an integrated management system that takes into account changes in weather, markets, regulations, and so on, not just at home, but abroad.

Technology will continue to transform agriculture, modifying farm machinery to achieve a precision of operation now unimagined. Tomorrow's farmers will have to be even more astute financial managers than they already are, and the computer will play an even greater role in their affairs. Information will be a key to success.

But certain characteristics of farming will not change. It will remain a family-dominated enterprise in which well-educated family members, particularly women, will integrate their advanced skills for a successful operation.

# STRATEGIES FOR THE
# TWENTY-FIRST CENTURY

## JOHN L. MERRILL

None of us can see far ahead clearly, but every person in a decisionmaking or decision-influencing role is obliged to try. I shall try to describe my thoughts and concerns for farm management in the twenty-first century.

The difficulty of such an effort is exemplified in the lives of my mother and father, now 79 years of age, who in their lifetimes have seen transportation change from the animal-drawn or water-borne modes to the fairly frequent, if not yet commonplace, space travel of the present. The burgeoning technology of recent decades undoubtedly will continue to accelerate geometrically in the years ahead as each new discovery leads to others. This technology at once holds the keys to our future and becomes a problem in itself as we tradition-formed and habit-bound humans scramble to acquire, assimilate, and apply it productively and thereby become victors with it rather than victims of it.

My remarks are organized in five areas, most of which overlap. I'll start with some holdover attitudes that affect our present and future, mention some ingredients essential if individuals and nations are to be successful in agriculture, describe management principles applicable now and for the future, relate my assumptions for times ahead, and, finally, discuss promising technology for future survival and success.

There are some long-standing attitudes in agriculture which will continue to affect us adversely as long as we hold them. The oldest and worst may be farming by habit and tradition, even though conditions have changed. A quick example is the 160-acre homestead law, which worked fine in humid Virginia but not so well in arid Nevada, where that much acreage would support one cow and no farming. Just behind is the persistent attitude that farming and ranching is a way of life rather than a business. The fact that farming certainly is a business and should be so conducted in no way dilutes it as a way of life but usually enhances and extends that way of life. The corollary holdover that anyone can ranch and nature will take care of him or her may have been true when land and labor were cheap and few other inputs were used. The cost of all inputs now demands experience, knowledge, and the ability to apply them well.

Another holdover is a preoccupation with input agriculture, which often only transforms inputs into other products rather than conducting efficient production and harvest from renewable resources with minimum or optimum inputs. Closely related is the emphasis on maximum levels of production rather than optimum levels of production for maximum net return. And finally, there is the myth, fostered by western movies, that farmers and ranchers are rugged individuals who go their own way rather than organizing for common cause. Even the most rugged pioneers were quick to circle the wagons or assemble at the fort when attacked.

Agriculture has been and will be led by nations and individuals to the extent they possess and efficiently use available resources. First are the basic renewable natural resources of climate, soil, water, plants, and animals. The United States and other nations of similar latitude, north and south, have been blessed with a built-in climatic advantage as have individual producers located in areas where resources are most productive and best balanced.

Second, and even more important in certain ways, are the profit-and-loss incentives of the free enterprise system which offer freedom and flexibility to thousands of individuals to make management decisions, as opposed to the limitations of central government land-use planning and management of agriculture. We must strongly oppose the mood of some that would move our system toward the centrally controlled systems of other countries we outproduce and are having to feed.

Third is a stable government that enhances the aforementioned freedoms and does not impede productivity and progress with excessive regulation or unpredictable policy changes. While we live in one world and operate in one world market, the United States must exert at least as much nationalistic spirit and enlightened self-interest as other nations do if we are to gain an equitable share in and return from the world market for agricultural products. To benefit more fully we must emphasize the export of value-added and finished products rather than raw materials.

Fourth is the generation, dissemination, and application of technology resulting from agricultural research to increase efficiency of production. Agricultural research is a legitimate function of government. It serves both producers and consumers in ways they are unable to serve themselves.

Fifth is adequate capital and credit to accomplish each step in the chain from producer to consumer, and sixth is an effective and efficient system for transportation, processing, safe storage, distribution, and marketing of agricultural products, both domestic and overseas. I repeat, the availability and efficient use of these resources will determine the success of both nations and individuals in agriculture in the future, as they have up to now.

Certain management principles that are applicable now are even more necessary for survival and success in the twenty-first century. First is the realization that each farm and ranch is an operating ecosystem and economic system in which almost every decision and action affects most, if not all, other parts of the system. Production and marketing should be integrated from the beginning by determining what products each operation is capable of producing with minimum inputs and for which there is a relatively stable market and a price level that will return a profit over costs. This mode of operation requires a careful inventory of all available resources, followed by a careful analysis of their capabilities and limitations, and a budget of possible alternative or multiple uses to discern the most profitable combination.

With careful planning most agricultural lands are suitable for simultaneous multiple uses which usually are compatible and often complementary while increasing offtake and revenue. These uses include multiple cropping and multiple use of crops, grazing by one or more species of domestic animals, yield of wood products, wildlife and recreational uses, mineral production, and both groundwater and watershed values in terms of maximum water absorption for plant growth and aquifer recharge with minimum runoff of clear water to reduce erosion and flooding damage.

A detailed program should be written for each aspect of the operation, preferably including measurable objectives and strategies to attain them. From these, an annual plan and a long-range plan of operation should be prepared with budget and cash flow projections. These programs should comprise an integrated management system to avoid single-shot applications of technology, which may not only be less helpful but even harmful. To be most useful, the operating plan must be flexible enough to accommodate frequent changes in weather, markets, and other circumstances and, whenever possible, use them to advantage.

As the plan is applied, careful observations should be made, and records should be kept and analyzed to make needed adjustments. To the greatest extent possible management knowledge should be used to replace purchased inputs, determine the optimum level of inputs for maximum return, and apply them in a timely manner for greatest efficiency.

To apply these principles in the twenty-first century requires some assumptions about the future operating climate. I assume that increased competition for available water, declining water tables, and increased pumping costs will reduce the present acreage of U.S. land that is irrigated and require that irrigation be used for higher-value crops. This will reduce the surplus of grain from these irrigated acres, which has so greatly affected animal production in the United States in recent years.

The ability of ruminant animals to convert forage inedible by humans into high-quality meat protein for people, plus the flexibility to use whatever quantities of grain may be profitable, will be even more valuable and valued. Cattle and sheep will continue to provide the buffer that allows grain farmers to produce near-maximum levels but when grain supplies are inadequate, to use reduced levels of grain or none at all. This characteristic is also supportive of swine and poultry, which are more dependent on grain. Few realize that in the present system of grain-fed beef, 80 percent of cattle's lifetime intake is from forage, which makes them convert grain to beef at 2 pounds for 1.

If the foregoing is true, then there is a massive job of revegetation to do in returning marginal irrigated farmland to productive grassland. Land values will increase in areas with rainfall adequate to produce good crops without irrigation, such as the Corn Belt, the Mississippi Delta, the Blacklands of Texas, and Alabama. The cattle-feeding industry will need to be more flexible to feed more cattle for a longer period when grain is readily available and inexpensive, and to feed less cattle for a shorter period when supplies are short and prices high. Producers and the meat trade will need to be more innovative in producing and processing beef desirable to consumers at a price they can afford. Research now going forward in flavor, tenderness, and restructuring will be helpful.

A generation ago, when irrigated grain became plentiful and longer feeding periods were standard practice, cattle accumulated in large plains feedyards, and the second generation of packers and packing plants moved there from the old rail centers. Cheap energy and transportation allowed shipment of feeder cattle and carcass beef literally from coast to coast and the importation of meat from distant countries. In the future decreased grain supplies and the resulting shorter feeding periods may make it more efficient and economical to finish cattle at or near their birthplace. The third generation of packing plants may be smaller and dispersed in the cattle-breeding and growing areas. Cost of transportation will make importation of beef to the United States from Australia and New Zealand less attractive, but their production may find a growing alternative market in the Orient.

Logically, there should be an increased export market for U.S.-produced food but only if the population of food-deficit countries continues to rise without concomitant increases in their own agricultural production, and if those nations are able to pay for food. U.S. taxpayers may no longer be able—and certainly are no longer willing—to do so. This scenario becomes possible if labor intensive industries can be developed to employ those large populations to generate goods and/or revenue for international trade.

Ideally, if foreign demand could improve the U.S. balance of payments and strengthen the dollar, and, if allowed by government policy, it could raise prices of agricultural products to a higher level, survival would be assured for efficient U.S. farmers and ranchers. U.S. consumers will need to pay a more equitable percentage of their disposable income for food—approximately 2 to 5 percent more than the current anomalous level—which is bankrupting American agriculture. A slight increase in price in the marketplace for producers is much cheaper and wiser than federal subsidies paid for by tax dollars and incurring further costs in bureaucratic overhead. In foreign affairs we must learn to use American food not as a capricious weapon of warfare but as a reliable instrument for mutual peace and prosperity as we exchange what we do best for the products of what others do best.

The structure of U.S. agriculture should not be rearranged according to any one person's or group's ideas and ideals. I already have indicated my belief that free enterprise and individual decisions are vital ingredients of strong agriculture. Government should not try to determine or encourage any "optimum" size economic farm or ranch. There always has been and always will be room for every size unit.

The small farm or ranch, not economically viable on its own, will continue to be successful in its own way—that is, supported by outside income. The most difficult path will be that of the middle-size unit dependent only on self-generated income. The large units with equity and adequate operating capital accumulated over several generations, or supported by substantial outside income, or incorporated with public funding, if properly managed can ride out ebbs in economic cycles and use them to enlarge their operations at the expense of those less well-capitalized. There should be freedom for upward and downward mobility among these three levels of operation.

Motivation is essential to the success of all three. For mid-size operations, capital or management ability may be the limiting factors. Management and motivation clearly are the major limiting factors for large operations and the main reason they have not posed an overriding threat to the others. Unless large operations are able to create and implement the personal, professional, and monetary incentives of working for oneself, their performance and profits will not be commensurate with their resources.

For many reasons I do not foresee unionization of farm and ranch labor. The purpose of unions is collective bargaining to improve working conditions and compensations when unfair advantage of workers has been taken. Most farm and ranch owners I know work side-by-side with their employees—often longer hours at worse jobs for less pay—are basically fair with and considerate of others, and leave little room for union improvement. In addition, the extreme variability in hours, seasons, and work to be performed makes prescribed hours and job descriptions difficult, if not impossible.

I do not foresee any widespread return to family gardens and self-sufficiency. There may be some value in pride of production and physical fitness, but better food can usually be purchased at less cost at the local supermarket. I do hope that for greater efficiency and economy, as well as reliability in the face of any possible disaster, our food supply and distribution system may be simplified, streamlined, and planned by cool logic rather than hidebound tradition and haphazard evolution.

What of future application of technology in production, marketing, and other aspects of farm and ranch management? A desirable first step in setting priorities for research is evaluating needs and limitations. Then each need becomes an opportunity for new knowledge to be generated.

With little knowledge of basic plant and animal physiology, we have come a long way toward increasing efficiency in energy conversion, harvests, and processing. From an even better understanding will arise the technology of the twenty-first century. We shall identify critical points in the life of the plant or animal so that we better comprehend the effects of and responses to natural events and the type, amount, and timing of inputs, and other management manipulations.

A completely different opportunity lies in more accurate forecasting of weather, in both the near and long term. Not only could severe losses in crops and livestock be avoided, but management plans could be altered to take advantage of foreseen wet or dry, hot or cold periods. More accurate weather forecasting will also lead to more accurate production and market forecasting. Remote sensing with infrared photography and other techniques to reflect biomass can assist not only in production estimates worldwide but also can be used to detect outbreaks of disease and pest damage for rapid response and minimum loss. Even today the passage of the earth resources technology satellite every 18 days allows a fairly accurate estimate of forage available for grazing.

Developments in electronic, sonic, and laser technology will provide telemetry that will increase accuracy and timely management response. Electronics can monitor functions of engines and other machinery; leveling, depth, and draft of tillage and earth-moving equipment; rate and depth of seeding; losses in combine harvesting; and temperature, water, and nutrient content of soils and plants.

Individual implants in animals coupled by electronics with computers might display location and movement of every animal in one's care, monitor its body functions, and print out or flash its individual identification number and location whenever any dysfunction is detected to allow treatment for minimum cost and loss. It may be possible to monitor individual milk production and rates of gain and analyze lean and fat content on a continuous basis in order to market on a specific objective standard. Intensive strip-grazing may be controlled by electronics or laser beams without the use of conventional fencing.

Computers will become even more important in accessing specific research and market information, maintaining accurate production and financial records, and analyzing alternative scenarios for improved management and marketing decisions. Greater knowledge and more accurate measures of specific characteristics of food products that affect value to consumers will allow efficient production and marketing of plant and animal products to exact specifications, such as protein or caloric content, lean-to-fat ratio, and tenderness. Electronic marketing based on these specifications may be possible from field, pasture, and pen to the consumer's shopping bag.

Improvements in processing will enhance nutritive value, palatability, and the shelf life of food products while simplifying and reducing the cost of transportation, storage, preparation, and serving. Irradiation is just one development with future promise.

These examples barely touch our future prospects in the technology of farm and ranch management. As always, the key to the future is the human element—the intelligence, imagination, initiative, industry, and integrity of people working for the benefit of themselves, their families, communities, nations, and all of humankind. And a single fact stands out: There can be no lasting peace when people are hungry.

Somehow leaders of nations must be made to understand the necessity for funding agricultural research. It is absolutely necessary if we are to feed, clothe, and house the people of the world. The Creator has invested in us all the capability and resources that are necessary. As Jonathan Swift said, "Whoever makes two ears of corn or two blades of grass to grow where only one grew before deserves better of mankind, and does more essential service to his country than the whole race of politicians put together."

# NEW APPROACHES TO FARM AND RANCH MANAGEMENT

## STEVEN T. SONKA

The farmer and the rancher are key economic agents in our society, and in the twenty-first century they will need special skills. Their economic survival will depend in great part on the effectiveness of their own decision-making ability.

## LOOKING BACK

The twenty-first century is only 17 years away. Today's college student who chooses agriculture as a career will be in the midst of what should be the most productive years of that career when the next century begins. Therefore we are considering what skills and tools those young people should be acquiring in their educational programs. This fact is particularly interesting to me because only two decades ago I was in the position those college students are in today. As a means to improve our perception of skills needed in the future let's focus on some of the changes that have occurred over these last two decades.

There have been many changes in the way food and fiber are produced in this country over the last 20 years. Today's average farm is considerably larger than its predecessor. With 430 acres, it is 23 percent larger than the average farm of 1964. The value of the products produced on that farm is also substantially greater. In 1969, for example, only 2 percent of the farms in the United States had gross annual sales exceeding $100,000. In 1981 the proportion of U.S. farms with sales exceeding $100,000 each year had grown to 12 percent. Although inflation was a major factor contributing to this change, even growth in annual sales due solely to inflation results in increased managerial information needs.

What factors have allowed this marked expansion in size of the farm unit? Technological innovations such as larger machinery and more effective chemicals have been important. I suggest that there is another equally important factor: the use of debt capital. In 1950 the entire value of the debt held by the agricultural sector was essentially equal to the annual net farm income of that year. In 1980, however, the farm sector's debt was $157 billion, 12 times greater than it was in 1950. Furthermore, that debt was eight times greater than the 1980 annual net farm income.

This relationship is not necessarily bad, but today's agricultural producer, in addition to being a skilled producer of food and fiber, must carefully and effectively manage the farm's financial assets and liabilities. If we believe that the use of debt capital will be a feature of farming in the twenty-first century, we must focus on the tools producers will need to accomplish this important and often frustrating task.

Another major change in our agricultural system is the increased role of the marketplace in determining prices for several of the important crop commodities. In the early 1970s the role of the export market became much more important to the producer. Improved marketing skills and the timely availability of information necessary to utilize those skills attained new prominence. Along with financial management, marketing management became a survival skill of the farmer and rancher.

## LOOKING AHEAD

Trying to predict what the farm unit of the year 2001 will look like is not easy. New technologies will change production practices, and economic and societal pressures will undoubtedly continue to influence the structure of the American farming industry. At a more basic level, however, patterns are emerging that are likely to continue into the future. In his book, *The Third Wave*, Alvin Toffler suggests that we are entering an "Information Age." The assumption which undergirds this prediction is that the prices of energy and raw materials will be relatively higher in the future than they were over the past two centuries. Information is a third, often overlooked, basic input to the production process. We can, and I believe will, continue to increasingly substitute information for more expensive raw materials and energy. This is why today's farmer and rancher are so interested in the small business computer as a tool to increase their management capabilities.

A related perception is that change itself will be a most important characteristic of the future. New production technologies will continue to become available. Changing market conditions will alter profit expectations. Fluctuating input prices—in particular, the price of capital now that rural credit markets are less well insulated from national money markets—will require flexible strategies for both operating and strategic decisions. These two phenomena, a marked change in relative prices and the need for continual reevaluation of management decisions, seem to be central factors defining the type of management attributes that will be required of the producer in the twenty-first century.

## WHAT PRODUCERS AND THEIR ADVISORS BELIEVE

In 1982 a major study of the future information needs of farmers and ranchers was conducted jointly by the University of Illinois at Urbana–Champaign and Arthur Andersen & Co., a public accounting firm. The primary research team was composed of Norman Carlson of Arthur Andersen & Co., Dr. Thomas Frey, Professor of Agricultural Finance of the University of Illinois, and myself. Our goals were to define the information-related management practices of today's producers and to determine what farmers, ranchers, agricultural lenders, and other farm advisors believe to be likely practices of tomorrow. The focus of the study was on those information practices most directly related to financial management activities.

The study results summarize the opinions of 535 individuals—269 agricultural producers, 179 agricultural leaders, and 87 agricultural consultants from the United States and Canada. Representatives of all major areas of agricultural production and regions are included. The respondents were asked to consider changes in management practices over the next decade, a shorter period than the focus of this book, but their responses provide confirmation of the trends suggested above.

### Financial Management

If financial management is to be effectively practiced in any business, the financial position and performance of that firm must be well-documented. To do so requires that the manager has accurate financial statements and be able to interpret them. In Table 1 the responses of the study participants regarding the importance of financial statements to agricultural producers in

Table 1. Future Importance of Financial Statements

| Statement | Proportion (%) of Each Respondent Group Predicting It Would Be Very Important to Prepare the Following Statements in 5 Years | | |
|---|---|---|---|
| | Producers | Lenders | Consultants |
| Balance sheet | 79 | 89 | 91 |
| Income statement | 79 | 89 | 80 |
| Projected cash flow statement | 82 | 89 | 91 |

the future are presented. A substantial majority of each group believes that in only 5 years it will be very important for farmers and ranchers to prepare these major financial statements. If correct, preparation of such statements should also be important as we enter the twenty-first century.

Of course, simply preparing financial statements does not mean they are accurate and useful. In nonfarm businesses verification of financial statements by outside evaluators is common. This is not currently true for farm businesses. The lender and consultant respondents in the study believe that this disparity will diminish in the future. For example, they predict that one producer in six will have to have outside verification of financial statements within only 3 years. This proportion is expected to increase to one farmer in four within 7 years.

## Accounting Methods

Preparation of accurate and verifiable financial statements requires accurate data. The farm's accounting system must capture and summarize those data if financial analysis is to be accomplished effectively. But American farmers are notorious for their inattention to this bookkeeping function. For example, the study respondents believed that over two-thirds of all commercial farmers (those with annual sales exceeding $100,000) currently do not maintain a separate checking account for their farm business. It is quite difficult to prepare accurate financial statements if personal and business transactions are mingled in one account.

The 269 producers responding to this study were nominated because they were regarded as being innovative information managers. Therefore their actions are likely to lead the agricultural industry as it adapts to the coming information age. They were asked to predict which accounting method is most likely to be used by farmers like themselves in 5 years. More than 60 percent of them responded that farm accounting is most likely to be accomplished by computer. Eleven percent of these farmers were already using this accounting method.

These producers expect future farm accounting systems not only to be computerized but also to be able to provide detailed managerial information. More than nine out of 10 expect these systems to determine costs and revenues per unit of production, determine costs and revenues by enterprise, and maintain a perpetual inventory for harvested crops. Leading producers believe that access to these types of information will be a necessity for the agricultural producer of tomorrow. Clearly, the single entry, manual approach to farm accounting will be hard pressed to provide that detailed information.

## Computer Use

As already noted, the study's farmer respondents were not selected as being representative of today's average producer. Their use of small computers indicates that they are leaders in innovation. Whereas it is expected that 2 to 3 percent of all farmers currently own a computer, 22 percent of the study respondents were already using them. The most common use of the computer was to assist in accounting and financial statement preparation activities.

The producer respondents foresee that computer use will be common on the farm of tomorrow. More than 80 percent believe that farmers such as themselves are likely to use a computer within 5 years in the following ways: to monitor receipts and expenditures, to maintain production records, to assist in forward planning, and to obtain marketing information. Of these four activities obtaining information is currently the least utilized. The potential for telecommunications and the electronic transfer of information seems particularly appealing to the farmer. Such innovations may reduce his geographic isolation and allow him to acquire quickly the information necessary to make decisions in tomorrow's environment.

## Management Consultants

How will the family farm compete in the future when financial decision-making and the interpretation of a myriad of data are critical management skills? One potential solution is the utilization of management consultants to provide specialized information and skills. The producers responding to this study already utilize such consultants. Producers and advisors believe that the use of such consultants is likely to increase. More than 90 percent of all the study participants believe that assistance in the following five areas will be important to producers in the future: accounting, financial analysis, estate planning, marketing, and use of computers. But the respondents noted that quality consulting assistance is not always available, particularly in the areas of financial analysis, marketing, and computer use.

Today's farmers and ranchers face vastly different problems from those of only a few decades ago. The increasing size of the farm firm and the growing use of debt capital mean that the producer must be as effective in the business management of the firm as in production management. Attention to costs and returns, financial decisionmaking, and marketing have become key survival skills.

The future promises an increasing rather than a decreasing importance of these factors. Change would seem to be the only constant feature of tomorrow's farm management process. Acquisition and utilization of information may become the critical factors determining whether the farmer of the twenty-first century is an effective decision maker. The small computer is emerging as a tool which can aid tomorrow's producer to manage the increasing information required. More importantly, business management skills which allow the effective utilization of information will become increasingly necessary. Among today's innovative producers these skills are starting to be developed.

I believe the independent family farm can prosper in this environment. Structures to accommodate the information needs of this business-oriented manager must develop. Both the public and the private sectors have key roles. Although the role of the public sector in providing recommendations for the typical producer may diminish, its role in providing education and unbiased analyses will be of paramount importance. The private sector's role is likely to increase, for routine provision of information and consultation seems essential if this business-oriented producer is to make timely and effective decisions in a continually changing world.

# MANAGING HUMAN RESOURCES

## JO ANN VOGEL

The key to the success of agriculture in the twenty-first century, I believe, will be in our ability to manage our human resources. One-family units, as viable production units, will no longer exist as they have in the past. Production agriculture will consist of several family units per operation. Therefore, the management of our working counterparts will be the key to success. Regardless of our ability to manage other portions of the operation, the management of labor will decide profitability. I will illustrate this theory by relating the past 25 years of our family operation.

Agriculture in the 1950s was regarded as an occupation that required very little management and gave a similar amount of prestige. I was born and raised on a small dairy farm which my father operated while also working in town. I never wanted to marry a farmer, and my husband-to-be had never considered becoming one.

When my husband proposed marriage 27 years ago he was employed by Western Electric, and I worked for Hamilton Industries. However, four weeks before our wedding, in June 1958, he became unemployed. After much discussion, we decided that we would invest his savings—$7000—and begin farming. We contributed $5000 to the farm purchase and kept $2000 for operating expenses. We borrowed the rest of the purchase price from his father and mother. The farm consisted of 17 cows, three heifers, and four calves; a minimum of machinery (in good condition); a barn with running water; and a house without. An outhouse had to suffice for the first few months, and the kitchen never had plumbing during the 11 years we lived there. The house didn't have central heating either, and when the west wind blew the water would freeze as it splashed against the sides of the bathtub.

To make this farm go we had to fine-tune our management. Our milk check was $280 every 2 weeks. That had to meet our farm and household expenses. Our first child, Gary, arrived in May 1959, and our second son, Marcus, was born 13 months later. In August 1962, a beautiful daughter, June, was born. During this period I milked twice daily, helped clean barns by hand, threw out and fed silage, chopped hay, cultivated corn, cooked, cleaned the house, canned, and sewed almost the entire wardrobe for the family. Pregnancy was never classified as a disability. The work had to be done regardless.

In 1962 a neighboring farm became available. We decided that if we were to continue farming we needed additional acreage. By this time we had paid off a considerable amount of our debt and we were able to seek financial funding at our local bank. In 1967 we purchased another neighboring farm, which provided 115 acres of cropland, needed buildings, a silo, and a new house. In August of that year our youngest daughter arrived, and when Julie was 6 months old we became involved in the Foster Children program. We welcomed three children from one family into our home. A year later we accepted another child. The next 9 years were very interesting, to say the least.

We had built our herd to 45 cows, and our butter fat average had advanced from 350 to 550 pounds per cow through the use of artificial insemination, the Dairy Herd Improvement Association, and a good management program. Our young stock numbered 55 head, and our yields doubled through soil-testing, fertilizing, and choosing the correct varieties of feed. We knew we could not rest on our laurels. We had to increase the herd to 80 cows to help meet our debt load. An addition to the barn was essential, along with additional milking machines. Through all the growing years the farm necessities always came first.

As a wife and also as a part of the management team I had to realize that management decisions are based on profits and the ability to keep the cash flowing. I always tried to maintain a delicate balance between farm and personal matters. Farm couples are together every day, and, as in any marriage, communication is essential to keep problems at a minimum. A good farm wife motivates and sustains her husband during the bad times, and she is a stabilizing influence during the good times. Hiram Drache, a farmer, author, and professor of economic history, maintains that the key to a farmer's success is his wife. "A successful farmer," he says, "will rank the six qualities of (his) success in this order: mate, motivation, management, money, marketing, and mechanization."

Yet the importance of the farm wife is hardly appreciated. In testimony given before the House Committee on Agriculture in March 1981, Fran Hill, professor of government at the University of Texas, stated, "Women are America's invisible farmers." We speak fondly of the family farm but use "farmer" as a male noun. Women's contributions to the operation of American farms have long been hidden by a screen of cultural myths about male gallantry and female delicacy. In common parlance men "farm" but women just "help," even when husband and wife are using identical tractors to plow the same field.

In our family, management decisions whether they relate to the purchase of a machine, changing rations for our cattle, crop rotation, or building structures, have been made without heated discussion. Since I have been involved in our operation from the very beginning I fully understand when new machinery must be purchased and why. We discuss the need and check our cash flow, including committed payments in the coming months and other equipment or major repairs that might need to be done. We compare prices of equipment, the availability of service proximity, and previous experience with the dealers. If financing is needed I will discuss it with our banker.

I have taken care of all the banking for the past 25 years. We discuss our finances together and decide how much will be deposited in each account. Each month I do all the bookkeeping and I fully understand why we cannot purchase new carpeting or whatever for our home. I readily learned that in order to succeed the business must come first. I think my husband and I work together very well because I tend to be liberal and he tends to be conservative and we always arrive at a middle-of-the-road decision. It has worked very well, and we have been happy and successful.

Women are vitally interested in the farm business in spite of obstacles to their involvement in custom and the law. Farm wives deserve a measure of public recognition and changes in public policies that do them active harm. For example, I am a strong advocate of reform of estate taxes to recognize a wife's contribution to the business.

Divorce is a problem on the farm as elsewhere—maybe a bigger one. In the last 35 years or so the agricultural community seemed to invade the divorce courts. In 1940, 56.5 percent of farm marriages ended in death and 43.5 percent ended in divorce. In 1979, 42.9 percent ended in death and 57.1 percent ended in divorce. In a recent article in *Successful Farming*, Jolene Brown, an Iowa management consultant, states, "Farm divorce often occurs because individuals are so involved with farm management that they forget the critical balance needed for successful living. We must learn to farm while respecting the other areas of our lives as well. With a farm marriage, respect is shown through such things as goal-setting, communication, and COMPROMISE."

Most farm families are often closely knit due to the fact that the children are with their parents a great deal of the time. For older children the choice of school activities are family problems affecting the farm work force. During their high school years our sons wanted to play football but with our setup we could not cope with additional labor costs. Instead, we installed a milk pipeline costing $10,000 so that we could handle the work load. Our sons would come home at 6:45 p.m., eat their dinner, work with us until 10:00 or 11:00 p.m., and then get up at 5:00 a.m. to do their schoolwork.

In our family we feel there is a need for higher education regardless of our children's choice of occupation. We paid them a salary for their farm work, and they saved it for college expenses. Guy graduated from the University of Wisconsin–Platteville. Marcus chose not to attend college but worked with us full time. Our daughter June is presently a senior at UW–Platteville in Business Administration, and our youngest daughter, a junior in high school, is planning to attend college. Our sons will take over the management of our farms in the near future.

In the last 25 years, U.S. farms have become large, complex operations with problems our ancestors did not have to cope with. A farm represents huge investments and financial risks frequently spanning several generations and families. We are now trying to decide if our operation will become a partnership or a corporation. Will the children become salaried employees with incentives, or will we have a rental agreement for a period of time while they adjust to the labor and stress involved in management? Management of human resources is a key to success in any operation, particularly a family operation.

I know from personal experience that management becomes very complicated when married sons enter the picture. They are young and ambitious and would like to apply their new-found knowledge. On the other hand it is very difficult for a father to hand over responsibilities that for 25 years have been his. Our sons married girls without farm backgrounds. It is difficult for these young women to adjust to the fact that farm dinners are not served at 5:00 p.m. on the dot. The day ends whenever the work is completed, even if that means baling hay until 2:00 a.m. because it looks like rain or washing clothes until 1:00 a.m. because the next day you are needed in the field.

An unhappy daughter-in-law can be a management problem, or even the demise of a son's farming ambition. I find that women who are not actively involved in farm management and farm work cannot cope with the long hours their husbands must work or possibly understand the financial decisions that must be made. In my family I have found that a frank discussion with a daughter-in-law is very effective. I "laid it on the line" in very understandable language and we both expressed our innermost thoughts. A better relationship resulted—in fact it could not be better.

In passing our farm operation on to our children we have decided to deal with them as if they were strangers. Every step of the business transaction will be legal and fully understood by each party, for their protection as well as ours. With luck they will be able to overcome adversity without ever losing their optimism.

Do I feel there is a future in the twenty-first century for agriculture? Yes, I do. The young farmers of tomorrow will face adversities that we cannot possibly envision. They will need a college education. Stress will become more intensive due to forces beyond their control. They will need to cope with the weather, environmental activists, animal welfare activists, new plant and animal diseases, a national conservation program, government programs or the lack of them, foreign trade, promotion, marketing, and many other variables. Agricultural financing is rapidly changing. Will the Farm Credit system as we know it today be available in the future?

Operations in the future will become larger production units. These units will require a complex business organization which in many instances will require the services of a consultant, not only for tax purposes but also for production advice. I feel that every home will own a computer during the next century and that the major portion of personal business as well as official business will be conducted at home on it.

Will the population centers of the United States change? Some experts believe this will be the case. Will Wisconsin become part of an "OPEC" of water states? Will the deserts revert back to whence they came due to the overutilization of irrigation in the rain-deficit states? Will it cause the aquifers to drop to levels so low that utilization becomes uneconomical? Or will we develop through genetic engineering plants that can survive the desert environment? At present, many varieties of plants and weeds not normally thought of as productive, such as the goldenrod and milkweed, are being studied for possible use. An experiment is taking place in the growth of plants without the use of soil. In the middle of a Lathan, Illinois, field there is a $1 million demonstration facility greenhouse in which plants are grown in black crystals. Production is year-round and 17 crops of lettuce are feasible. Other fruit and vegetable crops are being grown in the same manner with a computer totally regulating the environment. Through advances in plant and animal genetics, changes will come about that are beyond our imagining.

Fifty years ago the United States had 6.5 million farms that averaged 145 acres each and were operated by a farm work force that totaled 13 million. And each of those farm workers produced enough food and fiber for 11 people. Today, there are only 2.4 million farms in the United States and they average 430 acres each. And there are only 3.7 million in the farm work force. But each of those farm workers produces enough food and fiber for 78 people. All of which means that American agriculture is the world's largest commercial industry—an industry with assets exceeding $1 trillion and employing nearly 23 million people, or 22 percent of the total work force in the United States. Besides providing food and clothing each farm worker creates 5.2 nonfarm jobs for people who produce things farmers need and who process, transport, and merchandise the crops the American farmer produces. Farmers also are among our nation's most important taxpayers and consumers, paying more than $8.25 billion in various taxes in 1979. Farmers spent $131 billion on the goods and services they need to produce the commodities they raise. In grocery stores and supermarkets nearly $400 billion a year is spent on food and beverage products alone, yet it accounts for only 13.3 percent of the consumers' expendable income. In Brazil a family spends 41 percent, in the Soviet Union it is 31 percent, in Japan it is 21.5 percent, and in China 60 percent. We are very fortunate to live in a wonderful country in which food is so inexpensive.

Agriculture is truly America's heartbeat, and America's agribusiness keeps it beating.

# 12

# AGRICULTURAL INFORMATION DELIVERY SYSTEMS

JOHN R. SCHMIDT
THOMAS L. THOMPSON
LARRY R. WHITING
RICHARD KRUMME

Half a century ago the gasoline-powered tractor was still a new tool for the grower of row crops—and most farmers continued to use the horse. Similarly, today the microcomputer has been introduced as an information tool in farming, but most farmers still depend on printed matter and other conventional modes of communications. Undoubtedly, in the next century the computer will be in common use in farm management.

Limited computer networks have been in existence in various parts of the country for more than a decade. Used by agribusinesses, extension services, government agencies, universities, and some farmers, they are mostly conduits for factual information about weather, feeding programs, markets, and other news. Interactive systems in which information can flow from the farmers as well as to them are newer and surely represent the future for computers in agriculture since they offer capabilities to analyze complex problems for immediate plans of action. Up-to-date information will be quickly and widely available, not only for day-to-day farming decisions but also for financial management.

The technology of microcomputers is largely developed; in the future it will be refined and become cheaper and thus be available to more and more farmers. Software, however, is lacking, and many more programs need to be written and made available. But computers will not completely replace farm magazines, which will become more and more specialized and will continue to provide essential information at low cost.

## ON-FARM INFORMATION SYSTEMS

### JOHN R. SCHMIDT

The notion of farmers using information delivery systems and services typically implies the provision of information from off the farm to the farm manager at his or her location. The definition of information services for agriculture is much broader, however. It must include information flowing in the opposite direction—from the farm manager to a host of off-farm entities including marketing firms, input suppliers, and governmental agencies. There is yet another dimension to agricultural information, one which is largely internal to the farm as a business unit—information from the farm itself provided to the farm manager. Let us concentrate on this third on-farm dimension as we look to the twenty-first century.

The central theme is simple: Farmers in the twenty-first century will routinely apply sophisticated on-farm information systems based on microcomputer technology which is just now getting underway. Several observers have predicted that a majority of the commercial farms in the United States will be equipped with stand-alone computing capability through microcomputers by the end of this decade. However, some are skeptical about the adoption rate. They note that several national surveys completed within the past year estimate that only about 3 percent of our commercial farmers now own a microcomputer.

Dr. Earl Fuller of the University of Minnesota compares our current stage of on-farm microcomputer utilization with that of the tractor a half-century ago. He notes that by 1933 row-crop tractors had been available for about 5 years. But most farmers didn't own one; in fact, most were not yet convinced that the tractor could replace the horse. Even though the steam engine had been in common usage for 30 years, the gasoline tractor represented mechanical power of a different size and versatility. Apart from the plow, most of the tractor-drawn and PTO-powered (power take-off) farm implements had not been developed yet. Those that had were tailored to a specific tractor manufacturer and were incompatible with other makes. There were several deterrents—questions about reliability, confusion about makes and models, the need to develop new skills, and doubts about community acceptance. Many farmers saw at least one job where the tractor appeared to be highly desirable, but a comprehensive cost/benefit analysis was needed—a tenuous requirement because of the many uncertainties involved. Even for those convinced of the merits, depression farm prices had a major dampening effect. "Should I buy a tractor?" "When?" These doubtless were questions facing most farmers in 1933.

Compare this scene with that of today concerning on-farm computer acquisition. The parallels are striking. Microcomputer technology is about 5 years old. The Apple and Radio Shack models introduced in 1977 became generally available in 1978. Most farmers don't own a personal computer. As with the row-crop tractor, the microcomputer has a 30-year technology background going back to the early mainframe digital computers. Computer applications in agriculture can be traced to the 1952 work on feed blending by Dr. Fred Waugh of the U.S. Department of Agriculture. The micro differs greatly from the mainframe in size and versatility, like the tractor differed from its steam engine predecessor. The "implement equivalent" for the micro is its applications software. As with the tractor in 1933, these vital tools are only now emerging in acceptable form. And they, too, have limited transferability between microcomputer makes and models. Nothing akin to the standard three-point hitch has been developed for microcomputers. (Recall, though, that the three-point hitch did not become truly standard until after World War II.)

Most of the doubts and uncertainties cited above for tractors are now just as important for microcomputers. Although not nearly as severe, the current agricultural recession is likely dampening the rate of computer adoption at the moment. "Should I buy?" "If so, when?" These questions are prominent in farm discussion today just as they were a half-century ago. Only the topic has changed—from tractors to microcomputers.

Both the tractor and the computer represent sources of power on the farm. Tractor power is mechanical while the computer provides what we might best term "management power." Census data show that a majority of all U.S. farms (and presumably most of the commercial farms) had adopted the tractor by 1954. The on-farm microcomputer will be equally commonplace on the farms of the twenty-first century. They will power the management information systems needed to provide food and fiber for the generations to come.

What will these localized, on-farm information systems be like? They will have two major characteristics in common: (1) a dominant management orientation and (2) points of interface with computer-based information utilities and analytical services external to the farm. Beyond these commonalities the systems will differ by farm type since they will monitor and control the unique production processes and growth patterns of specific enterprises. These latter details can best be illustrated by example. Let's turn our time machine to the year 2000 and visit three U.S. farms: a Midwest corn/soybean farm, a Michigan apple orchard, and a Wisconsin dairy operation.

## A MIDWEST CORN/SOYBEAN FARM

The internal information system for this farm is relatively simple, centering on the growing cycle and field operations of the two major grain crops—corn and soybeans. At its heart is a spatially indexed data base which summarizes the important physical characteristics of this farm's cropland by its location; that is, the on-farm computer maintains a record of individual land parcels in units as small as a tenth of an acre. The basic data base information includes soil type, soil depth, slope, yield potential, and drainage characteristics. To this is added nutrient availability information from the latest soil tests together with fertilizer application information. Soil nutrients added through fertilizer are updated automatically, based on the quantity and analysis information that a computer program has "ordered" for each parcel. A microprocessor installed in the planting equipment determines which specific parcel is being traversed, finds the application rate which had been ordered, and controls the aperture accordingly. Seeding rates are similarly handled, as is the application of herbicides and insecticides.

An obvious factor of critical importance to crop production is the weather with moisture, wind, and temperature as key variables. An important component of the system then is a set of monitoring devices which can record this location-specific information and add it to the data base. Rainfall and soil conditions in most of the Midwest do not justify investment in supplemental irrigation equipment. Where this is not the case, the sensing information feeds into a computer model which in turn calculates the crop water needs and controls the timing and extent of irrigation. The latter determination draws on information services from off the farm as well as through localized weather forecasts—that is, a supplemental irrigation treatment otherwise needed may be reduced or deferred in cases where rainfall has been predicted for the area with a sufficient degree of probability.

All the information mentioned thus far—soil characteristics, soil nutrients, planting rates, herbicide treatment, weather experience, supplemental irrigation, and so on—becomes input to a computer model for yield prediction. These predictions coupled with weather forecast information, price forecasts, and an economic analysis of drying and storage costs will be used in a management-decision computer model designed to determine when harvest should begin. It will also be used to help in controlling the speed and internal settings of the harvest equipment. The latter locational adjustments are made under microprocessor control as well. As harvesting takes place the yield measurements are fed back into the data base. This directly updates the farm's data base for use next spring when the production cycle is repeated.

In short, the internal management information system provides for a circular flow in which inputs are controlled and the output results are monitored to give feedback for revising future inputs.

Is all this fantasy and wishful thinking? No. Many of the pieces are operable now and others are in various experimental stages. Dr. Larry Huggins, head of the Agricultural Engineering Department at Purdue University, states that there are two critical elements missing from the list of currently available technology in order for data base management concepts to impact field crop production. The first is an affordable system for automatic, continuous position tracking of field equipment. The second—a question of incompatibility—is a set of adequate industry standards to permit mass data collection, storage, and retrieval among subsystems from various vendors.

## A MICHIGAN APPLE ORCHARD

The internal management information system for our Michigan apple grower has some elements in common with the grain farm we just visited. It, too, has a central data base in which the physical attributes of the orchard are tied to location. Here the recording units are "blocks" of trees segregated on the basis of tree variety and planting/establishment date. As with the field grains, weather is extremely important. So similar rainfall, wind, and temperature monitoring devices feed their observations to the on-farm computer. Considerable effort goes into the monitoring of insect populations. This aspect may not be automated as readily, requiring specimen counts from insect traps. These field observations are recorded on portable keyboard devices with memory capacity for subsequent electronic transfer to the locational data base.

The most crucial management activity during the growing season revolves around the spray schedule in a perpetual battle against insect and disease damage to fruit. In a manner akin to the irrigation schedule for field crops, the spray schedule for this twenty-first century orchard is generated by a computer model. Many variables are used as input. They include past and current weather observations, weather forecast information from external sources, the kind and quantity of spray material used to date, as well as the insect population information.

The automated management information system deals with the question of harvest timing as well. But here the monitoring relies on a novel detection system. Have you ever walked through an apple orchard just before harvest? The aroma is keyed directly to the ripening stage. A sensing device for each block of trees "smells" this aroma and records the relevant chemical for transfer to the data base. Does this sound farfetched? Experimentation with this technique is underway at the Michigan Agricultural Experiment Station and early successes have been reported.

As with the grain farm, the production information is entered into the data base as harvesting takes place. In this case though, the information flow continues on to the grading and warehousing functions.

## A WISCONSIN DAIRY FARM

The internal management information system on our twenty-first century Wisconsin dairy farm centers on the individual cow as the basic production unit. Here the central data base is indexed to individual cows through a name/number identification system rather than to geographic location. The four major management areas are feeding, breeding, milking (production), and animal health. Each of these areas is incorporated in the data base, which also includes automatic updating of the milk production record information. The basic data base information on each cow begins with its date of birth, sire and dam information, date of first calving, and body weight. A computer model is used to predict production during the first lactation based on this information. Departures from this predicted level trigger such managerial responses as feeding adjustments, medical attention, or even animal replacement. The data base is updated when the animal comes into heat, is bred, receives clinical attention, undergoes a ration adjustment, is verified as pregnant, and is "dried off." All this information is used together with age-adjustment factors based on breed research data for a computer-generated production estimate for the second lactation. This establishes the monitoring and control cycle for individual animals—an all but impossible feat for the manager of a large herd without computer assistance.

For feeding purposes the control and feedback cycle will be set up on a daily rather than a lactational basis. That is, the 24-hour feed allowance will be computed for each cow, most likely based on her average production over the past few days or week. The allowance is typically based on the grain/protein supplement portion of the ration when forages are offered free choice. The allowance is for all feed when a total mixed ration scheme is followed. In either case, failure of the animal to consume some minimum portion of the allowance (say 80 percent) results in a special computer report suggesting the need for investigation. This is often the most effective kind of early warning system possible for detecting health problems; the cow loses its appetite even before milk production drops. Other causes of lowered feed consumption include the onset of estrus. Thus this kind of monitoring doubles as a heat detection device. This is valuable information, for timely breeding is of great economic consequence to the dairy farmer.

Not all cows drop their feed intake when in heat. The dairy information system therefore searches the data base daily for all animals whose pregnancy has not been verified. A computer message is generated listing any animal which may be coming in heat within the next few days. Typically, this message will be produced 55 days after calving and each 21 days thereafter.

Both the daily milk weights and feed-consumption weights are recorded automatically with this system. A signaling device on the neck chain of each cow emits a radio signal which can be interpreted by a microprocessor. Scales attached to both the milking and feeding equipment record the weights while the microprocessor transfers both the cow identification signal and the respective weight information to the data base.

In setting the daily feed allowances frequent use will be made of least-cost ration programs of the type now found primarily on college-sponsored mainframe timesharing systems. These linear programming models require more computing power than is common with today's on-farm microcomputer. The twenty-first century on-farm computer will have no such limitation, however. Recourse will probably be made to other mainframe-based information services though, such as sire selection programs. These programs contain the results of extensive progeny testing for the artificial insemination industry. When combined with the local data base information for a given cow, the genetic improvement from alternative matings can be predicted.

Computer technology will contribute to the internal information systems of twenty-first century farms by providing improved information for management. Each farm, however, has a financial transactions data base operating in parallel with the production-oriented data bases described here. The production and financial management information analyses will use the same computing system and be highly integrated. Developmental efforts in this area are now underway at Cornell University and elsewhere.

Finally, we must note that these twenty-first century farms are not islands unto themselves, operating in informational isolation. The use of computer-based information delivery systems from off the farm will be common—and not really very different from the situation of today. In my judgment, the more noteworthy informational aspect of the twenty-first century farm will be its internal management information and control system.

# INTERACTIVE COMPUTER-BASED INFORMATION SYSTEMS

## THOMAS L. THOMPSON

The impact of new technology on agriculture is much broader than the changes it is bringing to the farm and ranch. When evaluating this impact, we need to include all portions of the agricultural sector. Local agribusiness firms provide supplies and services to assist the farmers as they produce the food needed to feed the world. Financial institutions provide the capital for the producers to operate. Local agricultural management and consulting firms provide services to supplement skills that farmers do not have the time to perform. Government price support programs help stabilize the agricultural economy. The local school systems, community colleges, technical schools, and state universities provide education to train today's and tomorrow's agricultural producers.

All segments of the agricultural sector must work together to improve the efficiencies in the system. Probably the most vital component is the delivery and evaluation of current information. Today's agricultural producer can no longer rely on last year's data to schedule this year's activities. In addition to continuously evaluating changes in markets, government programs, weather, local and state laws, the producer also needs to incorporate new technology.

Interactive computer-based information systems are becoming an increasingly important tool for the delivery of up-to-date information to the agricultural sector. Today these systems provide the latest information and help evaluate the impact of this information on an individual farmer or rancher, a local supplier, a county, a state, or a country. In the future these systems will provide and evaluate much more information and they will also perform many of the routine tasks now performed manually.

Let us take a look at the use of interactive computer-based agricultural systems in 1983 and then project their use for the twenty-first century. After making these projections we will summarize what needs to be done and the implications of these changes on the agricultural sector of the world.

## INTERACTIVE COMPUTER USE IN 1983

For years batch processing on the mainframe computer was the only alternative for computer processing. About 10 years ago operating systems were introduced that allowed the user to interact with the mainframe computer. About that same time microcomputers came on the market but they required advanced computer skills to program and operate. During the last 10 years rapid advancements have been made not only in computer hardware but in the operating systems. The operating systems of today's microcomputers and interactive mainframe computers can be used by people who have little or no knowledge of programming, that is, these systems are now "user friendly."

Easy-to-use, powerful computers and on-line services are now affordable by small businesses and many individuals. Unfortunately, many potential users are not able to use this power because application software for their needs has not been developed. Users need to be involved in the development of software that will perform the tasks needed to meet their particular needs.

In the agricultural sector the computer is just starting to be put to effective use. Applications are being implemented on both mainframe computers and microcomputers. Generally, programs for the microcomputer are being developed to perform routine tasks and procedures that do not involve many changes and updates. The mainframe computer is being used to handle large data bases, perform analyses that require resources not available on the smaller computers, and as a repository of timely information and management models.

The first application software developed for agriculture was in the area of management models to perform detailed analyses or simulate the performance of a system. The first major application was a linear programming analysis that determined the least-cost combination of feed ingredients for a livestock ration according to preestablished nutritional specifications. This application continues to be a major activity on many agricultural computer systems. Other applications include simulating the performance of livestock, crops, and other agricultural operations.

## SERVICES OFFERED

By performing detailed analyses and simulations through management models, users can obtain answers for their particular situations and also evaluate the impact of new information on their enterprises. The size of the program and the frequency of use determine whether it will be run on a microcomputer or the mainframe.

Up-to-date information delivery through the electronic media is becoming important to all segments of the agricultural community. Many federal and state agencies can no longer provide free publications. Generally, these agencies found that less than 20 percent of their readers were willing to pay for the same information. Users indicate that they no longer want entire reports but only a few selected tables or sections—and they want them as soon as they are released to the public. Electronic information delivery is one way to meet this need. The information is immediately available to all users wherever they are located. This means the grain elevator manager in Cozad, Nebraska, can have "hot" U.S. Department of Agriculture information as quickly as the large grain companies who have previously received this information from expensive computerized services. Since the user pays to retrieve the information from the computer, the information supplier (e.g., USDA) can evaluate the real need for that information.

Communication is another major component of services currently being offered on interactive computer networks. A message can be sent easily and quickly to an individual or to many users, and it usually is less expensive than mailing a letter or calling on the phone. Electronic messages tend to be short and to the point—they do not talk about the weather or their families—and they eliminate the inefficient process of "telephone tag." The computer can also be used in a conference mode to discuss various topics and even to plan workshops and conferences. Conference participants can meet electronically every day. Without the expense and inconvenience of travel they simply contribute to and draw from the information pool.

The use of microcomputers will continue to increase in all segments of the agricultural sector. Routine, frequently performed tasks will represent the major use of microcomputer units. Generally, these tasks fall into four categories: spread-sheet analysis, word processing, data base management, and communications with larger computers. The micro has the ability to communicate with the large mainframe computers, collecting the latest updates and storing them on a floppy or a hard disk for later study, analysis, or printing. Likewise, some applications will require that data be uploaded to the larger computer systems.

## SCOPE OF COVERAGE

Although the number of users of available agricultural computer networks and microcomputers is increasing every day, it is small compared to the potential number of users. The agricultural leaders are just starting to see the real benefits of this new management tool. Let me describe a recent experience to illustrate the potential scope of agricultural computer networks:

On January 11, 1983, President Reagan announced details of the Payment-In-Kind (PIK) program. These details were entered into a news release on AGNET about noon (before the President finished his speech). This story was accessed 90 times (by midnight) the same day and a total of 397 times during January. The following week Professor Larry Bitney and Duane Jewell of the Agricultural Economics Department at the University of Nebraska modified an existing model that analyzes the returns to a farm operation under the various alternatives in the 1982 farm program. The modification incorporating the PIK option was released on January 19. Immediately, this program was used by many users all over the country. During the two remaining weeks of January and the month of February the program was accessed over 6000 times by more than 500 different users for a total connect time in excess of 2000 hours. This amounts to over 51 hours, or 146 users per day, for the 41-day period. The people that used this program included producers, bankers, elevator operators, local, state, and federal governments, and exporters from all over the United States.

This example illustrates the impact that electronic information delivery and analysis can have on the agricultural sector. The details of the program were announced on January 11 and the last day to sign up was March 11; thus the "life" or need for the program was limited to a period of approximately two months. The users of the program were able to evaluate easily those details of the PIK program applying to their operations without having to interpret all of the complicated government red tape. The university extension specialists were able to customize their analyses for thousands of producers effectively and efficiently.

## TIMESHARE PROVIDERS

In the late 1960s Michigan State University developed TELPLAN. A few years later Virginia Polytechnic Institute and State University developed CMN (Computerized Management Network). These early systems were based in the agricultural economics departments and each developed a library of programs geared mainly to extension applications. Both of these systems are still operational, but development of new programs has been limited.

While most land-grant universities have some agricultural software for use by their staff, a few universities have expanded access to their systems so that nonuniversity users can utilize their programs. AGNET, the largest such agricultural computer system, has many nonuniversity clients from the agricultural sector.

Several general purpose computer systems are available for information delivery, including Dow Jones, The Source, AgriStar, and Professional Farmer. Several suppliers are just starting to work with videotext units that combine graphics with information delivery.

## PROJECTIONS FOR THE TWENTY FIRST-CENTURY

The methods for information delivery and procedures for barter will change for the twenty-first century. Although it is difficult to project advances in technology and the impact of national and international policy on our society, I expect that much of our communications and information resources will be available via satellites through portable transmitter/receiver units. These units will have access to thousands of data bases and real-time information to assist the farmer in making management decisions.

The role of interactive computer networks will change significantly. New technology will provide for a rapid exchange of up-to-the-minute information (perhaps with a laser beam from a satellite), and computers will exchange information, data, and programs through the same message exchange media. These new capabilities will allow users to automate many of their shopping functions. Automated marketing, shopping, banking, forecasting, newspapers, and even private mail will be delivered through these new systems. In many areas employees will be able to work more effectively from their home offices.

Microcomputers and microprocessors will be built into most equipment and appliances. The microprocessor will monitor and control the operation of each process. For example, a microprocessor would be set up to monitor the comfort of each pig in a swine confinement building and control the environmental conditions of the building. The correct ration and amount of feed would be delivered to each pig based on its projected performance.

There will be no distinction between micro, mini, and mainframe. Powerful integrated processor chips will be customized for all kinds of specific purposes, and they will be networked together to perform the desired functions. There will be literally thousands of timeshare providers. All computers will be interactive and connected to numerous data bases through high-speed data networks. All information (data base) providers will compete to provide the best information—just as the TV networks compete for the best news coverage and programming.

## WHAT NEEDS TO BE DONE

Extensive development of application software will be the major task to prepare for the twenty-first century computer systems. That involves as a first step identification of the proper role of the computer in agriculture. Concurrently, extensive basic research needs to be conducted to provide the input for management models that can effectively assist with agriculture's day-to-day management. Agriculture's needs will not be limited by the availability of computer technology. Available technology has already surpassed our ability to use it effectively. We are hardware rich and software poor.

Policy is needed to develop the proper roles of the private and public sectors as they work together to serve the information needs for the twenty-first century. This issue has been addressed, and an overview of current technology provided, in a combined Congressional hearing and workshop in May 1982, which involved experts from the federal executive branch, state and local governments, universities, the agricultural sector, and the information industry.* As a follow-up, AGNET hosted a symposium in September 1983 on "Information Needs for Modern Agriculture."

# AGRICULTURAL INFORMATION UTILITIES

## LARRY R. WHITING

Sometime ago a futurist coined the term "information utility" to describe home or office computer-based information delivery systems. Although the label never caught on, for me it fosters an accurate image of a system that supplies information to multiple users through a network of computers in the same manner as a public electrical utility supplies electricity to consumers through a network of power lines.

Two types of computer-based information delivery systems are in use today: on-line and off-line. Both systems require the user to access a distant data base. The user—a farmer, rancher, agribusinessperson, and so on—must have a phone modem and either a remote ("dumb") terminal or a microcomputer. The modem permits use of the conventional telephone system to connect the terminal or microcomputer with the host computer containing the data. From this point on the systems have important differences.

With an off-line system, the user enters menu numbers for the specific information desired and then places a phone call to the host computer. In a few seconds or minutes, depending on the amount of information requested, the data base computer transmits ("downloads") the information to the user's terminal or microcomputer where it is stored in memory for either immediate or later use. At this point communication with the data base computer terminates, or goes "off-line." Consequently, an off-line system has one-way informa-

---

* For more information, see *Information Technology for Agricultural America*, prepared for the Subcommittee on Department Operations, Research, and Foreign Agriculture of the House Committee on Agriculture, December 1982.

tion flow from the data base to the user. The user can interact with the system only to the extent that certain types of information can be selected.

With an on-line system, the user remains connected with the host computer and can manipulate or influence the nature of the information that is to be provided. This feature is especially useful for problem-solving or decisional aids programs. For example, the user might request a dairy ration formulation program. Certain information such as feed ingredients on hand and level of milk production desired as well as a host of other variables might be entered by the user. The program would then calculate a balanced dairy ration given the variables specified by the user.

On-line systems can also accommodate educational programs based on principles of self-instruction or tutorial assistance. On-line computer networks can be used as one-way deliverers of information, but the converse is not true: off-line systems cannot be used for two-way communication.

## ON-LINE SYSTEMS

Michigan State University's TELPLAN was the first land-grant university, computer-based information utility and may even have been the first operation of this type in the nation. It was begun in 1966 with seed money from the Kellogg Foundation. The system became operational in 1968 when 40 programs were put on-line. TELPLAN was designed to provide computer-based, individualized problem-solving programs to county agents and state extension specialists. It was designed to be a low-cost, easy-to-use system—a system with which noncomputer-oriented persons would feel comfortable. In its very early days TELPLAN used touch-tone telephones to communicate with the host computer which had a voice response unit because hard-copy terminals were still very expensive. If the user did not have a touch-tone telephone, a touch-tone pad could be attached to the phone. The user would dial the computer after data were entered.

The high popularity of a dairy ration balancing program carried TELPLAN from Michigan into a number of dairy states. Today there are more than 90 programs available to TELPLAN users. And in 1983 the implementation in Michigan of COMNET, a statewide computer network with a computer in every county extension office, provides even greater accessibility for users to TELPLAN's problem-solving software.

Another very early computer network was the Computerized Management Network (CMN) at Virginia Polytechnic Institute and State University. This system was begun in 1972 as a pilot project of the Federal Extension Service and primary users were extension personnel. CMN now has users in nearly every state and in Canada.

Perhaps the largest and fastest growing university-related information utility is Nebraska's AGNET (Agricultural Computer Network) system. AGNET was launched in the mid-1970s at the Scottsbluff field station with two computer terminals linked to a computer at the University of Nebraska's Department of Administrative Services in Lincoln. Popularity increased rapidly, and in 1977 a $1.5 million grant from the Old West Regional Commission put the system in all the Commission's states: Nebraska, North Dakota, South Dakota, Montana, and Wyoming. Additional states also have come on board since then. Users are extension services, governmental agencies, agribusiness firms, and farmers. Farmers are accounting for an increasing percentage of the total as they rapidly adopt home computer technology. AGNET now has approximately 300 programs available.

In Indiana, Purdue University's FACTS (Fast Agricultural Communications Terminal System) consists of a computer in every county extension office, 10 area extension offices, and each campus agricultural department office. The central processing computer is housed on the Purdue campus and functions as a storage and forwarding system for the nearly 130 computers that form the network. Market and weather information and extension news releases, newsletters, and bulletins are disseminated from the state office to district and county offices. Each county office has a library of 50 problem-solving programs. The county microcomputers can also be used for word processing and other office management functions. Electronic mail is another feature of the FACTS system.

User fees or other costs to participate in a computer network system vary considerably. Generally, the user must pay any regular phone charges, a charge per hour when on-line with the computer, and a charge for storage of the user's data base. Rates also are generally lower for nighttime, weekend, and holiday use.

Michigan's COMNET and Purdue's FACTS systems were designed to be in-state computer networks, while systems such as Virginia's CMN and Nebraska's AGNET have attracted more external users. These four computer networks are the most active and advanced of the land-grant information systems.


## OFF-LINE SYSTEMS

As indicated earlier, off-line systems feature only one-way communication; that is, information is sent from a main computer to user terminals once the client specifies the information desired. The user cannot manipulate the information in the data base. Another term widely used to describe these kinds of systems is "videotext." One computer vendor, Radio Shack, a division of Tandy Corporation, has designed such an off-line system, called Videotex.

Between 1979 and 1981 the Cooperative Extension Service at the University of Kentucky in cooperation with the U.S. Department of Agriculture and the National Oceanic and Atmospheric Administration established a videotext experiment involving 200 farmers in two Kentucky counties. The purpose of the experiment was to test the feasibility of a computer-based information retrieval system for disseminating weather, market, and extension information statewide. By using a television set, telephone, and an interface unit that came to be known as the "Green Thumb Box," farmers could request information. Each of the two counties had its own computer, which would store information transmitted from a state computer.

The Green Thumb program was evaluated at the end of the 2-year experiment by the Institute for Communications Research at Stanford University. Market information was requested most often (55.5 percent) by the 200 users, and weather information was second choice (30.6 percent). Requests for extension information comprised less than 10 percent of the total system usage.

The evaluation concluded that the Green Thumb system was reasonably successful and that the overall design of the system was workable, but there were some trouble spots. The data bases were not always current. When users could not receive timely information, especially pertaining to weather or markets, usage fell sharply. Graphics were very limited and of poor quality. There were technical malfunctions because of environmental factors such as humidity and temperature fluctuations. Day-to-day management of the system was shared with other ongoing activities, so that the Green Thumb project did not always come first in terms of worker assignments and responsibilities.

A recommendation was made that a central system of data entry be set up so that updating responsibilities would be clearly delineated and there would be no redundancy of information. It was suggested that a central point of data entry might also enhance the timeliness of information. Another recommendation called for a single vendor to provide both hardware and software components giving the system greater internal compatibility.

Following the Green Thumb experiment, the University purchased a Videotex system from Radio Shack in order to determine its reliability, evaluate its updating techniques, and consider the feasibility of using it statewide. Unlike the Green Thumb experiment, Radio Shack's Videotex had been developed with its own software and hardware. Utilized were a TRS 80 MOD II microcomputer, a Ventel Auto Answer Modem, and a TRS 80 Communication Multiplexer with an 800 toll-free number to the University of Kentucky Agricultural Weather Center; the Weather Center would simulate a county computer center. The Weather Center also had a TRS 80 MOD II and a Ventel Auto Dial Modem so that weather and extension information could be updated at that location. A wire from the National Weather Service provided weather information, and the U.S. Department of Agriculture provided market information twice daily. The Videotex system was tested with 12 volunteer farmers in Daviess County.

The general conclusion from this limited test was that Videotex was generally reliable, but further developments were needed. Problems were encountered in getting reliable and timely market updates. A system is needed to coordinate outside sources of information. A typographical error in a market or weather report, for example, meant that the item had to be retransmitted from the information source after correction or that the Videotex system operator had to recreate the item and reinstate an update to the county computer. There were not enough information updates. As with the Green Thumb experiment, market information was in greatest demand followed by weather and extension information, respectively.

Perhaps the most significant conclusion was that the dissemination of extension information is not enhanced by a Videotex system because it has a lower demand in relation to perishable information such as weather and markets. The Kentucky Videotex test also prompted administrators to suggest that a two-way, interactive computer network may be more consistent with extension objectives because such a system would permit both problem-solving uses as well as straight dissemination of weather, market, and other agricultural information.

For nearly 2 years the Maryland Cooperative Extension Service has also been experimenting with Radio Shack's Videotex system. Maryland calls its system ESTEL (Extension Service Telecommunications). For a number of months the system was operated from the state office at College Park. At that time, system users were primarily extension field staff throughout the state and a few interested farmers (see Figure 1).

Market information originated from the Agricultural Marketing Service of the U.S. Department of Agriculture and consisted of commodity futures from Chicago and cash commodity prices from primary U.S. markets. This information was transmitted automatically by the USDA to the host computer in the state office. In addition, some local market information, such as livestock and cash grain prices, was contributed by the Maryland staff.

Weather information also was transmitted automatically several times a day from the National Weather Service. Reports ranged from county or zone forecasts to international weather and crop summaries. In addition, extension specialists in all program areas—agriculture, home economics, community resource development, and 4-H—contributed information to the system.

The Videotex system was run as a pilot program for 9 months in three counties on Maryland's lower Eastern Shore. In September 1982, Extension announced that ESTEL was ready to go public. A $25 annual user fee was established. To date there are 30 farmers on the system plus seven terminals in county extension offices which can be utilized by the public as well as extension

Figure 1.   Videotext Information Delivery System. A user of Maryland's ESTEL videotext information delivery system keys in menu numbers of information desired. Maryland
is piloting this computer network in 3 of its 24 county Extension offices and plans
to expand the system statewide if there is sufficient interest from farmers and other
Extension clientele.

faculty and staff. The user purchases a videotext terminal ($300). There are
no network charges to the user. In the original three-county area, all users
can access the system without dialing long distance.

Maryland extension administrators are allowing the system to expand into
counties where there is interest. In other words, there is no timetable as to
when the system will be in place statewide. Counties expressing no interest
will not be compelled to participate. It is thought that 18 county systems
could service the whole state without requiring any user to make long-distance
calls to access the system. Maryland has 23 counties and a city (Baltimore)
jurisdiction.

Counties must find their own funding to purchase videotext equipment: a
Radio Shack TRS 80 MOD II computer, a communication multiplexer, and
an auto answer phone modem. It is recommended that there be a second
MOD II computer so that local extension staff can input local information
into the ESTEL system. Total approximate cost for a county system ranges
from $11,000 to $15,000.

Although no formal user survey has been conducted in Maryland, user activity seems to be similar to the experiences of the Kentucky projects. Market information is most in demand, followed by weather and extension material. As with Kentucky Videotex users, Maryland clients also want more market and weather information and more updates.

Maryland also has occasional problems with automatically receiving information from government agencies. The only way to monitor the system from the state office is to call the county system to make sure information has been transmitted.

The above summaries do not encompass all of the on-line and off-line systems in operation across the nation's agricultural campuses, but they give a general notion of some of the innovations of computers and high technology.

In addition, a number of commercial firms are expanding into electronic information delivery. The Source is a large national computer network that offers a host of user services, from UPI news and stock market quotations to travel reservation services. Telidon is a large Canadian on-line computer network serving several hundred U.S. agricultural users and assisting with the start-up of a nearly identical system—Grassroots—in the United States. A large U.S. on-line computer network for agriculture is AgriStar, headquartered in Milwaukee, Wisconsin. Instant Update is a videotext system offered by Professional Farmers of America, an Iowa firm. Its major focus is farm market information. In Kansas the Harris newspaper chain has a statewide agricultural videotext network called Agritext.

Indeed, at some point private industry may move in and build upon the computer systems that have initially been developed in the public sector. Regardless of whether these systems are commercial or public enterprises or joint ventures, there are some quite obvious lessons to be learned from these early systems and some important implications for communications people.

## IMPLICATIONS AND CONCERNS

One of the important implications derived from the evaluation studies of the Kentucky videotext experiments and from firsthand observation of Maryland's ESTEL system is that most users want more information, particularly weather and market reports, more frequently updated. The underlying problem, I believe, is the failure of administrators to realize that computer networks are a communications medium, no different in intent from use of radio, television, or print media. Consequently, such systems should be managed as part of the traditional agricultural communications or information programs of the land-grant university. The best possible configuration of hardware can be assembled, but if information on the computer network is irrelevant, untimely, or simply in short supply, users will quickly become disenchanted.

Furthermore, the information must be unavailable, or not so easily available, from other media sources. For example, the National Weather Service prepares county weather forecasts and during the growing season releases agronomic weather reports. By and large, newspapers and local radio and television do not print or broadcast this specific information making it ideal for a videotext system. Videotext provides a unique opportunity to tailor information to very specific locations or even to very specific types of audiences or users.

A major conclusion from the Kentucky evaluations was that videotext systems do not enhance dissemination of extension subject matter. One could easily contend, however, that use of computer networks for that purpose has not been adequately tested. Improved graphics and presentation techniques, needs cited by Kentucky, might also increase usage of extension material.

Because profit margins in farming and ranching are so critically linked to weather and market conditions and because the chances for economic gain are so much greater in agriculture than in other extension program areas, one can understand the importance of planning and implementing videotext systems for farm use. But we need to be cognizant of the fact that extension and the land-grant system have responsibilities to other publics as well: home economics, community and resource development, 4-H and youth, energy, marine advisory programs, and many of their subgroups. A computer network could be just as useful to one of these as it is to agriculture.

Occasionally, the land-grant university has been chastized for devoting too much attention to large commercial farms and ignoring the needs of small or low-income farmers. We may need to consider equity implications: Will the small farmer be able to take advantage of computer technology to improve production and management practices? As prices continue to decline for microcomputers, software, terminals, and other technology, the potential for inequities seems to be less. A very exciting possibility of microcomputer technology is that any farmer, practically regardless of size of operation, may have immediate access to the same computer technology used by the largest business and industry. An accounting program, for example, will work essentially the same on a farmer's microcomputer as it will on a mainframe computer at the large company that manufactures the farm's machinery.

A major role for universities is to develop software that will be appropriate for the publics they serve. Universities also will need a "quality control" or peer review process for computer programs, just as they have for publication materials. University-produced programs must be useful and reliable.

Two-way, on-line, interactive computer networks will undoubtedly proliferate because of their problem-solving capabilities. One-way, off-line noninteractive systems such as Videotex are a satisfactory way to introduce the public to computer technology. Users, however, will soon outgrow these systems and seek the diversity of services that an on-line computer network offers. Furthermore, as more users relinquish the use of terminals for microcomputers, programs from on-line computer networks will be transmitted to the user for storage in his or her own microcomputer for use at leisure, decreasing the time for on-line connection and lowering communication costs.

Futurists indicate the world is entering the Information Age, the third such major historical era for humankind. What the plow was for the Agrarian Age and what the assembly line and standardized parts were for the Industrial Age, the computer will be for the Information Age. It has been said that 55 percent of the U.S. work force is already involved in "information processing," and the percentage is increasing at the expense of agriculture and manufacturing. Howard G. Diesslin, former Director of the Indiana Cooperative Extension Service and a major national proponent of computer technology for agriculture, has said that science was the great catalyst for the fantastic increases in agricultural production, but that agriculture is now entering an era when improved farm management will prompt equally great advances in agriculture. This will come about because of computer technology. The computer will greatly help farmers with the complexities of farm management. Computer networks—if you please, information utilities—will enhance the assistance that such technology can provide.

# THE FUTURE OF
# FARM MAGAZINES

## RICHARD KRUMME

I am not yet 40 years old and yet I find myself (1) worrying that electronic technology will pass me by; (2) defensive about the future of my livelihood—magazines; and (3) horror of horrors—resisting, or rationalizing, change.

More often than not change is perceived as a potential threat to the position and direction of an industry. And I readily confess to that thought. However, upon reflection one sees that "anticipatory" change—changing our products ahead of the competition to serve the reader better—has made *Successful Farming* and the Meredith Corporation more money than anything else. Besides, without change we journalists would have nothing to write about.

I do not know what is the wave of the future in agricultural information. I *do* know that enterprising publishers are looking into the new technologies which could broaden their publishing bases. I predict that farm magazines will be in the forefront of this technology if it should prove desirable.

It may surprise you to learn that farm magazines typically provide publishing leadership. They were well ahead of consumer magazines in collection of demographic data about readers. They were the first to provide regional and demographic editions to target audiences. And they were among the first to produce the enterprise verticals—publications that write about a specific topic only. Furthermore, farm magazine publishers were the first to experiment with electronic delivery of information. Many farmers have marketing wire services in their homes. Professional Farmers of America sells a service called Instant Update, and *Successful Farming* has a service called MarketCall. Also, *Successful Farming* has developed and sells software packages for farmers who have computers. We also started the first newsletter, *Farm Computer News*, for farmers with computers. Agricultural publishers still are, as they have long been, leaders when it comes to meeting the needs of customers.

So what do I believe farm magazines will be like in the next century? The truth is I don't know. I contend, however, that in one form or another magazines such as *Successful Farming* will still be around because we are so adaptive in serving the needs of both the reader and the advertiser. The trend toward specialization in print through highly targeted publications will continue. No other demographic group is so identified, so measured, as are farmers. There already are hundreds of magazine titles for farmers, from a more general farm news magazine such as *Farm Journal* to a general magazine with a business orientation such as *Successful Farming*, soybean and corn magazines, magazines for those who irrigate, and a magazine for those who irrigate corn.

The trend is toward targeting the high-income, upscale farmer, and many of us, including *Successful Farming*, have been quite successful in doing this. But other publishers have been equally successful doing just the opposite. For example, in 1968 when *Successful Farming* dropped its coverage of recipes and basic farm-homemaking information, a rival publisher started *Farm Wife News*. It was quickly successful. As many farm magazines began reducing the so-called farm lifestyle editorial, the same publisher launched a publication appropriately called *Farm and Ranch Living*.

Magazines such as *Successful Farming* tend to rise and fall with the relative health of agribusiness, that is, with the welfare of those who advertise their wares to farmers. Our subscription pricing structure does not require the reader to pay a full and fair price for the magazine. Instead, most farm magazines are subsidized by the advertisers. There are a number of successful controlled-circulation magazines, that is, free, which derive no revenue from readers. *Successful Farming* is the best, or worst, example, since we have both paid and controlled circulation. We have a circulation of 100,000—in the South—which is nonpaid. The economics of soliciting and procuring a low-cost subscription simply were not there, yet we wanted those readers in order to achieve a penetration into certain farm enterprises. A nameless rival publisher who thrives on doing just the opposite of what the old-line farm magazines do charges twice as much as *Successful Farming*. What does this mean? It means we in agripublishing are quite adaptive.

The trend toward print specialization offers an apparently limitless number of editorial and advertising format possibilities for the agricultural customer. But publishers ask whether the specialization trend can continue indefinitely. A recent *Wall Street Journal* article is entitled, "More Magazines Aim for Affluent Readers, But Some Worry That Shakeout is Coming." The article raises the concern that the special interest trend will create a "narrow, distorted picture of life as it is conveyed in print." At *Successful Farming*, as at other magazines, many of our lower-income readers resent the fact that we regularly portray large, late-model, expensive equipment with the farmers we feature. To some extent this is a distortion of reality.

Another challenge facing our industry is the development and proliferation of the home information system. Although the expense of such systems is something of a limiting factor at this time, the next 20 years will see more information consumers able to afford the initial purchase of such systems. As Alvin Toffler has suggested in *The Third Wave*, the electronic cottage is rapidly becoming a reality as an avocation, and more significantly, as a vocational possibility. The home-based computer system offers the user the opportunity to pursue leisure interests as well as management of financial operations. Many farm professionals (5 percent of our audience according to survey information) are employing this technology in the day-to-day management of their operations—from livestock inventories to projections about fertilizer costs.

With the increasing competition from cable, high technology, and computer forms, could the agricultural publication as we know it today become a relic of the past? Could the home computer lead to the ultimate phase-out of print in our society? I think not. I am inclined to liken computer delivery systems to television which could be a great source of information but most of us are watching "The Dukes of Hazzard." Likewise, computers in many of the homes I visit are used mainly to play *Star Wars* or *Pac Man*.

Nevertheless, I believe computers will revolutionize our businesses and our lives and perhaps even the way we receive information. I predict that this will not happen through home or business use of personal computers. It will happen as the miniaturized computers find their way into the machines and services we use in our everyday life: our automobiles, combines, feed grinders, ovens, telephones—even our bodies! I believe that farmers rather than routinely owning and using a personal computer will use information generated by someone else's computer. Several cooperatives have installed computers and will provide farmers with information about least-cost feeding programs. A major chemical company provides its large customers with free marketing information sent electronically. A machinery company will let farmers use its computer and software to assist in making buy, trade, or lease decisions. Lenders have installed personal computers which will show their customers the cash flow projections for a parcel of land the customer is contemplating buying. In other words, the information delivered by the computer will become either a premium or a sales tool.

Despite increasing competition from other media forms, the future of the agricultural publication lies in its stability as a traditional, time-tested source of information and in its proven ability to adapt. The magazine is nontransitory and this offers the consumer the opportunity of reading whenever and wherever he or she desires. Its low cost and secondary readership potential cannot be equaled by any other media form. Also, the diverse opportunities offered by advertising promotions and specific targeting to different farming operations ensure that waste and clutter will be kept to a minimum. Magazine CPMs—cost per thousand readers—have been increasing at a slower rate than that of any other media in the past 7 years, indicating efficiencies not present in other media.

The agricultural publishing industry and all publishing professionals in general have ample lead time to turn the electronic media forms into the handmaidens of magazines rather than combatants. We at *Successful Farming* expect to use the electronic media in harmony with our printed product. We expect the printed form to be the data base, if you will, and the springboard for any electronic delivery systems which we at *Successful Farming* would implement. We expect that other innovative publishers will do no less.

# 13

# MANAGING AGRICULTURAL TECHNOLOGY

JOHN T. CALDWELL

KENNETH R. KELLER

GEORGE HYATT, JR.

On a worldwide basis, and most dramatically in the United States, the march of agricultural progress is based on scientific knowledge and the systematic application of scientific principles. From Thomas Jefferson's identification of agriculture as a "science of the very first order" to the myriads of programs, grants, extension systems, and Acts of Congress that have come forth to nurture research and pass along the results to producers, our history clearly says that humans can command nature and that there can be an abundance of food. All that is required is the intelligent encouragement of science and the dissemination of its findings.

The world needs much more food just to maintain its current standard of living as it is. And the mere maintenance of that standard of living worldwide will by no means suffice as we enter the twenty-first century. Pressures of growing populations and rising aspirations make abundant food a vital determinant of peace and political stability.

Wherever they live, farmers will increase productivity when they are taught a better way to farm, when they have the resources to deploy new ideas, and when they perceive the rewards available from markets accessible to them. There must be a huge amount of technology transfer from the United States to farmers abroad, and it is the research done here that will develop the technology that we must transfer. As stated by Julian Simon in *The Ultimate Resource*, " . . . our cornucopia is the human mind and heart, and not a Santa Claus natural environment."

# SCIENCE AND AMERICAN AGRICULTURE

## JOHN T. CALDWELL

Once, most of the energy consumed by human beings was in feeding and clothing themselves. In American society today, however, only 3.1 percent of the work force is engaged primarily in producing food. This is a relatively new condition, and it is the result of knowledge. In no other arena of essential human activity has there occurred such dramatic progress.

When gathering was succeeded by planting and planting by cultivation, and when hunting was succeeded by husbandry, plateaus were reached that for tens of thousands of years determined the ratio of supply to population. As late as 1799 Malthus assumed that the food supply would be increased only by arithmetical increments, principally through added acreage being brought into cultivation. Of course, for many centuries people have made simple observations about sunshine, water, and soil that caused them to improve their practices in planting crops. At the same time they responded to the dictates of gods and spirits, to superstitions and folkways. Some farmers today still obey "signs" handed down from father to son. However, scientific agriculture based upon systematic knowledge is now the order of the day in the advanced Western world and has found its way in some measure to every continent. How did it come about?

In the thirteenth century Friar Roger Bacon set out to reform higher education, then dominantly monastical, by the introduction of a broad scientific program of secular studies. "He sought to show," writes J. B. Bury in *The Idea of Progress*, "that the studies to which he was devoted—mathematics, astronomy, physics, and chemistry—were indispensable to an intelligent study of theology and Scripture." (Roger Bacon may well have argued thusly to get the Pope on his side!) Bacon believed in the "close interconnection of all branches of knowledge," the "solidarity of the sciences." He could not escape, however, the medieval theory of the universe and remained focused on "things which lead to felicity in the next life."

So it remained for Sir Francis Bacon, "the first philosopher of the modern age," writing three centuries after Roger Bacon, to do two things of extraordinary relevance to every person today: to emphasize experimentation, "the direct interrogation of nature," as the key to the progress of science and to insist that the end of knowledge is utility. His "guiding star," writes Bury, was "the principle that the proper aim of knowledge is the amelioration of human life, to increase men's happiness and mitigate their sufferings and to establish the reign of man over nature." Francis Bacon was interested less in the next life than in the condition of human beings on this earth.

So the modern age of science was introduced into the Western world. In time it found its way to the fields and pastures. By the time of Malthus systematic experimentation and knowledge-building in agriculture had begun in Europe.

The first systematic research performed in Europe had to do with plant food and took place in Switzerland in 1804. Work on carbon, then nitrogen, then minor minerals followed in Germany, England, France, Russia, the Netherlands, and Denmark by the end of the century. As the work of individual scientists gained recognition, it encouraged other inquiring minds to investigate and publish, thus building a broader and stronger platform of knowledge.

The researchers of Europe did not escape the attention of eighteenth and nineteenth century Americans. Benjamin Franklin, more recognized abroad than at home as a leading scientific mind, was outspoken in his advocacy of science as a practical endeavor. Thomas Jefferson wrote that agriculture was "a science of the very first order." George Washington became an honorary member of The Philadelphia Society for the Promotion of Agriculture, which was founded in 1785. A similar organization, founded 6 years later in New York, announced a program of collecting information and conducting "experiments to seek improvement of exhausted land." Farm journalism and public fairs tackled the problem of persuading farmers on the merits of the scientifically produced facts pertaining to deep plowing, drainage, crop rotation, selective breeding, plant diseases, animal digestion, and fertilizers.

Some states appropriated funds to encourage agricultural production and marketing, and the federal government entered the picture in 1839 when Congress gave the Patent Office $1000 for the collection of farm statistics. The Patent Office "looked to the science of agriculture to make folklore yield to scientific analysis," and as historian David D. Van Tassel writes, "genius had stooped from its lofty height to lessen the farmer's burden." Historian Allan

Nevins, writing of "the scientific impulse" that characterized America's early nineteenth century, tells of the effect of a $500,000 grant by Englishman James Smithson to the U.S. government. John Adams wanted it to be used for "acquiring knowledge" as "the greatest benefit that can be conferred upon mankind." Specific proposals included one for a school of science, another for "an institution for the promotion of scientific agriculture with laboratories, workshops, and an expert staff." The Congress in 1846 settled on "The Smithsonian Institution" to serve as a museum, a research center, and a publisher of scientific reports.

As scientific knowledge in agriculture accumulated, its value and the need for more systematic investment became evident. In 1862 Congress passed two momentous acts: The Land-Grant College Act and the bill that created the United States Department of Agriculture. Both were aimed at bringing knowledge to bear on agricultural enterprise and, writes science historian A. Hunter Dupree, marked "a genuine turning point for science in government."

It is well-known that the Land-Grant Act of 1862 addressed explicitly the need for "education" in agriculture and the mechanic arts (engineering). It is less well-known that the 1862 Act also required the participating states to make an annual report "regarding the progress of each college, recording any improvements and experiments made, with their costs and results." These reports were to be sent to all the other land-grant colleges. A base was being laid for broader scientific endeavors, and field trials came more and more to be linked with laboratory work. The Hatch Act of 1887 established "agricultural experiment stations" at all the land-grant colleges and signalled a new era of federal–state cooperation in applying science to agriculture. The 1890 Land-Grant Act provided for an additional college in each state where the existing land-grant college was not open to blacks. The Smith–Lever Act in 1914 created the Agricultural Extension Service as a means of taking knowledge to the farm.

Research related to agriculture today covers an enormous spectrum of knowledge in the natural and social sciences. It involves fundamental interdisciplinary inquiries into the nature of matter and its behavior; the most refined applications of pure knowledge to manipulating plant and animal behavior in natural and human-influenced habitats; and the investigation of the relationships of natural and social factors in the production, distribution, and utilization of food. Our enterprise society has produced an incomparable structure of institutional, private, and governmental effort to support the individual intellect.

But even so it must be revitalized from time to time. At a recent conference at Winrock International in Arkansas, agricultural scientists were critical of the present state of the enterprise and pointed out that the researchers themselves are not sufficiently critical of the system; that federal "scientific and intellectual leadership is no longer visible;" and that institutional researchers are too insular and often pursuing political goals. The "system" supporting research needs fresh air. One way to open up to potentials would be to use competitive research.

Federal support for science, including the agricultural sciences, has experienced political manipulation. The requirement for geographical spread among the recipient institutions has reduced the effectiveness of expenditures. A preoccupation with socially oriented goals in increased federal grants to newly developing institutions has not increased the effectiveness of funds expended for agricultural science, but has spread them thinner.

The central mission of agricultural science is nothing less than the nutritional well-being of the human race, not transient and narrower goals. Yet we slide into neglect of research. We cut back funds. We act as if we already know enough. We tend to revert to solving the problems of the moment—a new plant disease, a new harvesting problem, a pest. We neglect to study how new fundamental discoveries on the nature of matter challenge all existing knowledge. We fail to study unfolding evidence of the wholeness of knowledge. We become bound to our existing ways and systems of funding and performing institutional research. We cut back on providing incentives to graduate study— the only path for bringing on new generations of scientists.

In other words, the scientific endeavor needs constant renewal, and we neglect the need. To put it another way we do not seem to grasp the meaning of past marvels of science. They tell us that the future is one of immeasurable possibility, that we never know enough nor have used enough of what is now known, that "out there" is a world of fact waiting to be uncovered and used.

One large component of the research enterprise in America is the private corporation, large and small. New products, new processes, and the creation of new markets are part and parcel of the enterprise system. Their contribution to what we know is not only large but indispensable. Let it be recognized, however, that the basic knowledge upon which their creativity is based is produced by institutions mostly, if not exclusively, in the public, nonprofit sector of our society—universities, foundations, and government. Any neglect of these seedbed institutions carries untold penalties for the producing and processing sectors of the food economy.

Politicians and budget cutters are not the only ones to blame. The scientists themselves, who now spend millions, must become pioneers again and combat more boldly the complacency bred by present abundance and the comfortable institutional structures that inhibit creativity as well as support it.

It would be derelict, however, not to salute the large foundations, especially the Rockefeller Foundation and the Ford Foundation, who have contributed so wisely and generously to the expansion of world food supply. Their efforts have been most dramatically exhibited in their pioneering support and continuing support of the great centers of production research in Mexico, Colombia, Peru, the Philippines, and India, notably in such basic foods as rice, wheat, and potatoes, but extending also into crops less universally utilized. These enterprises present stunning examples of how the U.S. government, foundations, other countries in the world, and indigenous government agencies in the host countries have worked to bring science to bear on the world food shortage.

The consciousness of decision makers must be aroused to a more realistic, imaginative, and adequate support of the knowledge enterprise that serves people's most fundamental physical need.

# GENERATING TECHNOLOGY

## KENNETH R. KELLER

The initiation of a research and development program for obtaining scientific information—in other words, generating technology—generally results from one or more major unresolved problems that are usually surrounded by complex forces. Many problems of the agricultural economy are difficult to resolve satisfactorily. Others yield a clear and concise understanding.

But Saul David Alinsky, the American sociologist, has said that "the world is deluged with panaceas, formulas, proposed laws, machineries, ways out and myriads of solutions. It is significant that every one of these deals with the structure of society, but none concerns that substance itself—the people. This despite the eternal faith that the solution always lies with the people." I am convinced that people are and always will be the most important element in generating technology.

Agricultural researchers, through scientific discoveries, have established a bond between producers and consumers. Plant and animal scientists have assured consumers adequate agricultural and forestry products and also sought to understand how new knowledge can be turned to beneficial uses. The results of their studies have made it possible for farmers to increase production and, indirectly, for new industries to develop. The research has been supported by state and federal funds, which are justified on the grounds that the welfare of agriculture is basic to the nation.

It is anticipated that, in the next 40 years, farmers must produce more food than was produced in the past 1600 years just to keep up with the long-term demand. The social, economic, and political problems which attend development are now as urgent as the problem of food production. Therefore, it would seem that population growth and unemployment be included in research objectives on a level commensurate with the following:

1. Development of new and improved plants and animals, management practices, and processing techniques that will enhance the use of agricultural commodities in domestic and foreign markets.

2. Development of management practices that will increase the productivity of crops and livestock.

3. Development of technology that will aid in the improvement of environmental quality.

4. Improvements in the safety and nutritive quality of foods and feeds.

Certainly, encouragement must be given to our farmers who supply abundant quantities of food at bargain prices. American families, on the average, spend only 17 percent of their income for food, compared to 27 percent in Germany, 32 percent in Japan, 45 percent in Communist countries, and 60 percent in developing countries. Our farmers manage a food and fiber system that employs 23 million people and contributes 20 percent to our gross national product.

The substantial research and development capabilities and educational resources of the United States can make a major contribution to the world capability to increase food availability and at the same time benefit domestic producers and consumers. The best research and development groups are, in general, those that are well financed and employ resourceful personnel. Adequate buildings, laboratories, greenhouses, and plot space, coupled with essential research tools and support personnel, are indispensable. The administrative staff must be sensitive to the needs of the scientists and strive to maintain and promote a creative scientific environment. Young and promising scientists need encouragement and guidance from their administrative leaders. An annual or periodic personal evaluation conference is in order. These initial years can be costly to the employer and efforts should be made to recognize deficient performance habits and to correct them.

In the beginning, federal funds were essentially the sole source of support for research and development programs. In time, state moneys and, later, grant funds from both public and private agencies became an integral part of the budgetary process. They have created the stability, continuity, and ample financial support necessary to attract and hold the outstanding scientists required for a successful research and development thrust.

The need for reliable technology has intensified gradually as farmers have come to realize that the many problems confronting the agricultural industry are complex and that well-trained people are needed to solve them. Thus the development of new technology is dependent upon a cooperative, multidisciplinary team approach.

The scientist is the foundation of the research and development program. Attributes such as scientific imagination, the power of discrimination of the essentials and nonessentials, confidence in data, honesty in observation, and fairness to associates and subordinates are essential. Enthusiasm for a project such that an individual will push forward in the face of seemingly insurmountable obstacles is of utmost importance. Organizing research by the project system provides a current record of the status of the studies and is of enormous value in their coordination and continuity, because it requires submission of an outline and its approval by peer scientists and administrators before any action is taken. Success depends upon the leaders and the peer reviewers for their scientific attitude, depth of motive, and conception of the problem and its requirements.

Annual estimated returns on investment in agricultural technology in developed and developing countries are summarized in Table 1. These rates of return are impressive, yet policy makers often lack these data in their decision making. An effective, organized, and sustained means of providing information on the results of agricultural research and development is essential.

A report entitled *World Food and Nutrition Study,* issued by the National Academy of Sciences, refers to "the substantial research and development capabilities and educational resources of the world capability to achieve markedly increased food availability and to minimize the age-old threats of hunger and malnutrition." The report says that "these efforts could benefit both the producers and consumers of food in our own country by lowering production costs, increasing agricultural profitability, and helping the rising cost of food." The report identifies the following 22 areas needing major attention: aquatic food sources, biological nitrogen fixation, farm production systems, fertilizer sources, food reserves, information systems, irrigation and water management, manage-

Table 1.    Estimates of the Rates of Return from Investment in Agricultural Technology

| Country | Commodity | Period | Percentage Annual Return |
|---------|-----------|--------|--------------------------|
| United States | Hybrid maize | 1940–1955 | 35–40 |
| United States | All research | 1949–1959 | 47 |
| Mexico | Wheat | 1943–1963 | 90 |
| Mexico | Maize | 1943–1963 | 35 |
| Bolivia | Rice | 1957–1964 | 79–96 |
| Colombia | Rice | 1957–1972 | 60–82 |
| Japan | Rice | 1930–1961 | 73–75 |

*Source:*   Adapted from S. Wortman and R. W. Cummings, Jr., *To Feed This World,* The Johns Hopkins University Press (1978).

ment of tropical soils, market expansion, national food policies and organizations, nutrition performance relations, nutrition intervention programs, pest management, photosynthesis, plant breeding and genetic manifestation, policies affecting nutrition, post-harvest losses, resistance to environmental stresses, role of dietary components, ruminant livestock, trade policy, and weather and climate.

In addition, I believe that more attention should be directed to developing technology pertinent to efficient production, marketing, distribution, and utilization of farm products for, in the words of the Hatch Act, the purpose of "improvement of the rural home and rural life and the maximum contribution of agriculture to the welfare of consumers." Cooperative studies by intra- and interdisciplinary teams is required. Multidisciplinary action may solve more problems than traditional approaches since it usually generates much interesting basic technology. Those who generate new technology must understand the framework within which the innovation will function, and the scientists must translate the results of their findings into terminology that is understood by the farmers.

The urgency to continue technological development for future higher yields is real, even though we are in a market flooded with grain surpluses. Estimates from the International Food Policy Research Institute pertinent to doubling agricultural production to meet consumption needs are presented in Table 2. It has been estimated that the entire world agricultural production must double in less than 20 years to meet expanding world food needs. Current genetic potential for various crops has been predicted at several times today's best average yields. Geneticists and plant physiologists estimate that the optimum theoretical genetic potential of the best corn hybrids will be in the 500 to 600 bushels per acre range. The projected U.S. average corn yield for 1982, which was a near-ideal corn growing year in most areas, was 113 bushels per

Table 2.   Estimated Number of Years for Agricultural Production to Double to Meet Consumption Needs

|  | Countries by Years | |
| 7–12 Years | 13–14 Years | 15–19 Years |
| --- | --- | --- |
| Algeria 7 | Bangladesh 13 | Mexico 15 |
| Costa Rica 8 | Colombia 13 | India 18 |
| Philippines 12 | Nigeria 13 | Brazil 19 |
|  | Morocco 14 |  |

*Source:*   International Food Policy Research Institute, Washington, D.C.

Table 3. Average Yield in Developing Countries vs. U.S. Average and World Record Yields (1975)

| Crop | Average Yield (metric tons per hectare) | | |
| --- | --- | --- | --- |
| | Developing Countries | United States | Reported World Record |
| Maize | 1.3 | 5.4 | 21.2 |
| Wheat | 1.3 | 2.1 | 14.5 |
| Soybeans | 1.6 | 1.9 | 7.4 |
| Sorghum | .9 | 3.3 | 21.5 |
| Potatoes | 9.1 | 26.6 | 94.2 |

*Source:* Adapted from S. Wortman and R. W. Cummings, Jr., *To Feed This World,* The John Hopkins University Press (1978).

acre—less than one-fourth of the estimated potential. Table 3 shows the average yield in developing countries versus the U.S. average and world record yields.

The FAO report emphasizes that three-quarters of the new food production needed through the year 2000 must result from increasing yields on currently cropped acreage. The key to meeting world food needs in the short term is to narrow the gap between today's average yields and currently attainable record yields.

According to Duke University botanist Paul J. Kramer, crop yield is controlled by the interaction between the genetic potentialities of crop plants and the environment in which they grow. Variations in the genotype and the environment, including weather and cultural practices, act through physiological processes to control growth.

David W. Dibbs, Southeast Director and Latin American Coordinator of the Potash and Phosphate Institute, has suggested a maximum-yield approach to meet expected world food needs and to help farmers' economic survival. Dibbs states that new research must focus more directly on maximizing yields. He further emphasizes that developing food production technology now and in the future depends upon the development of appropriate agronomic systems. He defines maximum-yield research as: "The study of one or more variables and their interactions in a multidisciplinary system that strives for the highest yield possible for the soil and climate of the research site." In their book, *To Feed This World: The Challenge and the Strategy,* Wortman and Cummings have put the maximum-yield research concept in perspective for the modern food researcher:

As a part of their agricultural research, biological scientists in each country must continuously test the limits of available technology as well as their ability to put together new technological components to achieve highest yields. Maximum yields initially may not be economically practical, but agricultural scientists should not, for supposed economic reasons, fail to attempt to raise the limits on productivity imposed by technology.

In reaching the yield goal simple interactions can be significant, such as varying the levels of two important mineral fertilizer components. Kramer implies that an important contribution will be made by complex interactions in which three or more variables are involved at various rates or combinations. As new technology is generated, the challenge will be to learn to manage the specific positive interactions and thus further refine the system for higher yields. Successes in maximum-yield research and development programs are already emerging in soybean, wheat, alfalfa, and other crops.

The Food and Agricultural Act of 1977, which put federal allocations to the land-grant colleges on a continuing basis, stipulated that a comprehensive program of agricultural research be developed. As a result, current research projects in agriculture are regional, national, and international in scope, and they involve numerous scientists and represent many disciplines. Multidisciplinary cooperative agricultural research is expanding and is expected to continue and become more prominent in the twenty-first century.

For example, scientists of North Carolina Agricultural and Technical State University and the North Carolina Agricultural Research Service of the State University have joined in the development of a comprehensive program of agricultural research which is fundamentally service-oriented and aimed at the citizens of the state. Scientists at each of the cooperating universities are directing their attention to both applied and basic research. Procedures for developing and coordinating the comprehensive multidisciplinary cooperative research program have been carefully defined, with special emphasis on communication, preliminary planning, project development, and project processing and initiation.

As cooperative research becomes more important and significant, other developments in the twenty-first century will not all be optimistic. The following are my projections for that time:

1. Scientists will continue to generate an abundance of technology, but they will do so with rising costs and increasing pressures on laboratories, greenhouses, controlled-environment facilities, land, energy, and personnel (technicians).
2. The relative cost of generating agricultural technology is likely to increase more rapidly than the cost of research in other fields.
3. Private sector moneys for support of research in land-grant universities will play a more important role in supplementing state and federal appropriations.

4.  The value obtained from developing new knowledge will have to be translated to the public and, specifically, to those who appropriate financial support for agricultural research.

5.  Research and development capacities of some nations will remain inadequate to meet domestic demands.

6.  Technology pertinent to cultural and management practices will continue to improve at low rates as problems become more complex.

7.  Multidisciplinary, cooperative, agricultural research will be pursued more intensely on regional, national, and international bases.

8.  Increased pressure will be exerted to pursue cooperative, comprehensive, multidisciplinary agricultural research among scientists of the land-grant colleges of 1862 and 1890.

9.  The increasing cost of producing food and fiber will stimulate the use of unconventional techniques not now in commercial practice due to the development and understanding of the growth processes and their interactions in both plants and animals.

10.  As population increases, the demand for generating new technology will intensify and exert greater competition for well-qualified and productive scientists.

The most important decision we ever make is when we employ people to devote their talents to gaining new knowledge. Administrators must refrain from prescribing the specific course scientific investigations should follow but should, rather, recognize the potential capabilities of the scientists they employ and provide for them encouragement, resources, and a creative environment.

The future for generating agricultural technology, therefore, is entirely dependent upon the minds, bodies, souls, and spirits of those who individually or collectively seek solutions of interest and value to all segments of our society. I challenge you to select employees with care, wisdom, knowledge, patience, discernment, and faithfulness to assure a rich and rewarding twenty-first century.


# TRANSFERRING TECHNOLOGY

## GEORGE HYATT, JR.


*The Global 2000 Report to the President,* released on July 24, 1980, said that we are headed for "worldwide poverty [and] food shortages by the end of the century." I disagree with this conclusion. Agriculture in the United

States has developed into a thriving industry during the past 80 years because through industry cooperation and our land-grant system we have learned how to develop new technology and to transfer it to those who can use it. We have much more to learn about technology transfer worldwide, but significant progress is being made in this tough area.

In *The Ultimate Resource*, Dr. Julian L. Simon states:

> I do not believe that Nature is limitlessly bountiful. I believe instead that the possibilities in the world are sufficiently great so that with the present state of knowledge, and with the additional knowledge that the human imagination and human enterprise will develop in the future, we and our descendants can manipulate the elements in such a fashion that we can have all the mineral raw materials that we need and desire at prices even smaller relative to other prices and to our total incomes. In short, our cornucopia is the human mind and heart, and not a Santa Claus natural environment. So has it been in the past, and therefore so it is likely to be in the future.

I would only add that there must be increasing support for research and education in agriculture and natural resources in our nation's land-grant universities.

There are many problems complicating the transfer of technology to farmers here and abroad. It is now the accepted view that farming is not merely a technological endeavor but a socioeconomic one as well. Years of research and study have been needed to bring us to the understanding that technology transfer, especially among small farmers with limited resources, is a very complicated process.

## FOUR REQUISITES

The late Sterling Wortman and R. W. Cummings, Jr. in *To Feed This World* state that farmers, regardless of size of landholding, generally will increase their productivity provided four requisites are met:

1. *An improved farming system.* A combination of material and practices that is clearly more productive and profitable, with an acceptable low level of risk, than the one he currently uses, must be available to the farmer.

2. *Instruction of farmers.* The farmer must be shown, on his own farm or nearby, how to put the practices into use, and he should understand why they are better.

3. *Supply of inputs.* The inputs required, and, if necessary, credit to finance their purchase, must be available to the farmer when and where he needs them, and at reasonable cost.

4. *Availability of markets.* The farmer must have access to a nearby market that can absorb increased supplies without excessive price drops.

Wortman and Cummings also point out that, if the above conditions are met simultaneously in any locality, chances are good that a high proportion of farmers will change in time. As I see it, strategies needed to bring about technology transfer domestically and in the developing world are somewhat similar, even though cultural differences greatly complicate the problem.

As I reflect on my experience as a dairy specialist in three states, a farm manager, and later director of the Agricultural Extension Service in North Carolina, I am convinced of the accuracy of the conclusions of Wortman and Cummings. My work in West Virginia in the 1940s was with many small dairy farmers producing a can or two of milk daily for manufacturing purposes. At the same time I worked with many large and emerging dairy farmers in West Virginia, Wisconsin, and in North Carolina. In western North Carolina we have had many years of joint work with the Tennessee Valley Authority in learning how to speed up the process of technology transfer to farmers developing a trellised tomato industry, a Christmas tree industry, and strawberry and dairy farms. To make progress—to transfer technology—the four conditions outlined by Wortman and Cummings had to be in evidence almost simultaneously.

Let's examine these four conditions in a little detail:

1. *The Profit Motive.* The farmer's attention must be attracted in the beginning by pointing out that changes in technology or production practices will give visibly higher yields of income—not just small increases, but yields that are 20 to 30 percent higher. Increases of 10 percent are just not enough incentive for many farmers, especially those of smaller means and resources. Small increases are difficult to demonstrate and to see with the eye.

To develop new and improved farming systems is a complex operation. If, for example, farmers are to use a new cropping system, they must know the best varieties, dates of planting, right amounts of fertilizer, when to irrigate, appropriate pesticides, and so on. They must also know the proper combinations of production practices and understand economic decisions that will affect them. More than just increasing yields is involved; many of these decisions must be drawn from on-farm tests in the locality.

2. *On-Farm Tests.* Through much research in recent years we are learning how better to instruct farmers to speed up technology transfer. Managers of large farms may only require new information coming from research centers, but the vast majority of farmers in the United States and throughout the world need much more. There has to be the understanding among our research scientists, extension workers, and farmers that new technology taken directly from the Experiment Station to the local farm is often not adaptable.

Robert Rhoades, working with the International Potato Center in Peru, points out that a problem must be defined by both farmer and scientist. In this way they can arrive at the widest possible consensus on the definition of the problem to be solved. Rhoades describes a commonly defined problem centered on seed potato storage and, more specifically, on how to reduce sprout elongation and maintain seed quality. The interdisciplinary team was able to move forward with on-station research guided by farm-level information. The potential solution stood a good chance of helping to bring about the technology transfer desired.

On-farm trials are an important facet of many state extension service programs. We use them extensively in North Carolina, especially with the tobacco program. These trials provide information from which scientists and others can develop recommendations for farmers of the area by allowing farmers to see the effects of inputs and practices on production and to see the payoff effect on the farms. Such trials attract the attention of farmers throughout the area, and they are also helpful to fertilizer distributors, bankers, and others in understanding the new technology and its need for transfer. Extension agents must demonstrate.

3. *Credit and Supplies.* Supplies or credit are often either not available when needed or are not available in the quantities desired, especially in other parts of the world. However, the present recession dramatizes the need for low-cost credit to farmers if they are to proceed with normal operations, especially if new technology is to be transferred and changes in the farming system are to result. On the other hand, the present tight money squeeze may speed up technology transfer for large operations where small increases are important at a high level of efficiency.

Farmers in many parts of the world depend a great deal on the use of their own land, labor, and seed to get the job done. To shift to a high-yield system they usually must make purchases. These supplies of seeds, fertilizers, pesticides, and so on must be available nearby. Farmers also need to be able to purchase such supplies in small amounts. Often a source of credit will be needed to make even small amounts of purchases possible.

4. *Markets.* The first question that should be asked is: What are the needs for the product? Another is: Are there people who want the product, and in what form? A particular geographical area may be ideal for growing trellised tomatoes, Christmas trees, green beans, strawberries, and so on, but where are the customers who will buy such products, and what and where are the facilities needed to process them? New technology for producing cheap eggs and poultry meat by flock owners in North Carolina was not of great value until the processor of poultry products and the transportation to distant markets became available. Then the industry expanded greatly, and the research at the experiment station could be used in an ever-expanding manner. The

experience with the swine industry in North Carolina in the past 20 years followed the same path. As the markets became available, there was an ever-greater need for transfer of the technology available. The conditions for the payoff were now in place.

In developing countries the availability of markets is a tough problem to solve. Many agricultural commodities are bulky and highly perishable. This means there must be purchasing points near each farm where the farmers can receive a fair price. In many cases this means a guaranteed price announced before the crop is grown. Governments must be prepared to honor such obligations when harvest time comes.

While examining the research extension programs in Ghana a few years ago, it was obvious to me that the new findings in research pouring from the Ghana Experiment Station dealing with increased cocoa production had great potential for use by the cocoa plantation growers if only a better means of curing, storing, and transporting the product to the markets could be developed before so much spoilage of the product took place. New technology was available, but not the kind that was actually needed at the time. The experiment stations were really not geared to the farmers' needs, and their personnel were critical of extension because yields per acre of cocoa production were not responding to the new technology.

## EXTENSION CREDIBILITY

Extension services have served Americans with information and educational programs for nearly 70 years. Their credibility has been built at a time when knowledge has grown, audiences have changed, and extension staff expertise has had to grow too.

In the early days of the U.S. extension experience it was learned that separating the regulatory and educational responsibilities was imperative. Extension educators could not be effective if they operated in a dual role as educator and enforcer of regulations. In such a setting technology transfer is made difficult, if not impossible.

Early in my career as a dairy specialist I was hired to work with dairy farmers on their farms to help increase efficiency of milk production by bringing to them the latest research information through their local county extension agents. My other hat was to spend several days a month in the milk processing plant to which the dairy farmers shipped their milk. To my employer, the university extension service, this seemed an excellent way to combine my previous experience as a milk inspector with my knowledge and training in dairy production. I objected to this arrangement, but I needed the job.

It turned out to be a disaster. My role in the plant was to reject milk for off-flavors, high bacteria count, sediment, and so on. Such rejected milk went directly back to the dairy farmers with my name on the rejection slip. Then it was my responsibility to go back to the farm and help solve the problem. It certainly doesn't require a degree in psychology to understand that the very client I was to help educationally developed a negative attitude toward me. Fortunately, new arrangements were made within a year, and my regulatory responsibilities were shifted to other personnel.

In many countries throughout the world the extension service is an educational arm of the Ministry of Agriculture, where many functions are often combined. The extension agent may be responsible for inspection, regulatory duties, and the dispensing of credit for the purchase of inputs. Such arrangements detract greatly from the main responsibility of effecting technology transfer. The credibility of the extension educator is largely destroyed if the role includes regulatory, purchasing, credit relationships, and other responsibilities.

Extension credibility is closely tied to the acquisition by its personnel of new knowledge, not only about agricultural technology but also about ways of diffusing it quickly to the user and of finding out what problems farmers are willing to consider. Extension personnel must be continually trained and updated if they are to be effective, a job that is both time-consuming and costly. Many extension agents are more successful in keeping up-to-date with new research than in getting it transferred to clients. R. J. Hildreth, Executive Secretary of the Farm Foundation, points out that "the extension professional field worker or specialist plays a somewhat different role than the other professionals with whom he works. Although he's an educator, he operates in an informal, non-certified way, quite different from the classroom teacher." I would say that he or she must be an interpreter.

According to Paul J. Leagans, Professor Emeritus, Cornell University, "the advantages of modern agricultural science and technology have not yet effectively reached at least a billion small farmers located throughout the world." He further points out that the data he and his co-workers collected "indicate that a great many additional farmers will use the new technologies when the conditions necessary for making innovative action understandable, possible, and favorable are readily available." Agents must be trained in such areas. Knowledge of new technology alone won't get the job done.

Extension administration in the United States has made it imperative that their personnel be given many opportunities for training. In recent years industry and the university have joined hands in sharing this cost. I was part of the joint arrangement between Philip Morris and North Carolina State University that provided financial support for our Extension Agent training programs. Such funds have not only helped bring some of our agents to the cutting edge of new knowledge but also have greatly improved the morale of the extension organization.

### THE MEDIA

Communication through a one-on-one situation is what most extension workers are limited to today. They look to the printed word, radio, television, computer-assisted instruction, telephone systems, and many combinations of these systems to bring information on new technology to the user. The extension educator must be trained in the use of all this new gadgetry. In the developing countries, where personal contact is still the most effective means for promoting any type of program, the huge numbers that must be contacted make it imperative that mass-media methods be employed in the most effective ways. William L. Carpenter comes right to the point when he says that "with such a variety from which to choose, today's educators may be more confused than comforted when they want to select communications media to use in their educational programs. But for the sake of their students' learning and their own professional competence, it will behoove them to understand their new media and learn to apply them."

Technology transfer will continue at an ever increasing pace when the profit motive is at a reasonable level. Supplies, credit, and markets are easily accessible, and the farmer becomes a partner with the scientist in guiding the research through on-farm tests. Extension credibility must be improved through constant training efforts so that extension personnel are well-informed about the new technology and the methods for transferring it to the user.

# IV
## FINANCING AGRICULTURE

# 14

# CAPITAL INVESTMENT
# AND THE BUSINESS
# OF AGRICULTURE

WALTER W. MINGER

By the dawn of the twenty-first century the politically powerful farmers in Japan and the European Economic Community will probably have seen much of their power eroded. Many of the younger generation will have left the farm and gone off to Tokyo, Brussels, Paris, Rome, Madrid, or other cities where bright lights and a supposedly easier and better life beckons. I believe that those left on the farm will lose political vigor with the advance of years. Consequently, I assume that in Japan and Europe it will be the interests of the urban consumer that will loom the most important in politicians' minds.

The problem of subsistence farming will still be with us, particularly in India, parts of Africa and China (both The Peoples' Republic of China and Taiwan), Pakistan, Bangladesh, Indonesia, the Philippines, Mexico, Nigeria, and a number of other less populated developing countries. By the year 2000 we will see hundreds of millions of subsistence level farmers and like numbers of landless people who suffer from low incomes and inadequate nutritional levels.

The present serious global depression is expected to continue into 1985, and it is anticipated that the world's economies will experience relatively sluggish economic growth afterward. Thus agribusiness companies are not building up liquidity and/or after-tax profits that would normally be used to upgrade productive capacity and improve the efficiency of the world's food system. Continued inflation, coupled with the need by management of agribusiness companies, particularly in the United States, to present the best picture possible to their shareholders, will lead to the general practice of setting aside too little in reserves to replace aging capital items in much of the world's food system.

In the less developed countries, the continuing lack of liquidity and foreign exchange earnings will force governments to husband cash. Agriculture will compete—unsuccessfully—with social programs required to forestall political unrest and uprisings. Therefore, roads, communications and agricultural extension systems, transportation systems, port facilities, water delivery systems, river and port ship channels, irrigation systems, and other sectors of the food system will be in need of major repair or replacement.

In the developed nations the vast investments needed to support relatively sophisticated food systems will also be victims of reduced cash flow to the government. Because of budget constraints, by the year 2000 there will have been a decline in the availability and quality of government-supplied services, such as statistics and market news. Budget cuts will also have hurt the operations of river locks and canals, inspection facilities, conservation and reclamation works, port facilities, railroad beds, and highway systems. We will see clearly that part of the United States' vaunted advantage as an efficient and low-cost producer of food and fiber was that the food system lived off capital.

Much of the equipment and many of the structures that house agribusinesses were already 30 to 40 years old in 1982; in the year 2000 they will be 50 to 60 years old. They will still be capable of being operated but most probably should be junked and replaced by more modern and efficient capital goods.

So, as we roll into the year 2000, we will see a global food system that still suffers from social problems, is strongly affected by politics, faces massive needs for capital to modernize a decaying plant, and yet remains one of the truly essential industries on the face of the globe. How will the basic resource—money—be used by agriculture during the next century to carry out its role of financing food and fiber for as many as 6 billion people scattered all over the planet Earth?

Agricultural financing cannot be discussed in a vacuum. The wise use of money is an important consideration in planning the overall food and fiber strategy. But capital is so interrelated with labor, land, management, and politics that we must consider some factors that would at first blush not seem to have much relationship to financing food production.

If we make a simplistic assumption that lenders will continue to lend and borrowers continue to repay, then there is not much change that is likely to occur in agricultural financing by the year 2000. But suppose there are further consolidation of farm units, continued increases in the level of capital employed, further vertical and horizontal integration, some continued inflation in costs, only modest increases in yield, and an extremely slow increase in commodity prices. Then we can see that farmers who borrow will represent a riskier class of debtor than those who borrowed 25 years previously.

Another factor that affects financing is the expectations for repayment. By the year 2000 the U.S. farmer will be dependent on the export markets for possibly 40 percent or more of annual sales. In 1982, with $140 billion in sales, the export value of agricultural products was down to $42 billion from the previous year's high of approximately $44 billion. Exports in 1982 represented 30 percent of farmers' sales. Greater reliance on the export markets is a necessity for U.S. farmers, but these markets have volatile prices and are uncertain in volume. As a result, there is additional uncertainty as to whether a farmer can repay his borrowed capital. While large-scale farming operations will be increasingly involved in direct export sales, the major share of sales of agricultural products into both export and domestic trade channels will continue to be made by cooperative and proprietary handlers, processors, and marketers.

Interest costs will remain high over the next two decades because of the U.S. government's need for massive amounts of credit to refinance maturing obligations and to provide funding for new government programs. The capital requirements of the private sector and the use of dollars in the global economy will combine with the needs of the public sector to maintain pressure on the capital markets. Since a farmer's capital needs are apt to become greater while his income grows at a slower pace, his financial position will be more highly leveraged than was that of his counterpart of 25 years ago. Recent historical performance has not permitted the average farmer to build up liquidity or increase his equity commensurate with the increase in the asset side of his balance sheet. He has resorted to heavier and heavier borrowing. Such a reliance on expensive debt may affect a farmer's ability to repay.

I forecast that an increased debt burden and the cost of carrying such heavy liabilities will force the use of options other than the traditional farm financing instruments. Many farm and agribusiness balance sheets will be restructured to reflect the interest of owners, investors, short-term lenders, trade creditors, medium- and long-term lenders, lessors, partners, and joint venturers. The year 2000 will see more farmers sharing their farm equities with nonfarm investors, and there will be greater use of limited partnerships and syndicated group capital.

This bringing together of a number of principals for common undertaking will be done for several reasons. One will be to share in the financial risks associated with the growing of food. A second will be to provide for a share of the profits in the production chain, probably most often via the cooperative organization route. Participating in the profits generated in a food production chain could also be undertaken by proprietary organizations involving the merger of a number of entities into a single-interest operation, such as a corporate holding company with a number of operating subsidiaries each responsible for a link in the food production process. The third reason for the merging of interests will be to compete better in the global markets for basic commodities. The fourth will be to provide for accumulating the resources necessary to develop and increase sales volume in newly recognized mass consumer markets.

Happily, there will be no shortage of individuals, organizations, or institutions, domestic or foreign, who are willing to finance the various aspects of agriculture. Agriculture offers something for everyone.

To lenders who are assured by assets, agriculture and agribusiness offer livestock, equipment, land, tangible inventories, accounts receivable, and other collateral for loans.

To those who are cash-flow lenders, farmers and agribusiness operate within annual cycles. In the United States a lender or a borrower can closely estimate the cost of bringing a crop to market, the yields, and even the price of the crop at harvest, since futures contracts or processor growing contracts are available for so many of our crops.

For investors, farmland has been a fine investment from a capital gains viewpoint. Farming operations have provided some tax shelter; even the research and development in certain agribusinesses can be funded through the use of limited partnerships.

For foreign concerns, the United States represents an excellent market. Its politics are reasonably stable; the economy is relatively open; its currency is the global medium of exchange; and its taxes, reporting requirements, and business constraints are probably less burdensome than in any other country. From a financial viewpoint, U.S. farmers are not currently highly leveraged. They have about $20 of debt for every $100 of assets (1983 estimate), although this leverage rate is above that enjoyed in the 1970s. The educational and literacy levels of U.S. farmers are high. The infrastructure of the food system, even with its great needs for replacement and upgrading, is probably the best in the world. Agribusiness is owned and operated privately, as are the farms, with the Commodity Credit Corporation being one of the few government-owned agribusinesses. From a unit cost basis finished foods and commodities, with only a few exceptions, are now produced as cheaply as anywhere else, although the decline in infrastructure efficiency and increasing costs will reduce the American competitive advantage in the future.

There will be some dramatic changes in the share of the market that lenders presently enjoy, and, since farmers will utilize more institutional sources of credit, the ways in which lenders participate with each other in the credits to production agriculture and agribusiness also will change. There will be many more lenders seeking to do business with the agricultural sector, and the lenders will represent many more sources of credit than are presently available to agribusiness. Not only will there be more diverse lenders, there also will be an increase in the number of financial instruments offered to borrowers. Some lenders will find that their loan appetite is for very short maturities, 30–60–90 days, or 1 year at the most. Others will take the medium- and longer-term maturities. Some will continue to do asset lending; others will make loans based mostly on estimates of cash flow. Some lenders will elect to take the deferred and later portions of a repayment schedule, while others will make the initial advances to the borrower and expect to be repaid from the first proceeds of sales made.

I believe that in the year 2000 we'll be seeing fewer but larger and more capital-intensive farm operations and agribusinesses. There will be a larger number of lenders, both domestic and foreign, all desiring access to the U.S. agricultural credit market. There will be more integration of various facets of the food business, vertically and horizontally. There will be more competition for the investor's and saver's money; that is, more activity in buying and selling funds in both the national and international money pools. We can expect continued volatility in commodity prices. There will be an even higher level of protectionism among the various trading competitors than in 1983 and also an increasing sophistication in the United States among people in production agriculture and agribusiness and in their use of computers and data. There will be a scarcity of top-flight agribusiness managers. More increasingly costly research and development will be done to build strong new markets. We will see continued activity by the government to support exports by the use of trade programs. There will be a growing perception that government and the private sector need to work together in ways that will present to the world a more united front, an America Incorporated. This will help to counter the long cooperation between government and the private sector that has faced the United States in the form of Japan Incorporated and the European Economic Community.

The organized futures markets will continue their strong growth in volume of contracts traded. Futures markets also will grow in volume because of the ever-increasing kinds of contracts that will be developed to permit almost any producer or user of an agricultural product to hedge, including the hedging of money costs. Much more of the energy used in agriculture will be produced in the United States from a variety of sources, but energy will continue to increase in cost. Given the productive capability of U.S. agriculture, the real need will be for highly paid management that is able to market a huge volume of foodstuffs and also able to use available resources wisely.

There will be an increase in the number of part-time farmers. The ownership of the large commercial farming operations that will produce most of the foodstuffs for domestic and export markets will become more and more a landed aristocracy, given the shift in tax laws. The limitations on water in the arid Western states will accelerate the shift of certain specialty crops to areas such as the Mississippi Valley and will intensify cropping programs in fertile but water-scarce areas. More and more we will see crops grown on the basis of competitive advantage, even though there will be strong efforts to protect some industries.

There will only be a few crops in the world for which American farmers will not be competitive. These are commodities such as cassava, palm oil, palm kernel oil, some tobaccos, some spices, rice, and tropical fruits. But the United States will continue to be dominant in wheat, corn, soybeans for meal, many of the seasonal fruits and vegetables, and cotton. The trend that began in the late 1970s in the United States toward more fresh fruit and vegetable consumption, as opposed to preserved fruits and vegetables, will continue. Because of growing competition, the margins of profit in agriculture will remain relatively tight, but these will be offset by some yield increases not only because of improved management in the field but also from continued improvements in seed stocks and also the beginning of benefits from the genetic engineering that began in the late 1970s.

American agriculture will have one disadvantage that will require major amounts of capital to overcome. Almost the entire logistics system in the United States has begun to wear out, and the whole system will need to be replaced. That includes many of the locks on the river transportation system and the barges that were originally purchased in the 1960s and 1970s. Our bridges and highways have begun to fatigue. The railroad beds for those tracks that serve agriculture are in need of renovation. Roads have deteriorated from the use of larger and heavier trucks and the lack of maintenance and repair.

The Farm Credit System will be a full-blown banking system by the year 2000. As a means of competing with the Farm Credit System various groups of banks in broad geographical areas have banded together to form, for example, Mid-America Bankers Service Co. (MABSCO) and Western Bankers Service Co. (WEBSCO).

MABSCO and WEBSCO were first organized in the 1980s by commercial banks grouped geographically in the Middle West and the Pacific Northwest. The function of both is to provide a means of buying member-originated farm loans at interest rates reasonably competitive with the Farm Credit System. The pool of loanable funds each of these organizations utilizes is made up in part from the equity contribution each member bank makes, plus the funds bought or borrowed from the investor market. The liabilities (loans) of both MABSCO and WEBSCO are supported by the notes farm borrowers sign when making their loans and by the member bank equity contributions which together provide the collateral for the pool of loanable funds.

Given the relatively recent development of alternative funding for agriculture, it is useful to consider factors that in the past kept agriculture from using such external sources of capital. There has been a traditional unwillingness on the part of farmers to pay going market rates for money. Family farms and ample foodstocks for the U.S. public have been felt to be national objectives. The United States has long supported a cheap food policy, and there has been a nostalgic attachment to the continuance of the family farm. Agriculture is often used as the best example of free enterprise.

However, the ability of agriculture to produce more than can be marketed at profitable prices has sometimes resulted in erosions of income and at other times provided a strong inducement for farmers and nonfarmers to invest in agriculture. The cheap food policy and the occasional restrictions that government puts on the marketplace in order to ensure adequate stocks of cheap food have supported the idea that farmers need the crutches of supports, target prices, and administered income levels that, hopefully, are on a par with nonfarm family income. As a result, until the latter part of the twentieth century farmers had a limited ability to pay during certain periods, and it was the general feeling that farmers could not compete with industry for funds. One result is that agencies such as the Farmers Home Administration and the Farm Credit System have used their agency status to obtain a pool of capital that could be loaned to farmers at rates lower than those paid by companies in the industrial sector.

In the latter part of the twentieth century, with the onset of a prolonged recession and the drying up of such low-cost money, agriculture began to realize that it was no longer insulated from the capital markets. So many of the lenders, including the Farm Credit System, began to use rates of interest that moved more or less with market rates of interest. Country banks were able to be reasonably competitive with the Farm Credit System only because their cost of money was reflective of local conditions. Local savers often invested funds in local institutions at rates far below the rates for comparable investments in the money centers.

As global interest rates moved up in the latter part of the twentieth century some banks were less able to handle the increasing credit needs of farmers. Local deposits began to dry up, and the local banks were unable to pay competitive rates to attract funds from outside the local area. Another factor was that small banks were not perceived by investors to have the kind of financial strength that would create confidence among investors and encourage them to deposit their money. The reason for the runoff in the availability of local funds was that local investors had become much more sophisticated and had access to a great number of alternative investment opportunities.

The traditional means of making farm loans larger than those permitted by the statutory lending limits of small banks was to lay off portions of the loans to their larger city correspondent banks. However, the traditional loan rates that small banks charged their borrowing customers no longer were attractive to the correspondent banks that had alternative uses of funds which would generate higher interest rates. This caused a rather dramatic shift of the agricultural loan market share from country banks to the Farm Credit System, which now has emerged as a dominant lender to U.S. agriculture.

Agricultural financing will change by the time we get into the twenty-first century for a number of reasons. One will be the ability of the system to transfer information. Going back a few decades, we saw the beginnings of the accumulation of data. As high-technology electronics was applied to the needs and problems of several sectors in the economy, massive amounts of information were developed. After that came the ability to select bits and pieces of information and analyze them in various combinations. We began to have available management information that could improve the decision-making process. It also was possible to make complicated calculations very quickly. Complex questions could now be asked with the expectation of getting a reasonably fast and often complex response to those questions. The whole data base will continue to expand, and analytical methods will improve further. The possession and use of high technology will become common by the year 2000.

The ability to transfer information will have an even greater impact on the style and manner of business. This change will have a major impact on lending to agriculture. All kinds of data will be available to an interested user such as the cost of production, commodity market information, produce movement, and the existence of global commodity gluts or shortages. Models have been developed to analyze a particular farmer's annual program, giving him schedules of information on costs, cash flow, and taxable income and providing a printout of pro forma profit-and-loss statements and balance sheets. These models can analyze the suitability of his cropping program or the feasibility of a proposed capital expense.

The availability of such analyses, augmented in some cases by additional information from consultants or staff personnel, will permit a number of lenders to make, or at least to consider making, agricultural loans or to offer financial services to the farm community (see Figure 1).

One constraint that kept some lenders out of agricultural lending in the past was the lack of credible data. Now there will be a plethora of data. The lender will not only have data pertinent to the growing, marketing, and financial aspects of the operation but also up-to-date reports on the physical condition of crops—their soil temperatures, moisture levels in the plants (or some index of moisture stress), nutrient levels, the presence of diseases or pests, the density of plant populations, and the degree of maturity of plantings.

The use of accessible data presented in a uniform fashion will have the

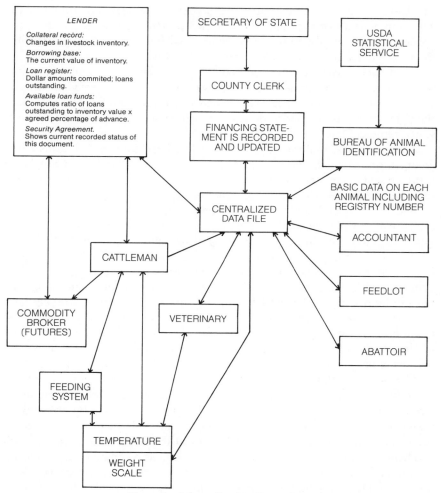

Figure 1. Information transfer system.

effect of standardizing the documentation, loan presentations, and credit analyses that are made by agricultural credit officers in processing a farm credit request. Only the personal comments included in the lending officer's presentation may be thought of as unique to each borrower, although each borrower's file will show different numbers relating to the borrower's cost, yields, income, acreage, and so on.

This uniformity of presentation will contribute to more participation in credit among various lenders. It will also, in my opinion, improve the examinations that banking regulators and internal auditors make of agricultural loan portfolios. The commonality of data and the standard presentation of information and documentation will also enhance an agricultural funding organization's ability to put collateral behind the sale of securities to the investing public.

Good data—easily available, up-to-date, and credible in a familiar format— will be helpful if agriculture is to get its share of funds from regional, national, and international money pools. The demand for capital will be great with both the public and private sectors, domestic and foreign, competing for a relatively scarce resource.

Good data and uniform formats will allow large numbers of loans to be grouped, or "batched," in a manner similar to the way blocks of FHA home loans are sold to investors. This ability to sell a large block of assets is particularly valuable to a lender who desires to maintain liquidity.

The problem of liquidity will encourage some of those who must make investments to replace capital items to finance them by resorting to leasing programs. Normally, when faced with growing capital needs and when after-tax profits do not keep pace with the financial demands of the farming enterprise, farmers have used long-term debt with farm real estate as collateral. This was a fine way to finance expansion as long as the increase in land values supported the loan and repayment looked feasible because of the level of after-tax farm incomes. But because some inflation will continue in the costs of farming and because the rather large annual increases in the price of farmland experienced in the 1970s will slow to a much lower percentage, the traditional way of borrowing on long-term assets will not be as readily available in the future.

This situation almost mandates that other forms of equity investment will be utilized, including syndicated groups, individual investors, or corporate investors acquiring a piece of the farm equity. This type of contribution to working capital is one that makes no demand on the farmer's cash flow. Increasing equity by foregoing the use of debt to expand or improve the enterprise will also help to keep the leverage factor down. As long as commodity prices remain volatile and are somewhat less apt to increase at a rate commensurate with the increase in production costs and off-farm costs, farmers have little recourse but to use sources of funds on an equity basis rather than trying to rely solely on debt funding.

By the year 2000 almost all the principals that have anything to do at all with financing, documentation, public record-keeping, or livestock registration and inspection will be tied together by a computer network and communications system that will permit the speedy transfer of information specific to a particular borrower. All the developments in improved transfer of information will bolster the technology of agriculture in the United States. When they are combined with the results from genetic improvement and the restoration of the infrastructure, the cost of producing basic commodities, processing those commodities, and marketing the finished product to the end user will also, of course, increase. However, yields should go up to some extent, and because of the ability to do a better job of overall financial management, U.S. farmers and agribusinesses should be able to maintain a reasonably strong competitive position with our offshore competitors.

# 15

# FINANCING THE STRUCTURE OF AGRICULTURE

MARVIN R. DUNCAN

DEAN W. HUGHES

MICHAEL BOEHLJE

RICHARD D. ROBBINS

Historically, the American farmer has depended heavily on debt as a means of financing farm operations. This will be true in the future as well but with a new ingredient. In the future, farming will be a riskier enterprise than it has ever been in the past. This fact has powerful implications for the way farmers borrow, who they borrow from, how they repay, and the methods they use to manage risk.

The Federal Government's involvement in agriculture, including its traditional role as a source or guarantor of loans, will continue for the foreseeable future, but the long-term trend is toward increased disengagement from the finances and affairs of farmers. At the same time, farmers will find themselves competing for loans and equity financing in an international setting. The velocity at which information moves will increase in the future, meaning that investors will have faster, more sophisticated ways of shifting from one investment to another.

In private financial markets the distinctions between bankers, brokers, and insurance companies are blurring as deregulation increasingly permits these institutions to offer fuller lines of financial services. This trend may benefit the larger, corporate farms. We will also see substantial specialization in financial services among some smaller banks and financial institutions that could benefit the smaller farmer. The prospect is that farm prices and incomes will rise sharply in the 1980s, which would enable farmers to compete for funds. Land prices also will rise, a boon to existing farmers but a barrier to those who wish to enter farming for the first time.

The successful farmers of the future will be those who focus on efficiency, cost controls, and risk management as well as production and marketing.

## PREPARING FOR THE TWENTY-FIRST CENTURY

MARVIN R. DUNCAN

DEAN W. HUGHES

For the purpose of analyzing the future of agricultural finance we assume that the U.S. and world economies will soon recover, but that they will not return to the ever-escalating periods of boom and bust that have characterized recent years. This assumption is admittedly optimistic, but it allows discussion of the future without presuming that there will be unpredictable cataclysms.

The Farm Credit System (FCS), the largest agricultural lender, held about 32 percent of all farm debt at the end of 1982. Chartered by Congress, which also provided its initial capital, FCS is a confederation of three banking networks, the Federal Land Banks, the Federal Intermediate Credit Banks, and the Bank for Cooperatives. All of the FCS banks are borrower-owned farm cooperatives. Federal Land Banks and Federal Land Bank Associations make loans to finance farm real estate. Federal Intermediate Credit Banks discount operating and intermediate-term loans from their owners—the local farmer-owned Production Credit Association—and from other financial institutions. The Bank for Cooperatives makes loans to farm cooperatives that supply production inputs to farmers and market farm products. The FCS raises funds for making loans by selling bonds in national money markets. Almost all loans in the system's portfolio carry a variable interest rate tied to the average cost of funds to each bank.

Depository institutions—national and state-chartered commercial banks and thrift institutions— are agriculture's second largest source of debt capital. These institutions accounted for 21 percent of farm debt at the end of 1982. While far less homogeneous than the FCS, depository institutions do share some characteristics. Deposits at the institutions are the main sources of the funds loaned to farmers. Most of the institutions, if not all, also lend to other borrowers besides farmers. And all are regulated by a complex set of rules provided by national and state laws.

However, there are substantial differences in the depository institutions lending to agriculture. Some institutions operate in unit-banking states. Others are free to establish branches anywhere in their county or state. Another form of ownership, the bank holding company, has become prevalent in recent years. While each bank in a bank holding company keeps its identity, all are owned by one company.

The government is also a major lender to agriculture. Loans from the Farmers Home Administration (FmHA), the Commodity Credit Corporation (CCC), and the Small Business Administration (SBA) accounted for 17 percent of all farm debt at the end of 1982. Each of these agencies has a national distribution system for providing loans to farmers and with each a farm loan generates either additional borrowing requirements or a contingent liability for the government. FmHA is the largest government lender to agriculture and makes loans for a variety of purposes, including relief from natural and economic disasters, aid to beginning farmers, rural housing, and rural development projects. CCC loans are made to farmers complying with farm commodity programs to provide flexibility in farm marketing decisions. SBA has never been a particularly significant lender to agriculture since it is primarily a lender on preferential terms to nonfarm businesses.

Other suppliers of farm loans include life insurance companies, individuals, and those who sell supplies and capital goods to farmers. Life insurance companies have traditionally made farm real estate loans, using the funds received from selling life insurance policies. More recently, they have also been investing the reserves of pension funds, which have added to the pool of credit for available agriculture. Seller-financed mortgages have also provided a source of debt funds to agriculture. Considerations of tax advantage, estate planning, and the limited availability of other kinds of loans have created the incentive for individuals selling farmland to finance part of the sale themselves. At the end of 1982, 29 percent of farm real estate debt was owed to individuals and others. Suppliers also provide credit to farmers either explicitly, as in the case of machinery dealers, or implicitly, as in the case of merchants who do not require payment until sometime after delivery.

Many factors determine the structure of the industry that provides loans to agriculture. Among these factors are technology, government regulations, and the international economy.

## CHANGES IN TECHNOLOGY

Rapid technological change has characterized the second half of the twentieth century. Nowhere has change been more significant than in the information industry. The fast growth of information systems can reasonably be expected to continue well into the next century. The speed with which resources can be transferred from one investment opportunity to another has increased as one result of an improved telecommunications network. When the speed of communications was coupled with very high interest rates and ceilings on bank interest rates, new financing instruments emerged to avoid the constraints imposed by financial regulation. Financial innovation continues, with Money Market Deposit and Super-Now accounts the most recent additions to an array of available financial instruments.

Improvements in information systems are also being incorporated into farm management. Computers with accounting software programs are improving farm record keeping. As management information systems improve, lenders to farmers will be better able to evaluate the potential risks and returns on their loans.

Revolutionary changes in information processing may also be changing the composition of industry in the United States. Investments will be required for the growth of high-technology industries and to bring high technology to industries now considered mature. Agriculture will have to compete directly with these uses of debt capital over the foreseeable future.

## CHANGES IN GOVERNMENTAL RULES AND REGULATIONS

The Depository Institution Deregulation and Monetary Control Act of 1980 (DIDMCA) signaled Congress' intent to remove many of the regulatory controls on financial institutions. For example, part of the Act progressively removes the ceiling on rates of interest on bank deposits over time. The benefits from these changes are that depositors receive a market-determined rate of interest and that lenders are able to attract loanable funds from a national market to meet their loan demand. These arguments are probably most convincing in the case of the small rural banks and thrift institutions that most often serve agriculture. Because most rural banks are small they have little ability to attract loanable funds from national financial markets. The payment of market interest rates is expected to resolve some of the fund-availability problems rural banks and thrift institutions have faced in the past and make them a more reliable source of financing.

DIDMCA is also having an effect on the structure of the banking industry. Many banks are moving rapidly to offer a full range of financial services, including brokerage service and insurance. At the same time other firms are beginning to offer financial services.

As deregulation has blurred the distinctions among them, financial institutions—banks and nonbanks alike—have moved aggressively to develop new marketing strategies across both geographic and product lines. Financial service companies have merged with brokerage firms. Banks have acquired interests in brokerage firms. Banks and thrift institutions have merged. These activities appear to be based on the assumption that marked change in the delivery of financial services has just begun.

Amendments to the Farm Credit Act were also passed in 1980, and studies were undertaken in 1982 to identify the possible effects of removing government-agency status from the Farm Credit System. The amendments opened up new activities to FCS, such as the financing of agricultural export sales and the sale of insurance. The 1982 study suggests there is little possibility that the FCS will lose agency status anytime soon because the loss would mean higher interest rates for farmers when they could ill afford them.

Farm policy will also change over the next 20 years. Specific changes cannot be forecast, but there are observable trends. Farmers may have to accept more business risk in the future or at least internalize the cost of such risk. Government programs that provide free insurance or floors under farm prices without capping price increases are ever more costly. These will almost certainly face increasing difficulties in Congress. Development of more efficient commodity futures markets and actuarially sound natural disaster insurance programs paves the way for more disengagement of the government from agriculture. Government intervention in the future may be concentrated increasingly on market development and on assistance limited to young farmers just starting out in agriculture.

## CHANGES IN INTERNATIONAL FINANCE

Anyone visiting the trading desk of a large bank late in the day cannot help but be impressed as the bank moves to meet its required reserves position by selecting from an array of funding sources that span the world. The Farm Credit System has also been marketing more of its debt instruments to foreign institutions—some 8 percent of all offerings in 1982. In projecting the future for agricultural lenders, therefore, it is impossible to ignore international financial conditions and the current problems in international financial markets.

Many developing and centrally planned economies are having difficulties repaying foreign debts. Even if these problems are short-run and can be resolved with acceptable economic growth, U.S. agricultural investment will have to compete for funds in an international setting. Domestic investment in agriculture will be compared not only to the profitability of other domestic investments but also with international investment opportunities. Importing countries will have to more rigorously prove their ability to repay. Thus growth in exports may not show the predictable gains they did in the 1970s. Given the risks of agricultural production and marketing, returns on assets will probably have to be higher. This can be done either by an increase in profitability in agriculture, a decline in the value of farm assets, or a combination of the two.

The combination of deregulation of financial markets, improved information processing, and international financial market integration will very likely bring about a homogenization of financial institutions. The distinction between thrift institutions and commercial banks is already blurred, and nonfinancial institutions are beginning to offer financial services. These financial services are expanding to include the sale of new assets and liabilities as well as financial counseling on how best to use these products in improving net worth, income, and cash flow. The fundability of funds, already apparent in businesses and households, is likely to spread to the sources of funds.

With the growing integration of financial markets probably forcing farmers to compete for funds on a national and international basis, better farm records and improved financial planning will be required. Farmers will have to pay market interest rates, and they will have to demonstrate management and repayment capacity equal to that of other investment opportunities.

Since farming may be riskier than in the past, successful farmers—those to whom credit is made readily available—will be farmers who master risk management. More skill in risk management will affect farm production, marketing, and financial management strategies, and likely emphasize a variety of strategies to reduce cash-flow risk and safeguard the capacity to repay loans. Such strategies could include hedging against fluctuations in prices and interest rates and the use of insurance to reduce the effect of unanticipated shortfalls in production.

## THE FARM CREDIT SYSTEM

To compete with other lenders offering a full menu of financial services, the FCS will likely increase coordination among its three banks. Merging of boards of directors and certain management functions appears likely. At the point of customer contact—Production Credit and Federal Fund Bank associations—a *de facto* merger of functions may be needed to provide full financial services to farm customers. Merger of the three banks at all levels is not impossible to contemplate. The FCS might ask Congress to expand its available services to match the expansion of other lenders. If so, it may begin to function more as a financial institution with national branches with its distinctiveness reduced. The change could raise more questions about the FCS's links to government. An evolutionary move away from agency status would then likely occur. In the process, the FCS may try to acquire offsetting advantages, such as the right to accept deposits or make loans to nonagricultural businesses.

The system is likely to make even more use of both private placement and international funding to raise funds for loans. If export financing becomes a more significant part of FCS's business, important experience in international financial markets will be gained. While the system will probably raise most of its funds in domestic markets, savings of other countries may be needed to serve U.S. agricultural credit demands adequately.

If the FCS prepares to market a more complete line of financial services, it will need to address at least two problems. First, there is customer reluctance to pay explicitly for many financial services. This reluctance may be reduced in the future as depository institutions also start charging explicit prices for services that previously were provided as implicit returns to low-interest deposits. Second, it is not clear that the human resources in the FCS are adequate to meet the challenge presented in providing a wide range of financial services to customers. The FCS will likely upgrade both the quality and diversity of the staff that provides the services. Additionally, the FCS may decide to contract with other institutions for access to many of the more specialized or sophisticated services.

## DEPOSITORY INSTITUTIONS

There is little doubt that competition for depositors is more intense. The number of competitors in lending markets is also growing. The structure of banking, therefore, is likely to change greatly in the next two decades. While the nature of the changes can be debated, it seems reasonable to expect a bimodal distribution of institutions offering financial services.

The distribution will almost certainly include a few very large organizations doing business on a national and international basis and serving as markets for large financial transactions. If interstate banking is permitted, many major financial institutions will extend their services across state boundaries by purchasing regional correspondent banks and bank holding companies. De facto interstate banking appears to be here already—with loan production offices and the national advertising of investment instruments. At the other end of the spectrum will be a large number of smaller organizations offering specialized financial services. Many of these will also merge into multibank holding companies to better serve their specialized customer base. This process is underway even in unit-banking states in the form of chain banking.

Many unit banks and small holding companies view themselves as targets for takeover by national banks or large regional bank holding companies. Some unit banks and holding companies have likely been formed in an effort to become a takeover target. Many of these bankers may be disappointed since expanding banks will be both sophisticated and cautious in their evaluation of takeover targets and will have limited interest in more than one outlet in a given rural market.

There is every reason to expect well-managed, aggressive unit banks to continue prospering in such an environment. This will be particularly true if these institutions view their function as conduits of loanable funds to and from broader financial markets. However, unit banks which are too small to service customers' needs or too complacent in their management posture are likely to disappear by way of merger, acquisition, and failure.

Such a distribution of financial institutions should service the needs of agriculture quite well. Studies of the future structure of farming suggest the same bimodal type of distribution appearing in agriculture. Large commercial farming operations might find their borrowing and investment needs met with ease through large financial institutions. At the same time smaller farms could still obtain financing from smaller financial institutions capable of also offering a broad range of financial services.

Depository institutions may have already taken account of some of the human resource requirements associated with delivery of a broader range of financial services through specialized personnel within the individual institutions or use of the correspondent bank network. It seems likely, however, they will also experience many of the same marketing and manpower problems as the FCS. Consequently, contractual arrangements with specialized firms for many of the financial services seem likely.

## GOVERNMENT AGENCIES

The main determinants of government lending to agriculture are political. The lending of the Commodity Credit Corporation is widely supported and can be expected to continue as a major source of short-term credit to farmers. Current political opinion seems to suggest a much more narrowly defined role for lending by the Small Business Administration and the Farmers Home Administration. As the lender of last resort to farmers, FmHA increasingly appears to be concentrating on loaning to those entering agriculture.

If in the long-run the government becomes less involved in agriculture, government lending may be reduced. This is not certain, however. If other commodity-specific programs are curtailed and some involvement is to be maintained, general credit programs may emerge as the principal means of government intervention.

Obviously, human resource demands will grow in government agencies as in private financial institutions. However, decisions on the quality and quantity of management resources available in government agencies will be determined primarily by political and budgetary pressures.

## OTHER SUPPLIERS OF FUNDS

Life insurance companies have become less important in financing agriculture over the past 30 years, but this trend could be reversed. Positive real rates of interest and greater emphasis on private retirement planning will probably lead to a rapid growth in pension fund and life insurance reserves. These funds will require investments with long maturities and relative safety. Over the past 3 years agricultural real estate loans have been shown to carry some risk, but the risks have been small compared with most other investments. Faced with a growing pool of funds, portfolio managers of life insurance companies could begin adding to their investments in farm real estate loans.

A number of nondeposit financial firms have been formed in recent years to provide specialized credit services to the agricultural sector. These firms typically raise their funds in national financial markets through sales of bankers' acceptances and commercial paper. Generally, the firms concentrate their lending activities on a few narrowly defined customers such as large commercial feedlots or large and well-capitalized crop producers. Since these firms have been efficient in generating loanable funds and since they utilize highly skilled people to serve the credit needs of selected customers, they have grown rapidly and been quite profitable. The increasing specialization in agricultural production and the profitability of such firms suggests that more will be formed. It seems unlikely, however, that such firms will provide credit to more than limited and very select segments of U.S. agriculture.

Farm real estate loans by individuals might become less important as rural financial markets lose some of their imperfections. Faced with more alternative uses for their funds, individual lenders could hesitate to invest them in relatively illiquid farm mortgages. It is unlikely, however, that this form of financing will disappear altogether. Capital gains taxes will still provide incentives for sellers to spread their gains over a number of years, and a secondary market in farm real estate mortgages could improve their attractiveness not only to individual investors but also to institutions. In fact, there are some indications that financial service firms are already exploring the viability of such a secondary market.

As more nonfinancial companies enter the business of providing financial services, specialized production credit and merchant credit may be made more readily available to farmers. Manufacturers will continue to offer financing packages as an inducement to buy. These firms will be more financially sophisticated than in the past, however, and may require higher returns on their financing. Convenience has probably played an important role in the demand for this source of farm financing. In the future, however, farmers will have better tools for financial management. They may use this form of financing less if it is not priced competitively.

Leasing of farm equipment will probably become more important in farmers' overall financial strategy. Not only does lease financing provide an alternative to debt financing, but it also provides increased financial flexibility. Such flexibility can be used to control the cost of debt and maintenance costs and to protect against the technological obsolescence of equipment. Of course, growth in leasing will depend to a significant degree on the tax laws since one of the main advantages of leasing is its transfer of depreciation allowances and investment credits to a party with a higher tax rate. In any case, farm equipment manufacturers will be an important source of lease financing. To compete with traditional, nonfinancial organizations, many financial institutions will likely include leasing of farm equipment as one of their services.

Meeting the demands for agricultural credit into the next century will be a challenge for U.S. financial institutions. The higher risks that may be associated with farming will require more sophisticated management by farmers and lenders, better farm financial records, and higher returns on farm loans. Adjustments to farm asset values and profitability will be needed for the agricultural sector to compete with other national and international investment opportunities.

Institutions that now lend to agriculture will adapt to a new financial environment. Financial institutions will probably become much more homogeneous, all offering most of the same services. To be competitive, the Farm Credit System will need new authorities and increased integration with its component banks. Depository institutions can be expected to offer a broad range of financial services and to build national systems for delivering their services either by purchase of other institutions or by marketing arrangements. A viable role for small banking groups or even unit banks is likely, however. Not only will they continue to better understand local needs, but they will be better positioned to serve those needs by linking into national networks providing financial services. There will still be government lenders, but the scope of their activities will be more narrowly defined.

Overall, the changes taking place in financial markets will probably help both the farm and nonfarm economies. In all likelihood the institutions that emerge over the next 20 years will serve agriculture well.

# FINANCIAL NEEDS OF AGRICULTURE IN THE NEXT CENTURY

## MICHAEL BOEHLJE

### THE NEW FINANCIAL ENVIRONMENT

Significant changes occurred in the financial environment for agriculture in the last decade. During the 1970s the higher income that farmers earned suggested that they could service more debt, so farmers added to their debt load at an unprecedented rate. For example, from 1975 to 1980 the debt load for the agricultural sector increased at an average rate of 14.4 percent per year. In essence, farmers were borrowing twice as much money in 1980 as they were in 1975. But their incomes did not increase as rapidly as the growth in the debt load during the 1970s, and the problem has obviously become much more severe in recent years with declining incomes.

During the 1960s the debt-to-income ratio in agriculture averaged 2–3 to 1: In other words, if farmers had used all their income to service debt, they could have paid off that debt in 2 or 3 years. In the 1970s the ratio almost doubled, rising to 4–5 to 1. In 1983 the ratio is approximately 10 to 1. This rising debt-to-income ratio means that farmers must commit more of their future income to debt servicing. Thus less income will be available for expansion and reinvestment in the farm or for an improved standard of living.

In addition to a larger debt load, interest rates have risen dramatically in recent years in both real and nominal terms. The relative importance of interest as a percentage of all cash expenses has doubled since the mid-1970s so farmers have not only a higher debt load and annual principal payments, but their interest payments have also increased dramatically. Higher interest rates and interest costs will probably continue during most of the decade of the 1980s, although we would not expect interest rates to remain at their recent high levels indefinitely. In fact, that is another change in the financial environment for agriculture: Interest rates will be quite volatile in the future, thus increasing the financial risk that the borrower faces.

Farmers also restructured their balance sheets during the 1970s, which reduced their liquidity and ability to service debt. Many farmers converted current assets in the form of grain and livestock inventories into intermediate- and long-term assets such as machinery, equipment, or land. These assets are much less liquid and are not readily available to service debt. Furthermore, a substantial proportion of the increased debt load incurred was of a shorter-term nature, including short-term land contracts and short-term or even 1-year notes for capital items such as machinery and equipment. This reduced liquidity compounds the problem of debt servicing. In essence, farmers are trying to pay higher principal and interest payments with less cash and liquid assets.

When farmers encountered problems paying off their growing debt load in the past, it was not uncommon to refinance to solve the debt servicing problem. With very low or even negative real rates of interest, monetizing the capital gain in land and other assets through refinancing was a relatively low-cost way for farmers to solve their liquidity problems. Lenders were willing to "play the game" because appreciation in land values was considered to be a solid base for borrowing.

Monetizing capital gain to cover debt-servicing problems will not be as easily accomplished in the future because nominal and real interest rates are expected to be higher and capital gains will not be as predictable. Consequently, both borrower and lender will be less willing to use this traditional safety valve as a means of solving liquidity problems.

Another change in the environment is in the area of risk. Farm incomes were more volatile in the 1970s than in the 1960s, but the downside risk in the agricultural sector was severely limited because of growing demand for agricultural commodities, elimination of burdensome surpluses, and government assistance such as the ASCS disaster and the FmHA Emergency Loan programs. These programs provided funds or low-interest-rate loans to those who encountered financial problems. In the future, we expect that incomes will remain volatile and that the downside risk in agriculture will be substantially greater than it was in the earlier decade.

A recent USDA study focused on likely credit needs and problems in the 1980s. Highlights include the following:

1. Farm production expenses will more than double by 1990. Funds needed to finance annual farm production expenses could increase by more than $200 billion over the next 10 years, compared with about $134 billion in total farm production expenses in 1980. Most of the additional funds will have to be borrowed, although there are expected to be some innovations in equity financing.

2. Farm sector debt, which increased from $12 billion in 1950 to an estimated $158 billion in 1980, could be about $600 billion by the end of the decade. However, asset values in farm businesses are expected to rise to over $3 trillion and the ratio of debts to asset values will not be significantly higher than the 16 or 17 percent rate of recent years.

3. Competition for loan funds will remain strong. Agriculture will remain competitive and will be able to attract its fair share of funds. Farm

prices and incomes should begin to rise strongly by the middle of the decade, increasing the ability of farmers to compete for production and investment funds.

4. Land prices will likely increase rapidly, especially in the latter half of the decade. This will increase the wealth of landowners but will also increase the difficulty of getting started in farming, especially for those having no other source of income to subsidize the beginning years. The added wealth of existing landowners, combined with tax advantages, will enable them to outbid other would-be buyers and thus continue the trend to fewer and larger farms. Higher land prices also greatly increase the flow of debt funds needed simply to refinance the ownership of land, generally into the hands of fewer and fewer owners.

Similar trends are expected to continue into the twenty-first century.

The dynamic financial future of agriculture suggests a number of issues that merit further discussion and analysis.

## INTEREST RATE RISK

The increased risk in agriculture means that careful financial planning and risk management have become more important than ever. The risks originating in financial markets—manifested as unanticipated adverse changes in the cost and availability of farm loans—are becoming more important. With the increased dependence of rural banks on money market certificates and other deposits tied to national money market conditions, interest rate variability is reflected in the cost of funds for farm borrowers. And with the variable rate interest plan, the Farm Credit System has also transferred in part the risk of interest rate changes to their borrowers. Thus farmers are subject to more fluctuations in the cost of debt, and with increased utilization of debt in comparison to the income stream the risk of repayment is substantially increased.

A key issue in the agricultural financial markets today is who can best handle interest rate risk, borrower or lender, and what methods might be used to reduce the risk arising from interest rate fluctuations. Could lenders use interest rate futures to protect themselves and their customers against rate fluctuations? What about offering fixed-rate loans at a premium rate to the borrower to reduce his risk but also compensate the lender for accepting this risk? What about matching the maturity structure of assets and liabilities? And how can variable length amortization schedules, which result in adjustments in the duration of the loan rather than in the amount of the annual payments when interest rates change, be used to reduce the financial consequences of interest rate fluctuations? The policy issue of investment behavior and supply response to interest rate fluctuations also merits investigation.

## NEW FINANCIAL INSTRUMENTS AND TERMS

Changes in the national and international financial environment, including higher and more volatile interest rates, suggest that innovation in lending terms and instruments is needed in the future. Variable interest rates that reflect changes in money market conditions have become more popular. Delayed or variable payment schedules may be used to adjust the repayment schedule to repayment ability. Some lenders may offer options for the borrower to delay a payment (e.g., in 1 year out of 10) to reduce the risk of default in years of low-cash income. Joint financing arrangements may be a possibility, such as a joint Federal Land Bank installment land contract agreement to guarantee long-term financing without losing the tax benefits of contract sales. Other types of arrangements may be utilized whereby the lender receives a percentage of the increase in value during the life of the loan agreement (an "equity kicker") in addition to receiving the annual payment.

A form of personal loan that may be used increasingly to finance farm firms is the family debenture. As noted later, this financing arrangement can be used along with the corporate structure to facilitate the intergenerational transfer of the farm firm. If such arrangements are used to encourage nonfarm heirs to leave their investment in the corporation, intrafamily debt will be substituted for external debt which will affect the amount of institutional credit needed to refinance each new generation of farm ownership.

## LEASE VERSUS DEBT FINANCING

Just as the primary source of equity capital for agriculture has been internally generated savings, the primary nonequity source has been debt financing. But with recent passage of the safe-harbor and finance lease provisions of the tax law, leasing of agricultural assets may become more predominant. To evaluate the lease versus purchase option, the present cost of the lease payments must be compared to the present cost of depreciation, interest, and other ownership outlays. The tax treatment of the various costs is extremely important in making this comparison, and because of the full tax deductibility of qualified lease payments the leasing option will typically become more attractive compared to the purchase option for those in higher tax brackets. Furthermore, some captive lease companies are providing attractive lease terms in an attempt to promote sales of machinery and equipment.

Lease financing has at least two advantages compared to current debt financing arrangements for capital items. First, whereas a large majority of machinery, equipment, and facility loans for agriculture are now made on a variable rate basis, a majority of the leases are fixed rate. Consequently, the farmers know what the annual payments will be, and they are not subjected to the risk of fluctuating interest rates. Second, it is becoming increasingly difficult to obtain term financing for many capital purchases; many lenders prefer to write annual renewable notes rather than provide a 3- or 5-year term commitment. With a lease arrangement the farmer knows he or she has a specific number of years over which to make the lease payments. Thus leasing allows the farmer to obtain fixed-rate term financing which is increasingly difficult to negotiate with conventional credit institutions.

Some people have argued that leases involve a smaller financial commitment on the part of the farm operator compared to loan terms and that the down payment on a loan may present a significant cash flow burden to the business whereas a down payment is not required with a lease. However, most leasing arrangements require the initial lease payment to be made in advance, and in some cases this first payment may be no smaller than the down payment with conventional loan terms.

## ENTRY INTO AGRICULTURE

Concern has been expressed in recent years about the opportunities for and impediments to entry into agriculture. This issue is not only of concern to those who are attempting to enter farming, it has also become a policy concern at both state and national levels, with particular emphasis on the implications of entry into agriculture for the structure and control of the farming sector and the future of the family farm.

Programs to help beginning farmers should consider the control of resources, subsidized credit, and risk. Some have proposed to assist the beginning farmer to attain control of resources, particularly land, through a program of government acquisition of real property. This property is then leased to the qualified beginning farmer for a period of years with an option to buy at the end of the lease period. An alternative strategy that might have some potential in the area of resource control, particularly with respect to real estate, is to provide an incentive for retiring farmers to rent their land on a long-term lease to qualified beginning farmers rather than to sell the land to an established farmer or even to a government agency. The incentive might be in the form of a tax credit for rent received under such an arrangement. This arrangement might also include for the tenant a first option to buy after a designated period of time at a market or other designated price.

A program to encourage seller financing to beginning farmers with an installment contract arrangement might be developed with tax incentives. For example, recent amendments to the Minnesota Farm Security Act exempt from state income taxation all interest received on an installment sale of land to qualified beginning farmers participating in the program. Such a program might encourage retiring farmers to sell their real estate at a lower price or on longer contract terms than would be offered to an established farm operator.

## INTERGENERATIONAL TRANSFERS AND NONFARM EQUITY

With the rapid growth in the equity base of agriculture a key concern of farmers and their lending institutions is the issue of maintaining that equity base during the process of intergenerational transfers. The estate tax reforms of recent years certainly reduce the potential equity drain from agriculture to pay estate taxes, but they do little to solve the potentially more serious problem of equity outflows to compensate nonfarm heirs during the intergenerational transfer process. With growth in farm size and the desire to maintain these larger farm units as "going concerns" beyond a single generation, the problem will become more acute.

Certainly the legal structure is available to encourage business continuity during the intergenerational transfer process (in particular, the corporation or even a properly structured partnership can be used), but additional innovation in financial markets and instruments may be necessary to encourage the nonfarm heirs to maintain their equity interest in the farm firm. Using a combination of debentures and stock (common, preferred, noncumulative voting, convertible preferred, etc.) in the capital structure of the corporation, with the nonfarm heirs receiving debentures that generate a competitive rate of return and have a specified liquidation value, holds promise, but other innovations may be needed. Furthermore, the credit demands that will exist in the future to finance intergenerational transfers will be substantial and will significantly influence the debt-equity structure of many farm firms.

A related issue is that of nonfarm sources of equity capital for farmers. Many people abhor the infusion of nonfarm equity capital into agriculture, whether it be in the form of a large corporate entity buying and operating a farm or a local businessman, banker, or doctor buying farm real estate. Yet others claim that such nonfarm investment is beneficial to beginning farmers by maintaining the rental market for farmland, thus enabling such farmers to acquire a land base, obtain economies of size, and share risk with other equity investors.

The issues of nonfarm equity capital in farming can probably be most rationally analyzed if one focuses on the terms of trade and incentives provided to farmers and nonfarmers to acquire farm assets. Public policy should attempt to eliminate any unique incentives that nonfarmers may have in comparison to farmers in purchasing agricultural assets—both groups should at a minimum have equal tax and other incentives for such purchases. Furthermore, policy should be structured to maintain a balance between the rights of owners and the rights of renters of agricultural assets. If "reasonable" terms of trade are maintained and artificial incentives eliminated, the issue of who owns the asset becomes less crucial, although it is by no means eliminated entirely.

## ASSET ADJUSTMENTS AND LIQUIDATION

The recent financial pressure faced by farmers has resulted in the stark reality of asset adjustment and liquidation. Although we might suspect that this problem is short-run in nature and that improvements in the income of farms in the future will reduce the concern about liquidations, the problem may be more persistent than we realize. The reason is the dramatic restructuring and reduced liquidity of farmers' balance sheets. Whereas during the decade of the 1970s refinancing or restructuring the liabilities was the commonly accepted procedure to solve liquidity problems and cover cash flow deficits, the 1980s and 1990s may be a period where restructuring of assets may be a more common method to solve these problems.

The choice of which assets to liquidate is a difficult one at best. Some of the major considerations that should be evaluated in this choice include: (1) the cash generated after secured debt, (2) the discount or liquidation loss, (3) the loss of earning power of the asset, (4) the interest rate on secured debt, (5) taxes and tax recapture provisions, (6) legal constraints and obligations, and (7) the debt reduction schedule on secured assets including balloon payments. In many cases, because of the low cash flow generated by real estate compared to the very high cash subsidy required to maintain ownership of real estate (particularly with current high interest rates), farmland may be the prime asset to sell if the firm must liquidate part of its asset base.

Liquidations and other forms of asset adjustments, particularly if such actions involve land or machinery and equipment, are not popular in rural communities. Yet for some operators such adjustments may be necessary for survival. Shrinking the business to a size where it can carry the debt load may be essential for some farm operations. Borrowers and lenders alike must become more aware of the conditions and circumstances that require asset and/or liability adjustments and what the potential impacts of those adjustments will be on the long-run growth and survival of the firm.

## SURVIVAL STRATEGIES

The increased price and production risk in the agricultural sector of the late 1970s and early 1980s combined with higher debt utilization have increased the operating and financial risks in most farm businesses. The environment in which agriculture is operating has changed dramatically, yet many farmers have not changed their management strategies to match this new environment.

There is reason to believe that farmers' incomes will improve in the future, but the focus should be on managing the operation to maximize the probability of still being in business when incomes do improve. This philosophy suggests that the focus should be on survival strategies. Some of these strategies include: (1) understanding the difference between short-run cash flow and long-run profits; (2) locking in profits if possible; (3) emphasizing cost control, not growth and volume; (4) adjusting land rental arrangements to reduce risk and cash expenses; (5) reducing cash and capital outlays; (6) practicing preventive maintenance on machinery rather than replacing machinery; (7) maintaining an adequate insurance program; (8) refinancing to lengthen loan repayment schedules; (9) using sale-lease backs to generate cash; and (10) considering partial liquidation, if necessary.

During the 1970s much of the effort in managing the farm business focused on growth and expansion and the use of leverage to accomplish the expansion goals. In the future farmers must focus on efficiency, cost control, and the risk as well as the rewards of borrowing money. The successful farm managers of the future must not only be good production and marketing specialists, they must also have an understanding of financial and risk management.

# FINANCING THE SMALL FARM

## RICHARD D. ROBBINS

Because of the continual decline in the number of small farms more attention is now being placed upon the problems faced by the small farmer. While there is no commonly accepted definition of a small farm, perhaps the one area of agreement is that it should not be based solely on the amount of land under cultivation. Most definitions include the values of sales and require that a substantial portion of total income come from the farm.

The Food and Agriculture Act of 1977 defines a small farm as having $20,000 or less in annual sales and establishes that the owner's principle source of income must be from farming and that he must have less than $5000 nonfarm income. The U.S. Department of Agriculture's definition of a small farm is one that is operated with family labor, has total family income below the medium non-metropolitan family income in that particular state, and shows a substantial portion of income as coming from farming.

## TRENDS IN THE NUMBER OF SMALL FARMS

Historically, there has been a sharp decline in the farm population and the number of farms in the United States. No section of the country has escaped the decline, and this trend has continued for over 40 years. However, the 1978 Census of Agriculture did show slight gains in the number of farms. Data from that census show that from 1910 to 1940 the number of farms remained somewhat constant. Since then, the number of farms has declined, at least until 1974, while the average size of farms has risen. By 1978 only 2.5 million farms existed in the United States (Table 1) and farm population was less than 4 percent of the total U.S. population.

The amount of land being farmed has remained constant at around 1 billion acres. Thus average farm size has risen and now averages around 430 acres in the United States.

Almost 38 percent of U.S. farms had sales of less than $2500 in 1969 while 25 percent were in this group in 1978. Seventy-two percent of U.S. farms fell into the group with sales of less than $20,000, while 64 percent were in this group in 1978. Thus a significant number of U.S. farms still are in the small farm group.

A higher proportion of the farmers in the Southeast fall into the small farm group. Nearly 88 percent of this group had sales of less than $20,000 in 1969, while 77 percent fell into that group in 1978. While the number of small farms has declined in the Southeast, they make up a much larger share of the farms.

Another interesting point is that the Southern farms generally make up 40 percent of total U.S. farms and account for a significant share of total U.S. small farms. Nearly half of farms with sales less than $20,000 are in the South. Thus while small farm problems are not limited to the South, their impact is more heavily felt in that region.

## CHARACTERISTICS OF THE SMALL FARM

Only 53 percent of all farmers list farming as their principal occupation; nearly half are engaged in another principal occupation, with farming left as

Table 1. Number of Farms by Sales and Year for the United States and Southeast

| Value of Agricultural Products Sold | United States | | | Southeast | | |
|---|---|---|---|---|---|---|
| | 1969 | 1974 | 1978 | 1969 | 1974 | 1978 |
| $100,000 or over | — | 152,599 | 222,682 | — | 45,589 | 67,142 |
| 40,000 to 99,999 | 221,690 | 324,310 | 363,383 | 63,984 | 73,869 | 84,356 |
| 20,000 to 39,999 | 330,992 | 321,771 | 306,112 | 77,045 | 75,700 | 85,714 |
| 10,000 to 19,999 | 395,472 | 310,011 | 309,594 | 103,793 | 99,563 | 116,277 |
| 5,000 to 9,999 | 390,425 | 296,373 | 331,042 | 148,076 | 124,044 | 154,876 |
| 2,500 to 4,999 | 360,033 | 257,263 | 331,874 | 171,452 | 129,913 | 174,803 |
| Less than 2,500 | 1,031,638 | 649,448 | 611,653 | 597,049 | 380,817 | 331,520 |
| Total[a] | 2,730,250 | 2,311,775 | 2,476,340 | 1,161,399 | 929,495 | 1,014,688 |

[a] Abnormal farms excluded for 1974 and 1978.

Source: U.S. Department of Commerce Census of Agriculture, 1969, 1974, 1978, Summary and State Data.

a part-time activity. A substantial number in both groups are over 65 years old. About 11 percent of those who list their principal occupation as farming are over 65, while 5 percent of those in other principal occupations are over 65 (Table 2). Thus one-sixth of all farmers are beyond the usual retirement age and are likely drawing social security and other retirement benefits.

Whether older operators have their principal occupation as farming, or some other principal occupation, they are concentrated in the small farm group. Ninety percent of those operators over 65 who list other principal occupations and 75 percent of operators who are principally farming are small farmers. However, for those in other occupations, only 11 percent are 65 while over 38 percent of those who are principally farmers are 65 or older. Apparently, those who have other occupations retire totally, while the farmers continue to work after retirement age.

The percent of small farmers who are principally farmers is low: 65 percent have other principle occupations. Many are treating agriculture sales as a supplemental source of income. Seventy percent of those in other occupations work 200 days or more off the farm while only 5.4 percent of the more intensive farmers work off the farm 200 days or more. Even though 11 percent of the small farmers in other occupations do not work off the farm, over half of them are over 65 and 74 percent are over 55 years of age.

A slightly different pattern holds true for the South. Almost two-thirds of the small farmers in the South are in other principal occupations, a much larger rate than the nation as a whole. The farmer in the South also works off the farm: 42 percent reported off-farm work exceeding 200 days per year.

## FINANCING SMALL FARMS

Small farmers can be divided into three groups: (1) those over 65 who have largely retired; (2) those under 65 who have full- or near full-time employment off the farm; and (3) those who are full-time farmers with limited nonfarm employment. The type of credit policies appropriate for each group may be quite different. A retired farmer drawing pension and/or social security benefits is likely to use some of the funds for operating expenses. He is unlikely to be expanding the farm business and therefore has little need for capital to buy and make other large improvements. Only if he is in partnership with a younger person is he likely to make such commitments.

The other two groups of farmers are more likely to have major credit needs for operating expenses, machinery, and real estate. While the over-65 group may have problems and should not be ignored, the latter two groups are the ones for which intermediate and long-term credit will be important.

Separation of farmers into the two groups is useful in analyzing borrowing

Table 2. Summary by Age and Principal Occupation of Operator

| Value of Agriculture Product Sold | Total Farming and Other Occupation | | Principal Occupation Farming | | | | Other Principal Occupation | | | |
|---|---|---|---|---|---|---|---|---|---|---|
| | | | Total Farming | | 65 yr and Over | | Total Other Occupation | | 65 yr and Over | |
| | No. | Percent | No. | Percent | No. | Percent | No. | Percent | No. | Percent |
| $500,000 or more | 17,976 | 0.7 | 15,949 | 1.2 | 1,364 | 0.5 | 2,027 | 0.2 | 211 | 0.2 |
| 200,000 to 499,999 | 62,780 | 2.5 | 57,461 | 4.3 | 3,626 | 1.3 | 5,319 | 0.5 | 493 | 0.4 |
| 100,000 to 199,999 | 141,926 | 5.7 | 130,551 | 9.9 | 7,574 | 2.7 | 11,375 | 1.0 | 1,024 | 0.8 |
| 40,000 to 99,999 | 363,383 | 14.7 | 325,108 | 24.5 | 24,745 | 8.6 | 38,275 | 3.3 | 3,144 | 2.6 |
| 20,000 to 39,999 | 306,112 | 12.4 | 232,455 | 17.5 | 34,826 | 12.2 | 73,657 | 6.4 | 5,820 | 4.8 |
| 10,000 to 19,999 | 309,594 | 12.5 | 184,592 | 13.9 | 47,393 | 16.6 | 125,002 | 10.9 | 10,512 | 8.7 |
| 5,000 to 9,999 | 331,042 | 13.4 | 135,762 | 10.4 | 52,113 | 18.2 | 195,280 | 17.0 | 20,022 | 16.5 |
| 2,500 to 4,999 | 331,874 | 13.4 | 108,226 | 8.2 | 48,909 | 17.1 | 223,648 | 19.4 | 24,661 | 20.3 |
| Less than 2,500 | 611,653 | 24.7 | 135,249 | 10.1 | 65,549 | 22.9 | 476,404 | 41.4 | 55,527 | 45.7 |
| Total | 2,476,340 | 100 | 1,325,353 | 100 | 286,099 | 100.1 | 1,150,987 | 100.1 | 121,414 | 100 |

Source: U.S. Department of Commerce Census of Agriculture, 1978, Summary and State Data.

habits over time. Each group receives similar treatment in loan requests. Borrowing levels for both depend upon the size of their net worth, and they receive similar amounts when they apply for loans. However, there appears to be some difference in the terms received. Small, part-time farmers in the 1950s received a larger percent of loans with unsecured notes than did small, full-time farmers. Furthermore, 28 percent of the full-time farmers used cosigners, while only 11 percent of the part-time farmers used cosigners. Also, fewer real estate loans were made to full-time small farmers. One could conclude that part-time farmers with supplemental income sources received better terms than full-time farmers.

The situation has changed little since then. In a survey conducted at North Carolina A&T State University in 1982, we found that lenders still look for cosigners in a larger percent of their loans to small farmers. Furthermore, the small farm group as a whole receive smaller loans in proportion to their assets: 75 percent of value for equipment loans for small farmers and 85 percent of value for large farmers. For all loans, less than 20 percent go to farmers with sales of less than $40,000 annually. The repayment terms for intermediate credit vary, with larger farmers receiving longer payback periods. Loans for machinery and equipment have 3-year payback periods for small farmers, but large farmers were permitted 4 years to repay.

The problems with loans varied by the size of the loan. Lenders reported overly optimistic income projections and weak financial positions as major problems with small farmers. Regardless of size, lack of adequate records and inadequate collateral are problems in analyzing loan requests.

## SOURCES OF CREDIT

Farmers receive their credit from many sources, dependent primarily upon the type of loan. The source is not related to the size of the farm. Commercial banks, Production Credit Associations, Federal Land Banks, and the Farmers Home Administration remain the major sources of credit for farmers. In addition, a few loans are made by insurance companies and private individuals. In general, insurance companies lend for real estate purchases. Although private individuals make some non-real-estate loans, the largest proportion is in real estate.

The major change in the source of loans over time has been in the proportion of the debt held by lending institutions. Historically, commercial banks have been a major source of loans to agriculture. Recently, the share of loans held by banks has been falling while the share held by government agencies has been rising. Production credit (non-real-estate) loans have declined slightly. However, the share of real estate loans with the Federal Land Bank has risen sharply.

The largest share of the farm debt is owned by larger farmers. In 1983 those with sales of over $100,000 held 56 percent of the debt, while those with less than $40,000 in sales held only 21 percent. Furthermore, the debt-to-asset ratio is higher for the large operator than for the small operator (28 percent to 18 percent).

Farmers in our study are similar to those of the nation: 34 percent held loans from the commercial banks, 23 percent with the Federal Land Banks, 21 percent with production credit associations, and 10 percent with the Farmers Home Administration.

## FUTURE NEEDS OF SMALL FARMERS

Capital requirements in agriculture will continue to grow in the future. As more farmers seek to expand they will need funds for operating expenses and to purchase land and machinery. Technological changes that have been characteristic of agriculture historically have required farmers to become more dependent upon capital inputs.

The small farmer will be at a distinct disadvantage in competing for funds. This disadvantage occurs because lenders give more favorable terms to the larger farmer and because small farmers have fewer resources and are limited in collateral and therefore in the amount of funds they can obtain. In addition, there appears to be some evidence that lenders show preference to large farmers in making lending decisions. Many lenders believe that small farmers are inefficient and lack management skills. The cost of servicing the loans is likely to be higher per dollar loaned to the small farmer. Since larger farmers are preferred, the lender often works with them in preparing financial statements and projecting profits and losses. The small farmer often does not receive this kind of assistance. Thus the small farmer could be adversely affected in his ability to access needed resources and services.

Another factor that could limit access to credit is a self-imposed limitation on borrowing. Some individuals feel that debt is "bad" and are reluctant to borrow. Thus some farmers may not be taking advantage of the credit that may be available to them.

The overall economic conditions of the country play an important role in access to credit. With the fluctuation of interest rates in the recent past and low prices for agricultural products, many commercial banks and other lenders may be reluctant to enter into long-term, fixed-rate loans or even into the agriculture loan market at all. Those that do may show preference for operating loans because of their self-liquidating nature and the shorter term and collateral associated with those loans.

With high interest rates the farmer may not be able to acquire land, machinery, and other assets needed to expand operation or to replace existing assets. The returns to agriculture have generally been low. Combined with low agricultural prices in the recent past many farmers cannot acquire capital to operate.

Despite these problems farmers are planning major purchases in the future. In our survey we found 40 percent were planning to purchase machinery and equipment, make land improvements, and construct barns and buildings. However, only 12 percent reported plans to purchase land. It is interesting to note that they plan to change the source of their loans. A larger proportion said that they will seek loans from the commercial banks and production credit associations. Fewer expect to use government sources. Perhaps the emphasis of this administration on the role of the private sector and recent changes in the agricultural support program have led farmers to look to private sources. Farmers may feel that less funds will be available from FmHA.

With an increase in demand for funds from the private sector, the credit problems of small farmers are likely to intensify. Many farmers have viewed FmHA and government sources as lenders of last resort. If they now turn to the private sector and the private sector does not or cannot provide these funds, more small farmers will be forced out of business.

## OTHER AVAILABLE ALTERNATIVES

One alternative for the small farmer is to consider a cooperative, which would provide inputs and assist in marketing. Properly organized, credit could be available for inputs—seeds, fertilizers, chemicals, and so on—and advances made for crops during marketing. There appears to be some willingness for small farmers to form co-ops. One survey found that 90 percent of the farmers would participate.

The potential for a successful marketing co-op depends, however, upon the product. Generally, livestock co-ops have not been feasible. Given the limited land available to small farmers, most have not provided sufficient volume for the co-op to break even.

The survey found that establishing co-ops for purchasing inputs and marketing (primarily grains) was possible. Sufficient volume could be generated for small farmers to make the input co-ops profitable. Farmers benefit through lower prices, credit for inputs, and a share in any profits. Credit is provided by establishing an open line of credit to members, based upon their needs and determined at the time of joining the co-op.

While sufficient volume for a marketing co-op might be generated, market forces would be more pronounced. The co-op's estimated profit margin for grain crops would be small or negative during periods of low prices brought on by large grain surpluses. However, during other years, and should programs such as PIK continue during surplus years, profits may average enough for the co-op to prove feasible.

A separate co-op providing only financial services may not be feasible. A co-op organized to serve these needs would be faced with a large number of small loans with high administrative costs. The absence of large farmers who would reduce overall administrative costs and provide a volume business would lead to higher interest charges. The small farmer may be able to get competitive rates at existing financial institutions should he decide to borrow. Perhaps the best alternative is to provide credit through the input co-op, charging a premium for deferred payment of inputs. Some of the administration costs could then be covered out of the wholesale–retail price differential. Thus the small farmer could obtain inputs in a timely manner and get the benefit of good marketing strategies while minimizing his cash-flow problems.

Another possibility for small farmers could be contract farming. The North Carolina broiler industry provides a model. Farm input suppliers, processors, and/or distributors could expand resource-provider contracts. The farmer receives inputs (except labor, utilities, and fixed-cost items) and a guaranteed market. The resource provider receives a market for his inputs and some control over production and management decisions. Risks from production and marketing are shared by both parties. Profits for both parties may be increased because of better coordination of production and marketing activities. Of course, the farmer loses some management control, and the resource provider may have many small farmers to service. However, these problems were overcome in the broiler industry and could be overcome in other areas.

A third alternative involves the raising of capital through the sale of stock. And it is possible that a move toward corporate farming may grow in the future. Historically, farmers entered agriculture primarily by inheritance or purchasing land from relatives. Many therefore obtained farms with small outlays or at least below-market value. As inheritance laws change, and without migration from agriculture, these options may be limited in the future. The issuance of stock would provide an alternative for acquiring capital.

However, this alternative may be limited for the small farmer. Investors are likely to view small farmers, as many lenders do as undercapitalized, inefficient, and lacking in managerial skills. They are not likely to invest in small-scale agriculture. Thus the potential for corporate agriculture is likely to be concentrated among very large farms.

The recent trends in small farms are expected to continue into the future. A growing percentage of small farmers will seek nonfarm employment. Many part-time farmers will likely use nonfarm income to assist in paying operating expenses, thereby reducing dependence on short-term loans. The number of small farms is likely to stabilize as part-time farming increases. Even though their share of market will be small, it will be significant for selected crops. Those who remain as full-time farmers will have substantial credit needs. In order to stay reasonably competitive they must acquire modern technology and seek innovative ways of financing their operations. Their primary credit needs will be for intermediate- and long-term needs.

The specific policies that should be addressed depend upon the small farmer's principal occupation. The part-time small farmer is likely to fulfill credit needs through private sources who see nonfarm income as a part of the farmers' ability to repay farm loans. Those who are full-time farmers will suffer from lack of access to operating capital. Perhaps governmental sources could be directed specifically toward the needs of small full-time farmers.

Small farmers are likely to exist well into the twenty-first century. They are likely to have problems that will have to be addressed. Government policy could be directed to the protection and development of the small farm sector as the private sector may favor the large commercial farm.

# V
# THE WORLD'S EXPECTATIONS FOR AGRICULTURE

# 16

# AGRICULTURE
# AND U.S. TRADE POLICY

## D. GALE JOHNSON

A little more than three decades ago, I wrote a book entitled *Agriculture and Trade: A Study of Inconsistent Policies* (1950). Except for a number of details and changes in dates, a book on the same topic would be as timely today as when it was written at the time the General Agreement on Tariffs and Trade (GATT) was negotiated and a new framework for trade liberalization created. The general structure of such a book today would differ little from the 1950 version and, unfortunately, the same general conclusions would be put forward.

The basic source of the inconsistency between agricultural and trade policies now and then is that particular agricultural price and income programs require control over international trade as an adjunct to carrying out the domestic programs. When governments intervene in agricultural product markets through price supports or commodity loans, minimizing the costs to the public treasury requires that imports be restricted. If domestic output is larger than can be sold in the market at the artificially established price, export subsidies are required so that stocks will not accumulate to the point where they become so burdensome that they must be disposed of.

In 1950 I wrote as follows: "The group of programs that constitute what may be called agricultural price programs has required the insulation of domestic markets from the impact of alternative sources of supply. These programs have driven a wedge between domestic and foreign prices of the same product. They have operated on the premise that the way to improve the income of the American farmer is to raise prices. No one nation by itself, with only minor exceptions, can afford to take the actions required to raise the level of prices for a product in world markets. To do so, as the United States tried with cotton in the middle thirties, leads to increased production in other nations and the loss of a large share of the export market."

What was written about the interrelations between certain types of domestic farm programs, namely, those that attempt to achieve their objectives by controlling market prices, applies with equal force today not only to the United States but to nearly every nation in the world. The majority of the interventions in the international trade of agricultural products, whether limitations on imports or exports or subsidization or taxing of exports, are undertaken as a consequence of domestic agricultural and food policies. This is the most important reason why it has been so difficult to successfully negotiate reductions in agricultural trade barriers.

To negotiate about agricultural trade is very often to negotiate about domestic agricultural programs and policies. Examples are the level of dairy price supports in the United States and Canada, the target prices established by the European Economic Community, the high domestic prices for beef, rice, and wheat in Japan, or the low price of bread in Egypt. So far no country has been willing or able to engage in serious negotiations about the domestic agricultural programs and policies that require trade barriers and interventions.

My emphasis here is upon the United States and the continuing conflicts between its agricultural and international trade policies. Put simply, the United States professes its support of a liberal trade policy and has had the major leadership role in the world in pressing for more liberal trade for nearly a half-century. While the conflicts between domestic farm programs and a liberal trade policy are now less numerous and striking than in 1950, many of the conflicts have persisted and others have emerged in a more subtle form. It is time, in my opinion, that we take a serious look at ourselves—take a good look in the mirror—and reflect upon what we would have to do if we really practiced what we preach.

## TRADE AND AGRICULTURAL POLICIES

Officials of the U.S. government have long complained about the restrictions affecting agricultural trade imposed by most of our major trading partners. Currently, we are pressing Japan and the European Community for significant modifications of their agricultural and trade policies. The modifications would reduce the level of protection of agriculture and thus expand the opportunities for exports of U.S. farm products. The reduction in the level of protection may affect either the amounts imported or the quantities exported by particular countries.

Over the past quarter-century there has been substantial progress toward the liberalization of trade in manufactured products, and this is true even with the recent protectionist moves of many industrial countries. However, none of the major rounds of GATT negotiations has resulted in a significant reduction in the barriers to imports or in uneconomic maintenance of agricultural exports. One important reason is that such barriers and interventions result from the character of domestic agricultural policies and programs. In other words, trade regulations result from the domestic decisions that determine national prices, production, and consumption. When pressed to reduce their regulation of agricultural trade, nations often retort that such pressure is an illegitimate interference with their domestic policies. Trade regulations generally cannot be reduced except by changing some component of an agricultural program, such as the level of price support.

The complaints made by the United States have had minimal effect in reducing restrictions that impede the flow of trade in agricultural products or the value or quantity of our farm exports. Why is this so? Perhaps the most important reason has been noted, namely, that trade interventions are generally extensions of domestic agricultural policies. But there is at least one other important reason. It is that the United States continues to use means of trade intervention that are of the same type as those we argue should be reduced or abandoned by other nations. Claude Villain, an often shrill spokesman for the Common Agricultural Policy of the European Economic Community, can say with much truth and only some exaggeration:

> . . . let me say how amazed and even envious I am when I review the panoply of instruments which a liberal country such as the U.S. possesses to support its exports!
>
> To mention only some, there are measures of trade promotion, export credit, export guarantees and insurance, food aid under P.L. 480, the DISC tax arrangements, government-to-government agreements particularly for milk products, the drawback system for sugar, and so on.
>
> I will simply say that if you compare the budget spending on agriculture with the value added of the agricultural industry, you will find that in 1976 to 1978 the ratio in the EEC was 39.2% and in the USA it was 37.6%. That is nothing like the gap which some critics would have you believe.
>
> Initially, the 1982/83 U.S. Farm Budget envisaged large reductions. However, substantial revisions have since been made which increase foreseeable outlays by some $6.8 billion. When I look at the Farm Bill, I see that the target price

system is continued, that the milk support program, despite some changes is still likely to lead to a big surplus, and that a new sugar support program is being introduced. I always thought that the U.S. was the great defender of free competition, but I must say that all these target prices and support programs have a familiar European ring to them . . .

. . . I must say that I think the free-trading United States is armed with a remarkable weapon in the form of the GATT waiver. This is the exemption which since 1955 has allowed you to ignore certain rules of the GATT; and I note that the U.S. authorities have invented an impressive number of other measures, including import quotas, supplementary taxes, domestic price rules, and marketing orders.

Mr. Villain, as well as other apologists for the Common Agricultural Policy (CAP), compares only the budget costs of the farm policies and ignores the costs imposed upon consumers due to the high prices resulting from the CAP. The U.S. Department of Agriculture has estimated that the EC consumers in 1979–1980 paid nearly $29 billion more for food products than they would have paid if EC prices for farm products had been at the prevailing world market levels. I have calculated that the excess consumer costs of the CAP for 1979–1980 were somewhat more than $30 billion. Allan E. Buckwell and his associates estimated consumer costs at $34.5 billion for 1980.

Furthermore, I have been able to duplicate the estimate of budget expenditures for agriculture in the United States of approximately $20 billion for 1976–1978 only by assuming that the $8 billion annual expenditures on food stamps and other nutrition programs were included. Such inclusion is hardly reasonable since the purpose of these efforts is to expand demand for farm products and to reduce costs to consumers. If such items are included in budget expenditures, then it would be appropriate to reduce the U.S. excess consumer costs by approximately the indicated amount. Our estimate of excess consumer costs in the United States for 1979–1980, not including any adjustment for the food stamp plan, is approximately $9 billion. Obviously, this is much less than the excess consumer costs of the CAP.

As 1982 came to a close, the United States was threatening to engage in an export subsidy war with the European Economic Community in an effort to counteract presumed losses of exports of grains and chickens resulting from the high levels of export restitutions, or subsidies, paid by the EEC. On this point I am in complete agreement with Villain, who responded to a statement by U.S. Secretary of Agriculture John Block to the effect that the United States was ". . . going to do battle with the EEC wherever and whenever it is necessary." Said Villain, "Make no mistake about it: If there is economic warfare between the United States and Europe, there will be no winner. Both of us will be losers."

What Villain did not add is that there would be other losers, namely, such competing exporters as Brazil and Thailand, which do not have the enormous financial resources available to officials of the world's great economies. I must admit that I can think of nothing that would be more effective in strengthening the position of the supporters of the Common Agricultural Policy than an export subsidy war. Such a war would only confirm the view that the United States intervenes in agricultural trade to the same degree as does the EEC.

## WHAT IF THERE WERE FREE TRADE?

On occasion, I have a dream. In that dream I see consternation on Capitol Hill and on Independence and Pennsylvania Avenues in Washington when the European Community and the Japanese offer to eliminate all economic barriers to trade in agricultural products except for a modest *ad valorem* tariff. The consternation is great indeed. The U.S. government has made no plans to respond to an offer of a significant liberalization of trade in agricultural products. There is no difference on this point, either in my dream or in the real world, between Democrats and Republicans. There has been a nearly universal unwillingness to even consider what the consequences of such liberalization might be.

When I awake from my dream, I remember that there does exist a report done in the early 1970s at the request of the Council on International Economic Policy as part of the preparation of a proposed round of multilateral trade negotiations. The report, which had been prepared as an internal document to assist in the determination of a negotiating position, was leaked by the late Hubert H. Humphrey, presumably in an effort to embarrass the Nixon Administration.

In the report, which is generally identified as the Flanigan Report, the unimaginable had been asked: What would be some of the economic effects, as well as the effects on trade, of a significant liberalization of agricultural trade by the United States and the other industrial countries? At the request of then White House staffer Peter Flanigan, a more detailed study was prepared to estimate the effects of trade liberalization upon employment in agriculture and net farm income in the European Community and Japan. The work was nearly completed in 1973, but the report was never published. It had been decided in the Office of the Secretary of Agriculture that there would be adverse foreign reactions to the results.

As far as I know there has been no subsequent work along the same line undertaken by the U.S. government. Nor has there been any systematic effort to discover how U.S. farm programs could be made consistent over a defined period of time with a liberal trade policy. Trade is now so important to U.S.

agriculture that it would seem appropriate for some responsible component of government to consider how the United States could accept substantial and universal liberalization of agricultural trade. This may seem too utopian to consider, but there is much ground that could be covered that would improve our position as a supporter of liberal trade in agricultural products. As things now stand, it is rather easy to characterize the U.S. position as one favoring trade liberalization for products that we export but also as one that is highly restrictive for imported products that are competitive with our own production. This is something of a caricature—but not very much.

## GATT PROVISIONS FOR AGRICULTURE

The General Agreement on Tariffs and Trade permits special forms of protection for agriculture. These provisions were included at the insistence of the United States. One of these is Article XVI, which deals with the issue of agricultural subsidies, both domestic and international. Under this Article countries are not supposed to use subsidies "including any form of income or price support, which operates directly or indirectly to increase exports—or to reduce imports of any product." No form of direct or indirect subsidy is to be used to acquire "more than an equitable share of world exports, account being taken of the shares . . . during a previous representative period, and any special factors. . . ." The loose nature of the provisions with respect to the use of export subsidies plagues GATT to this day: Apparently the Tokyo Round trade talks completed in 1979 accomplished nothing with respect to subsidies and countervailing duties, and the discussions to implement the code envisaged by the agreements have gotten nowhere in 3 years.

Article XI bravely provides that "no prohibitions or restrictions other than duties, taxes or other charges" shall be "instituted or maintained on imports or exports." Specific reference was made to the banning of quotas and licenses and similar measures. But, alas, the second paragraph of the same Article included a clause at the insistence of the United States that quantitative import restrictions could be used where necessary for the enforcement of governmental measures that (1) restrict domestic output or marketings, or (2) remove a temporary surplus from the market by making it available to certain groups of domestic consumers at less than current market prices, or (3) restrict the output of an animal product directly dependent on the imported product.

## U.S. FARM PROGRAMS AND GATT

The United States has a number of farm programs that clearly violate the letter or spirit of GATT. Some of the provisions have no significant economic effect, but yet are retained in spite of their violation of GATT Article XI. Other provisions that violate GATT provisions do have significant economic effects. It is relatively easy to see why foreign officials react with skepticism when we loudly proclaim our support for liberal trade in agricultural products. Simultaneously, we have a GATT waiver so that we can maintain import quotas on dairy products even though our dairy price support programs are not designed to "restrict domestic output or marketings" of milk. Quite the contrary: The programs are designed in such a way that milk production is encouraged.

While it is true that the United States has fewer quantitative restrictions on imports of agricultural products than any other industrial country, save perhaps Australia, it is not particularly helpful to be reminded that:

> And some would argue that the Section 22 legislation and the waiver that grandfathers it into GATT is proof positive that agriculture is basically protectionistic. But Section 22 provides for border protection mainly in the form of quotas not to protect agriculture per se, but rather to protect the ability of the government to carry out effectively the mandated agricultural price support programs without undue cost to the Treasury. In recent years Section 22 has been invoked for dairy products, sugar, peanuts, and cotton but not for the main traded commodities—wheat, feed grains, rice, or soybeans.

These sentences are from what is an otherwise excellent paper on U.S. agricultural trade policies and the efforts to ward off protectionist moves by Fred A. Mangum, Jr. and George E. Rossmiller, the Director and Deputy Director, respectively, of the U.S. Foreign Agricultural Service. What I find objectionable, but probably realistic, about the quoted words is that the import quotas imposed under Section 22 are used to hold down Treasury costs and not to protect agriculture. This means that either we have import quotas that violate GATT principles, for the convenience of the bureaucrats, as was true for at least three of the four cases cited, or we are trying to reduce Treasury costs to make the farm price support programs more acceptable to American voters. The latter reason can just as readily be given to justify the variable levies and export subsidies used by the European Community and the numerous import restraints imposed by the Japanese to maintain their somewhat peculiar internal price structure for agriculture.

Our most blatant violation of GATT principles is the dairy price support program. We use import quotas and we use export subsidies to dispose of part of the mountains of butter and dry skim milk that we acquire in our efforts to maintain dairy product prices at levels significantly above those of the world's low-cost producers. We often cloak our subsidized exports by calling them food aid, but in doing so we fool no one except possibly ourselves. In nearly three decades we have done nothing to bring our dairy programs into conformity with our GATT obligations: We have apparently considered it preferable to continue to take advantage of the GATT waiver. This has been done in spite of the substantial impairment of our negotiating position for liberalizing agricultural trade that results from the existence of the waiver.

The weakness of our commitment to GATT principles and liberal trade in agricultural products can be seen by considering some of the consequences of our dairy price support programs. From 1960 until 1975 our dairy price supports did not encourage increased milk production. Over that period milk production declined by 6 percent. That is admittedly not very much, but it is nevertheless a decline. However, in response to sharp increases in milk price supports in early 1977 milk production increased by 15 percent through 1981 and continued to increase further in 1982. In 1981 the U.S. government had to purchase a tenth of total milk production to meet its price support commitments. While a significant fraction of the milk acquired by the government was disposed of domestically, in 1981 we exported the equivalent of 1.4 million metric tons of milk while importing just 1 million tons. This is not the pattern that would have resulted from a liberal trade regime. With liberal trade we would be a significant net importer of dairy products.

In recent years our farm prices for dairy products have been equal to or higher than those received by farmers in the EEC. In 1980 U.S. farmers received $28.60 for 100 kilograms of milk sold to all plants; the average price received by farmers in the Netherlands was $28.70 and in France $26.85. In 1981, when the value of the dollar was much higher than in 1980, milk prices received by U.S. farmers were 25 percent higher than prices received by Common Market farmers, when local currencies were converted to dollars. At the same time the U.S. dairy farmer had access to significantly cheaper feed than did his Common Market counterpart.

It is not possible to argue that our return to a sugar program dependent upon high price supports and import quotas is consistent with GATT principles, since the current program has no output control features. Imports of sugar are being adversely affected by the high support prices for sugar and the encouragement these high prices give to the increased production and use of high fructose sugar from corn.

U.S. imports of sugar have declined from nearly 6 million short tons in the early 1970s to 3 million short tons in 1982. Over the same period of time sugar production in the United States has remained approximately constant. The decline in imports of sugar has been due to increased domestic production and consumption of high fructose sugar, displacing about 20 pounds of imported sugar per capita between 1970 and 1980. If present trends continue, we will be a minor sugar importer by the end of the present decade, primarily due to a price policy that encourages the production and consumption of a substitute for cane and beet sugar. Has anyone bothered to ask if such a reduction of sugar imports is in the interest of U.S. agriculture? If so, I do not know who has considered this question.

The programs for major export commodities—wheat, feed grains, and cotton—have features that bring U.S. policies into conflict with GATT provisions on the subsidization of output. These programs contain a number of subsidies. Currently, the most important ones are the deficiency payments and the subsidies on the storage of grains under the Farmer Held Reserve Program. Until recently, the disaster payments program represented a clear subsidy to farm production by making payments when yields were below some specified level. By reducing risk and increasing the mean level of income, the program encouraged a higher level of output than would have otherwise occurred. The disaster payments program has been replaced by an insurance program, but it must be noted that the insurance program involves a subsidy. That is because premium income will be substantially less than the cost of claims.

The European Community permits free or nearly free entry of a number of nongrain feed materials, including corn gluten feed. Such imports have displaced several million tons of grains, perhaps as many as 14 million tons, in the EEC livestock rations: In 1980 corn gluten feed imports were 2.6 million tons, up from 0.7 million tons in 1974. The United States now exports 90 percent of its corn gluten feed production of about 3 million tons compared to just one-third a decade ago. One of the important products of corn gluten meal is high fructose sugar; a much less important product is ethanol for use as a motor fuel. Each of these end products is now heavily subsidized, the one by the high domestic price of sugar and the other by the forgiveness of federal and other gasoline taxes on gasohol. In addition, we have had significant credit subsidies and guarantees for the construction of production facilities for ethanol. Small amounts of corn, claimed to be off-grade, have been sold by the Commodity Credit Corporation at subsidized prices for the production of alcohol. While the corn may be off-grade, the corn gluten meal produced from it is not.

One might ask if anyone in a policy position in the U.S. Department of Agriculture or in the White House or a member of one of the relevant committees of Congress ever questioned whether the substantial indirect subsidies of such grain substitutes endanger the duty-free status of this high-quality feeding material. It is more than a little curious that we are satisfying two very small special interest groups, namely, domestic sugar producers who number no more than 15,000 and only a handful of firms that produce high fructose sugar.

Moreover, the benefit to corn farmers from the high fructose sugar subsidy and the export of less than 3 million tons of corn gluten feed to the European Community is very small. With the United States producing 250 million tons of feed grains, the increase in grain processed for high fructose sugar and for ethanol, even if this were a net increase, would have little effect upon the United States and the international market price for corn and other feed grains. And the EEC imports of corn gluten feed do not represent a significant net increase in demand. Thus far these imports appear to have had little effect on EEC feed grain production and consequently the increased imports have been largely offset by larger EEC grain exports.

It is highly likely that most of the gains from the expanded production and exports of corn gluten feed go to the firms in the wet-milling industry. In the early 1970s corn gluten feed was priced at approximately 10 percent below the price of corn. In the United States from 1979 through 1981 corn gluten feed was priced as much as 12 percent more than corn and in one year only 2 percent less. The higher relative price of corn gluten feed compared to corn could not have had a significant effect upon the price farmers received for their corn.

Both the United States and the European Community have had difficulties in disposing of their respective "butter mountains." The European Community, in spite of significant resentment of its citizens, sells some part of its mountain to the Soviet Union at bargain basement prices. People who pay about $2.50 per pound for butter have some reason to resent a sale to the USSR at about a third of that price. Because the United States complains so loudly about the EC's use of export subsidies (dumping) to dispose of its agricultural surpluses, there is some sensitivity in Washington about how our actions will be received.

In September 1981 the officials of the U.S. Department of Agriculture came up with a solution to minimize the adverse reactions to a substantial export of butter at a highly subsidized price. The solution, were it a mechanical invention, would have rivaled any Rube Goldberg flight of the imagination. New Zealand is the world's lowest-cost producer of dairy products and the world's largest exporter. And yet, the United States sold 100,000 tons of butter to New Zealand at a price that was less than half our wholesale price at the time. The loss was in excess of $155 million. The butter was disposed of in this way rather than selling it directly into the world market for two main reasons:

1.  A major buyer of butter in the world market at the time was the Soviet Union, and as Richard Smith of the USDA noted ". . . to sell butter at subsidized prices to the Soviet Union would be to send the wrong signal to the Kremlin during a very tense period in foreign relations."

2.  The United States is said to be very conscious of its commitments "to open and nondiscriminatory world trade and our obligations under the General Agreement on Tariffs and Trade (GATT)." About this time the United States had been challenging ". . . Japan's use of subsidized rice exports and the European Community's increased use of subsidies on exports of wheat, flour and poultry, and we are gaining support in these efforts from third countries. To unload 220 million pounds of subsidized butter on the market could call into question the sincerity of these U.S. efforts to restrain the use of subsidies by other nations."

I do not believe that the USDA was able to disprove the famous dictum: "A rose by any other name is still a rose." And so an export subsidy by any other name is still an export subsidy, even when the sale is among friends.

## TRADE AND U.S. AGRICULTURE

The prosperity of U.S. agriculture depends upon export outlets for its production. The production from nearly two-fifths of all cropland is exported. We export more than one-half of the output of wheat, cotton, and soybeans and over one-third of the output of feed grains. If these markets were lost or significantly reduced, American agriculture would face a long period of adjustment during which returns to farmers would be adversely affected.

It is not possible to dispute the view that U.S. agriculture has much more to gain from liberalizing agricultural trade than from maintaining our restraints on imports and from direct and indirect output subsidies. Yet there has been almost no effort made to modify our farm programs to make them more consistent with the GATT principles for governing trade in agricultural products.

During the 1950s and 1960s U.S. agricultural policy was modified by reducing the level of price supports and the need for governmental intervention, such as export subsidies. Over an extended period of time price supports were reduced to levels that were generally below market prices. The market was permitted to allocate the available supplies among the competing market outlets, including the allocation between the domestic and foreign markets. These changes occurred, however, exclusively for those products for which the United States had a comparative advantage, namely, the major export crops of cotton, wheat, and feed grains. Soybeans had generally not been burdened with high price supports, except in 1968 when election politics were stronger than economic rationality.

The changes in domestic farm programs were associated with a rapid growth in exports of farm products. The United States did not have significant net exports of agricultural products until the early 1970s, but even this small excess was a change from our being a net importer as late as 1956. While the policy changes were effective in achieving an expansion of exports, we did nothing to liberalize trade in certain products where we had severe trade restraints, such as for dairy, peanuts, and sugar, until well into the 1970s and not at all for dairy products. And in the mid 1960s, when we were preparing for the Kennedy Round of trade negotiations, our dedication to liberal trade was revealed by the imposition of quotas on beef and veal imports.

Import quotas and export subsidies continued to be applied to dairy products. Our sugar program remained an affront to liberal trade until 1974 when the sugar program was abolished, but only for a relatively brief period. Our current sugar program is at least as restrictive as the program of the 1960s and early 1970s. While the peanut program was modified during the 1970s by introducing a two-price system that reduced the degree of subsidization, it still required import quotas to maintain the high domestic price for edible uses of peanuts. The notification to the peanut program that provided for pricing marginal peanut production at approximately the market price has not resulted in a decline in peanut production. This may be due in part to the very high price paid for the so-called quota peanuts.

With the possible exception of the peanut program, we have made no effort to liberalize trade in the farm products that we would import under a liberal trade regime. We remain in significant violation of GATT principles, or would if we had not used our muscle to obtain a waiver. Some 15 years ago, as one of the members of the National Advisory Commission on Food and Fiber, I presented rough outlines of adjustment programs that would permit eliminating most if not all quantitative restrictions on imports as well as doing away with export subsidies (1967). I believe I made a reasonable case that for peanuts, sugar, manufactured dairy products, rice, wheat, feed grains, wool, cotton, and tobacco it would be possible to eliminate the trade disruptive aspects of the then existing commodity programs and, where necessary, to compensate fully all those who would lose from the required program modifications. I do not argue that my proposals were the most appropriate ones that could be devised. What I now say is that I have not seen any alternative proposals that would accomplish the same objectives, including the reduction or elimination of interventions in trade of agricultural products.

The costs of price supports and their associated trade restrictions to consumers and taxpayers for sugar, peanuts, and dairy products are so large and the net income benefits so small that it would be possible to compensate all who would lose from trade liberalization from the savings to consumers and taxpayers for a relatively brief period of time, perhaps no more than 2 to 4 years. I have made no recent estimates of the costs and net benefits, but in an earlier time I did so. For sugar, for the early 1970s, I estimated that full compensation for loss of income to producers (present value at 10 percent interest) plus full compensation for the capital loss to processors plus 1 full year's wages to all employees engaged in sugar production and processing would cost no more than the savings realized in 4 years from abolishing the sugar program (1974).

How far we are from having a consistent and coherent liberal trade policy for the nation and agriculture was recently revealed by the unsuccessful efforts of the U.S. government to negotiate textile quotas with the People's Republic of China. The failure to reach an agreement acceptable to both sides resulted in China taking retaliatory measures. It may well be that those retaliatory measures will hurt China as much as they will threaten U.S. agriculture. But this makes little difference. Another distortion in world trade in farm products has been created as China turns to other and presumably higher-cost sources for its cotton, soybeans, and synthetic fibers.

In many ways our national policy for the textile industry parallels our dairy program: For more than a quarter-century major components of the textile industry have been protected by agreements intended to disguise our violation of the GATT principles that oppose quantitative restrictions on imports. Adjustments in textiles that should by now have been completed if we were attempting to adhere to a liberal trade policy have not been made, nor is there any evidence that the industry has been given signals that make it clear that adjustments were required. And why should it have adjusted if it could rely upon continued protection?

But leaving aside the debacle of the textile quota negotiations, it remains true that the textile industry is protected at the expense of the American farmer and, I might add, the American consumer. The American consumers who bear most of the costs of our textile trade policies are those with lower incomes, since it is low-priced textiles that the quotas are primarily designed to exclude.

American agriculture has much to lose from rising protectionism in the world. While the spread of protectionism is due, in part, to the current worldwide recession, it is not obvious that all of the forces in the United States that are pushing protectionist measures will be dissipated by even a strong recovery and prosperity. American farmers should not assume that a return to high levels of employment will be enough to prevent further erosion in our liberal trade policies.

Farmers and those who represent them and support them must be aggressive in resisting the current moves to increased protection and must not relax their vigilance in the expectation that recovery will assure the abolition of some of the recent departures from the principles of freer trade. Some of the industries that are now complaining so strongly about import competition have acquiesced to cost structures that may well have destroyed the comparative advantage such industries may have at one time possessed. If so, it is absolutely essential that no effort be spared to fight the rising tide of protectionism: It cannot be assumed that the tide will recede automatically.

# 17

# DEVELOPED AND DEVELOPING COUNTRIES

## W. ARTHUR LEWIS

Two decades of extraordinary growth preceded the recession of 1974. World industrial production grew at around 6 percent per year compared with 3.5 percent before the First World War. World agricultural output, which previously grew at about 2 percent per year, grew in the 1950s and 1960s at 3 percent per year, or 50 percent faster. The rate of growth of world trade was most spectacular—it more than doubled! The 1950s and 1960s will go down in the record books as our most prosperous decades since the Industrial Revolution began two centuries ago.

This extraordinary growth was not confined to the richer countries of the world; the developing countries shared in it too. These poorer countries also grew at unprecedented rates, significantly faster than the developed countries had grown in their comparable phases of development. They grew faster, for example, than Britain, France, or the United States. In expanding so fast the poor countries also ended arguments about their economic viability. They showed that the money they invest has the same yield as money invested in the richer countries: In both groups the ratio of investment to yield is about 4 to 1.

The benefits of growth were not spread equally throughout these economies, but they were not confined to the upper and middle classes either. The number of children in primary school has multiplied by four since 1950. The infant mortality rates have fallen from 300 per thousand to 60 per thousand—the reason why is that the population is exploding in these countries. Thousands of villages that used to be isolated in rainy seasons now have all-weather roads with a constant stream of trucks and buses. If one measures progress as the villagers measure it, in terms of hospital beds, village water supplies, bicycles, and radios, for example, their progress is not in doubt.

However, almost any generalization you make about these developing countries is bound to be wrong because of their immense variety of performance. About one-third of these countries did not manage to sustain a growth rate of 1 percent per person annually. At the other end of the spectrum about one-third grew continuously at over 2.5 percent per person per year. Most of the countries that did badly fell into one of two groups. Either their rainfall provides an inadequate or unreliable basis for agriculture, as in Niger or Senegal, or else they experienced internal war or other breakdowns of internal security, as in Ghana or Mozambique. Lack of rain and instability of government are the two most prominent causes of slow growth.

Two elements created and sustained the rapid pace of the developing countries, taken as a whole. The first was the great increase of imports of agricultural commodities into the industrial countries, such as rubber, tea, and cotton. And the second was the rapid increase of production of manufactured goods.

The growth of agricultural exports was not surprising. This has always responded to the prosperity of the industrial countries, which create the demand for agricultural commodities. If the industrial countries grew twice as fast as before, their agricultural imports also grew twice as fast. The growth of agricultural exports from the less developed countries is facilitated by a high elasticity of supply. The commodities exported are less than 20 percent of agricultural output, so volume can be expanded without sharply rising costs. Besides, there were heavy expenditures during the 1950s and 1960s on transport facilities, especially roads, bridges, ports, and motor vehicles, so transport costs fell relative to output costs. Nevertheless, despite the rapid increase in demand, the terms of trade for agricultural products from the tropics actually worsened a little.

Agricultural exports from the less developed countries would have fared even better but for the restrictions which some developing countries put on imports, mainly to protect their own production. Sugar is the most important example, but in some countries meat and fruit are also restricted. Quite commonly a higher tariff is imposed on refined products than on raw commodities in order to discourage the developing countries from refining their own production. The World Bank has estimated that in 1977 the developing countries could have sold some $20 billion more of agricultural and manufactured commodities but for restrictions by the richer countries. These countries must allow more space in world trade for the developing countries if the latter are to enjoy their full development potential.

During the 1950s and 1960s export commodities were growing rapidly, but the situation with domestic food production was quite different. Production of domestic foodstuffs barely kept pace with population growth and in a number of countries actually fell behind. To be sure, the effects of the green revolution were powerfully in evidence during the 1970s, but the new technology was still rather limited to wheat and irrigated rice and not available to rain-fed rice, to maize, to coarse grains, or to root crops. It was therefore also limited geographically, showing its most spectacular results in parts of Asia.

In areas where people are already not adequately fed, the rapid growth of population is disquieting. This is at its worst for the several hundred million people who live in arid lands bordering the great deserts of Asia and Africa. We need some breakthroughs in dry farming to double or treble current yields per acre. Temporary progress is possible through better cultivation, control of livestock numbers, planting of trees, and conservation of water and soil, but these measures are easier to talk about than to make effective. Permitting some of the people of these countries to move to more favorable climates would be helpful, but this is a formidable task given the numbers involved and the hostility with which the migrants would be received. Land reform would also be helpful. In the end the only permanent remedy is a sharp decline of the birth rate, but this also seems a distant prospect.

The feeding of these hundreds of millions on arid lands may become an issue. If the tropical peoples cannot feed themselves, they will move to get food from the temperate lands and may find themselves competing with China and with Russia for the temperate surpluses. Given the precipitous decline in the number of people working in temperate agriculture, such surpluses may be rather small even where fertile land is available, unless low-wage immigrants become a major source of farm labor, as is already the case in California. Presumably, the countries needing food would insist that the temperate lands be fully cultivated one way or another and would resist the imposition of export quotas by the temperate countries. In order to pay for food, these countries would have to be allowed to sell manufactured goods in temperate markets and to buy food at reasonable terms of trade.

The situation is grave. We are depending on our scientists to come up with more technological breakthroughs, not necessarily just in the arid zones but also in rain-fed agriculture, since the main point is to generate surpluses— never mind where. Because of the lack of technological possibilities, I would guess that the chances of running into a world food shortage are greater over the next 20 years than they may be in the twenty-first century, but no technique exists for predicting such matters with precision.

In manufacturing the less developed countries have concentrated on production for the domestic market of goods that were previously imported. This process is known as import substitution. Given a backlog of demand, output can grow quickly. Thus the average rate of growth of manufacturing in the 1950s and 1960s has risen faster in the poor than in the rich countries. Some of the poor countries were growing at 10 percent per year compared with only 5 percent per year for industrial countries as a whole.

Such a pace is sustainable in the early stages of import substitution, but in about 20 years the country in question will be producing at home all the easiest kinds of manufactures, especially light consumer goods, and it will lose steam unless it can find some new way to sustain demand at 10 percent. Latin America had completed import substitution as early as 1955, and by 1960 Asia had done the same.

A few of the bolder spirits then launched into exporting manufactured goods to the industrial countries. Contrary to expectation this turned out to be relatively easy, especially as world trade in manufactures was growing by leaps and bounds. This solution was adopted by one developing country after another with such speed that, whereas in 1950 manufactures were only 10 percent of the exports of developing countries, today manufactures are about half the exports from these countries, excluding oil.

Looking just at rates of growth, this would seem to be an ideal situation. Both world trade in manufactures and the exports from poor countries were growing at 10 percent per year, so the poor countries' share of world trade in manufactures was constant. It was quite small, only 7 percent of world trade, and you might think that the ability to keep such a small share constant would be beyond challenge.

But it did not work out that way. The poor countries began by concentrating on low-wage goods, where their comparative advantage lay, and were soon pouring cotton textiles and clothing into Europe and North America. Factories there were closed, and people lost their jobs. The developed countries retaliated with a whole series of restrictions on imports of manufactured goods from the developing countries. These barriers began with textiles, but as industries in the developing countries turned to other commodities restrictions spread to those other commodities too.

There was a clear confrontation between North and South, as the rich and the poor countries now came to be called, in the matter of trade. The South was shocked by the open discrimination. On the one hand, the OECD countries were negotiating at GATT a whole series of agreements reducing tariffs and quotas on trade with each other, while on the other hand, they were unilaterally imposing restrictions on the trade of the South. The South protested against this discrimination, but this got them nowhere.

Next, the developing countries of the South brought out the economics textbooks and proved that it is not in a country's interest to keep out imports.

Specifically, they said, to bring in low-wage imports releases labor and resources locally to be transferred to high-wage industries that can export their goods to pay for the imports. So there is a net gain.

It does not happen like that, said the North. What happens is that local labor becomes unemployed. Some people do get new jobs, but most of the people in the domestic, low-wage industries are old people or married women tied to the towns where the family lives and they can not be transferred. Therefore it does not pay to import low-wage goods if the labor and resources in competing industries locally just become unemployed. What you have is a fall in total production, which is not in the national interest.

This line of argument did not make sense in Western Europe in the 1960s when Europe was extremely short of unskilled labor and was importing such labor from Southern Europe. Germany, for example, did not use this argument. On the contrary, by 1973 Germans were talking about the desirability of importing low-wage goods instead of importing unskilled labor, a movement that causes problems of social integration. The Europeans restricted textile imports because of pressure from their own textile interests rather than because of sophisticated economic theory.

The American case was different because the bombardment by imports was not confined to low-wage goods from the less developed countries. It was happening all along the line to every kind of U.S. industry, light and heavy. The high-wage industries in the United States were under equal bombardment from other industrial powers. Where were the people from the low-wage industries to go? Could they go into automobiles? The industry was on the verge of bankruptcy. It was the same in the steel industry. The arguments of the South presupposed that the high-wage sector of the economy in the United States was expanding rapidly and could therefore take in displaced people from the low-wage sector. If this was not so, then imports would cause a fall in employment and output.

The proximate cause of America's troubles was easy to see. In the 1950s and 1960s, imports of manufactured goods into the United States increased at an annual rate of 13 percent, doubling every 6 years. Exports of manufactures were rising only at 6 percent per year, the difference between 13 and 6 percent having a deflationary effect on the market. The contrast with continental Europe was clear. Under this bombardment U.S. industry became relatively unprofitable. Its investment ratio was the lowest among industrial countries, and the industrial growth rate was among the slowest. Deindustrialization was taking place. This is a term we use when the labor force in manufacturing grows less rapidly than the labor force as a whole. It is something we expect of older countries as they consume fewer commodities and more services. In the American case consumption of commodities was maintained, but it was met from imports not from domestic production. This produced a drag on the labor market.

Instead of the reservoir of cheap labor emptying, as was happening in continental Europe, it was filling up. One little index was that the ratio of women's wages to men's was declining: from 63 percent in 1956 to 58 percent in 1969 and 57 percent in 1974.

The position of the United States in the world economy was changing. From the end of the First World War to the end of the Second World War the United States was the leading economy in the world. It pioneered new commodities, demonstrating not only their efficiency but also their attraction for mass markets. Industrial output per worker was twice that of Western Europeans. The United States could pay its workers wages 50 to 100 percent higher and still undersell all other industrial countries. This state of dominance ended in the 1950s. Germany, France, Sweden, and Japan reconstructed their industries after the war and wiped out most of the gap, leaving the high-wage U.S. industries stranded.

The rich countries must make space in world trade for tropical agricultural exports; restrictions are unjust and offensive. The same is true for manufactured goods. The industrial countries must make more space for the developing countries in the world trade in manufactures, of which they have only 7 percent. Some of the developing countries that are relatively short of agricultural resources, notably India, Egypt, and Indonesia, will be driven by comparative advantage—like Britain and Japan—to import agricultural products in return for manufactures. Currently, their bill for oil equals about 4 percent of their gross domestic product. They cannot pay for this indefinitely by increasing debt by 4 percent of GDP each year. Neither can they pay by exporting more primary products to the industrial countries since the inelasticity of prices would reduce rather than increase their earnings. In theory the problem could be solved by the developing countries cutting imports other than oil by 4 percent of GDP. They could do this by buying more manufactures from each other and less from the industrial countries. This also counts as making more space for developing countries since it means that the industrial countries surrender part of their markets for manufactures in less developed countries in favor of other LDCs.

This question of access to markets is one of the issues in the demand for a new international economic order which developing countries have been promoting since 1974 through the United Nations agencies. The list of topics on this agenda is rather long. It includes, apart from the trade issues, a code of behavior for multinational corporations, a multiplication of foreign aid, a redistribution of voting power in the World Bank and the International Monetary Fund, and much more.

Much of the argument derives from a basic disagreement as to the obligations of rich countries to poor countries. The industrial countries are surprised at the magnitude of the claims made against them, especially in those industrial countries that had no colonies to speak of, such as the United States and Germany, and which cannot therefore be charged with past neglect or exploitation.

Our moral obligation, the rich countries say, is to trade with you fairly; to pay prices for your produce determined by freely competitive markets. If we do that, we have fulfilled our major obligation. Oh, no, say the poor countries, what you call free competitive markets yield unjust prices. There is nothing sacred about market prices or competition. These prices exploit us.

How? The argument runs like this. The competitive market fixes the price of a tropical crop at a level equal to what the peasant farmer could earn if he produced something else. This is how the market works. Now the "something else" that the farmer might otherwise produce is food: The price of the export crop is determined by the amount of food the tropical farmers could otherwise produce. Since tropical food production technology is still rather backward, the amount of food the tropical farmer could produce is rather small, and the price he would get for export crops—tea, rubber, coffee, or whatever— would therefore also be rather small. Neither is his position improved by becoming more proficient in cultivating the export crops. The more productive he is the lower the price he receives since the price is tied to the alternative amount of food that could be produced, which remains the same. Farmers gain if they raise their productivity in food crops, but they are no better off for raising their productivity in export crops.

There is nothing sacred about a market price determined in this way. To pay according to alternative cost is to exploit the weak. The tropical farmers are not paid for their skill in producing what they offer for export—such skill does not benefit them—but are penalized for their lack of skill in producing what they are not offering for sale, namely, food.

This differs from what we pay each other within a developed economy. There, the principle is equal pay for equal work. If cocoa and steel were produced inside the same economy, we would expect the cocoa farmer and the steel worker to have equal incomes if each were of average working capacity. This would come about through competition in the labor market, bringing equal payment for equal work irrespective of what the work was used to produce. Such competition, however, would imply that cocoa farmers can emigrate to the developed countries. It is only because we assume away the possiblility of migration that we abandon the principle of equal work and are left with the principle of alternative cost.

Developing countries consider that they are exploited by market forces; the prices of their products are too low. This is the background to the demands they have been making since 1974 for what they call a new international economic order. Supposing that one accepted the argument: How would one remedy the situation?

The most obvious way would be to create a mechanism for raising the prices of tropical exports. Many agreements have been tried, such as international commodity price-fixing with buffer stocks, as in tin, or without buffer stocks, as in sugar. But the only resounding success has been in oil, and even this arrangement now threatens to come apart. The trouble is that when you raise the price of a commodity the supply increases and the demand contracts, so a surplus develops. Different producers are affected by the surplus in different degrees, and the worst hit tend to withdraw from the agreement. So these arrangements tend to be unstable.

A more straightforward way of helping the poorer countries, whether you accept their argument about prices or not, is simply to give them money. We have been doing this quite successfully since the 1950s both in bilateral programs and also through international institutions, of which the one handling the most money is the World Bank. It is fashionable to sneer at foreign aid, but there is no justification for doing so. As I said earlier, the ratio of investment to yield has turned out to be the same in developed and developing countries: It is 4 to 1. This means that money spent in the poorer countries is as effective as money spent in the richer countries. Of course, waste is to be found in all human affairs anywhere in the world, so one can easily produce a string of true stories about waste in Africa or Asia or for that matter in Washington. But this should not prevent us from recognizing the achievements of foreign aid.

The principle of foreign aid is not in dispute. The developed countries now devote an average of one-third of 1 percent of their national incomes to foreign aid. However, they all indicated at the beginning of the 1970s, with the sole exception of the United States, that they would aim at raising the average to seven-tenths of 1 percent, slightly more than doubling it. The U.S. Congress is opposed to raising foreign aid, and though the U.S. Administration is much embarrassed by this, no progress has been made for more than a decade. The American position is devastating, not just because of the absence of American money but also because other nations that favor foreign aid are reluctant to move too far ahead of the United States. Some positive movement by the United States would therefore be doubly effective.

Finally, a word about prospects. We had come through a period of extraordinary prosperity in the 20 years up to 1973. Since then, we have had a severe recession. This is not surprising since there is a severe recession in the U.S. economy about once every 18 years—the commencing dates are 1872, 1892, 1906, 1929, 1956, and 1973. This latest recession was running true to form until 1979 when instead of moving upward it moved downward under the pressure of the new monetary policies adopted in that year. Whether these policies will change the typical shape of recessions remains to be seen.

These severe recessions last about 10 years. Whenever we are in such a recession and we try to assess the future, we come out with gloomy forecasts. Profits are low, unemployment is high, productivity is low, a number of ancient industries are doomed—and it looks as if the economic system will never again deliver a fast rate of growth. This view has always proved false, and I see no reason for the long-run gloom that pervades such forecasting today. My guess is that the world economy as a whole will grow at reasonable speed over the next two decades, although a little more slowly than from 1950 to 1970 when there was a large backlog of innovations. Rapid growth is what we all need; the farmer, to give him better terms of trade; the urban worker, to give him employment; women and minorities, to create more jobs worth having; and the United States, to facilitate its adjustment to a changing world economy. Rapid growth is what we need; whether we shall get it is anybody's guess.

# 18

# A JAPANESE PERSPECTIVE

KENZO HEMMI

According to a projection by the Economic Planning Agency of the Government of Japan, Japan's share of the world GNP, which has increased from 3 to 10 percent in the past 20 years, will increase to 12 percent in the year 2000 (Table 1). Although Japan's share of the world GNP will not increase in the coming 20 years as rapidly as in the past, the Japanese people are gradually preparing themselves for world affairs. They think that the coming 18 years will be a shifting period when the country moves from its state of isolation to become a full member of the international community.

At the same time, there will be profound demographic changes within the Japanese population. Assuming that the future is an uninterrupted extension of past trends, the proportion of Japanese population aged 65 years and older is expected to increase from 9.6 percent in 1980 to 15.6 percent by the year 2000 and to 21.3 percent by the year 2025. There will be a significant shortage of youth in the labor force in these coming years. For this reason and others, the rate of growth of the Japanese economy is not expected to be as high in the future.

The public relations section of the Prime Minister's Secretariat has conducted a series of opinion polls about food consumption, food problems, and agriculture. The last poll was conducted in September 1980. The results of the opinion poll dealing with future food consumption are as follows (see Tables 2 and 3): About one-half of the respondents thought that present levels of rice and

*319*

Table 1.   Distribution of World GNP by Country

|  |  | 1960 | 1978 | 2000 |
|---|---|---|---|---|
| Developed Countries | Japan | 3 | 10 | 12 |
| (DC) | United States | 33 | 22 | 20 |
|  | Other DC | 26 | 31 | 26 |
|  | (Subtotal) | (62) | (63) | (58) |
| Less developed | NICS | 3 | 4 | 7 |
| countries (LDC) | Other LDC | 11 | 11 | 13 |
|  | (Subtotal) | (14) | (15) | (20) |
| Centrally planned | U.S.S.R. | 15 | 13 | 12 |
| economics (CPE) | Other CPE | 9 | 9 | 10 |
|  | (Subtotal) | (24) | (22) | (22) |
| World total |  | 100 | 100 | 100 |

*Source:* Keizai Shingikai Choki Tembo Iinkai Hokoku, Keizai Kikakucho, *Nisennen no Nihon-Kokusaika, Koreika, Seijukuka ni Sonaete* (Japan in 2,000—Preparation of Internationalization, Aging and Maturing of the Economy), 1982, p. 32.

wheat consumption will continue in the future. More than one-quarter of the respondents thought that rice consumption will increase. Those respondents who thought that wheat (bread and noodle) consumption would increase represented only 17 percent of the total. Corresponding figures in the August 1978 poll were 41 percent, 23 percent, and 26 percent, respectively. It would seem that rice has become more attractive to the Japanese people while wheat has become less attractive. However, estimates from the Food Balance Sheets of the Ministry of Agriculture, Forestry and Fisheries reveal that per capita con-

Table 2.   What Do You Think of Future Consumption of Rice and Wheat?[a]

|  | No. of Respondents | Present Consumption Pattern Will Continue | Wheat Consumption Will Increase | Rice-Based Food Consumption Will Have a Favorable Turn | Don't Know | Total |
|---|---|---|---|---|---|---|
| Total | 2494 | 48%(41%) | 17%(26%) | 26%(23%) | 9%(10%) | 100% |
| 20–34 years | 720 | 53  (41) | 18  (29) | 23  (22) | 6  (8) | 100 |
| 35–44 years | 578 | 49  (43) | 22  (28) | 25  (22) | 4  (7) | 100 |
| 45–54 years | 502 | 44  (41) | 14  (24) | 32  (26) | 10  (9) | 100 |
| 55 years and more | 694 | 44  (38) | 15  (24) | 27  (20) | 14  (18) | 100 |

[a] Figures in parentheses are the results of an August 1978 poll.

Table 3. Do You Want to Increase or Decrease Consumption of Beef and Other Products?

| | | Beef | Pork | Chicken Meat | Eggs | Processed Meats[a] | Fresh Milk | Milk Products | Sardine and Mackerel | Tuna and Skipjack | Cultivated[b] Fishes | Crustaceans | Squid and Octopus | Shell Fish | Fish Products[c] | Frozen Foods[d] | Prepared Foods[e] |
|---|---|---|---|---|---|---|---|---|---|---|---|---|---|---|---|---|---|
| (A) Want to Increase | Total (%) | 30 | 10 | 13 | 13 | 7 | 32 | 22 | 8 | 12 | 12 | 18 | 10 | 13 | 3 | 3 | 4 |
| | 20–34 years | 39 | 14 | 16 | 17 | 9 | 43 | 33 | 8 | 11 | 13 | 20 | 12 | 15 | 4 | 4 | 6 |
| | 35–44 years | 37 | 13 | 16 | 15 | 8 | 35 | 28 | 9 | 14 | 12 | 20 | 9 | 14 | 3 | 3 | 5 |
| | 45–54 years | 30 | 10 | 12 | 11 | 5 | 27 | 18 | 10 | 13 | 16 | 21 | 11 | 12 | 5 | 4 | 4 |
| | 55 years and more | 15 | 4 | 8 | 8 | 5 | 20 | 9 | 6 | 12 | 9 | 13 | 8 | 9 | 2 | 2 | 3 |
| (B) Want to Decrease | Total (%) | 5 | 10 | 4 | 3 | 12 | 2 | 4 | 5 | 2 | 4 | 3 | 4 | 4 | 10 | 14 | 10 |
| | 20–34 years | 3 | 8 | 5 | 3 | 12 | 1 | 2 | 7 | 3 | 4 | 2 | 3 | 4 | 12 | 17 | 13 |
| | 35–44 years | 4 | 10 | 3 | 3 | 10 | 2 | 2 | 4 | 1 | 3 | 2 | 4 | 3 | 10 | 13 | 13 |
| | 45–54 years | 5 | 11 | 4 | 4 | 11 | 2 | 4 | 5 | 2 | 5 | 4 | 3 | 5 | 11 | 15 | 11 |
| | 55 years and more | 7 | 13 | 6 | 4 | 12 | 2 | 7 | 5 | 2 | 5 | 5 | 6 | 4 | 7 | 13 | 8 |
| (A) − (B) | Total (%) | 26 | 0 | 9 | 10 | 5 | 30 | 18 | 3 | 10 | 8 | 15 | 6 | 9 | 7 | 11 | 6 |
| | 20–34 years | 36 | 6 | 11 | 14 | 3 | 42 | 31 | 1 | 8 | 9 | 18 | 9 | 11 | 8 | 13 | 7 |
| | 35–44 years | 33 | 3 | 13 | 12 | 2 | 33 | 26 | 5 | 13 | 9 | 18 | 5 | 11 | 7 | 10 | 3 |
| | 45–54 years | 25 | 1 | 8 | 7 | 6 | 25 | 14 | 5 | 11 | 11 | 17 | 8 | 7 | 6 | 11 | 7 |
| | 55 years and more | 8 | 9 | 2 | 4 | 7 | 18 | 2 | 1 | 10 | 4 | 8 | 2 | 5 | 5 | 11 | 5 |

[a] Processed meats include ham, sausages, and others.
[b] Cultivated fishes include all cultured fishes such as eel, rainbow trout, and young yellow tail.
[c] Fish products include kamaboko, chikuwa, canned fish, and others.
[d] Frozen foods include such frozen foods as hamburger, spring roll, and others.
[e] Prepared foods include such foods as tempra, croquette, and others.

sumption of rice declined at an annual rate of 3.2 percent from 1965 to 1970, 1.5 percent from 1970 to 1975, and 2.2 percent from 1975 to 1980. Wheat consumption, on the other hand, increased 1.2 percent, 0.4 percent, and 0.5 percent, respectively, over the same periods. Therefore the figures in Table 2 are on the surface contradictory to the actual trends of rice and wheat consumption.

A key to the contradiction may be found in the changes in the composition of the Japanese population: the shift toward an older population, decreasing size of average household, and a more urbanized population. Table 3 tells us that the Japanese people want to increase their consumption of beef, milk and milk products, chicken, eggs, and high-quality seafoods. They want to decrease the consumption of processed, frozen, or prepared foods. It should be noted that livestock products are more attractive among young Japanese than older Japanese. It is clear that in each food category in Table 3, A added to B is far less than 100 percent. Although the percentage of those who simply responded "Don't Know" or "Never Have Tasted" is not shown in Table 3, it is clear that more than half of the respondents did not feel the need to increase or decrease consumption of the items in the future. In regard to eggs, chicken, sardines and mackerel, tuna and skipjack, squid and octopus, and fish products, about 80 percent of the respondents replied that they didn't want any change in consumption. As for beef, fresh milk, milk products, and frozen foods, about 60 percent of the respondents replied that they wanted no change in consumption.

Apparently, the Japanese people are generally satisfied with their present pattern and level of food consumption. Consumption of animal products will increase gradually while consumption of starchy staples will decline. It is questionable that the consumption of wheat will continue to increase in the future.

Another important aspect of the Japanese view, which was made clear by the public opinion polls, is that Japanese are very pessimistic about the future supply of food (see Table 4). In 1975 they were unusually pessimistic: 58 percent of the respondents answered that Japan's food situation would worsen in the next 10 years. If we consider the fact that 1975 was the last year of the world grain crisis, we can understand this pessimism. However, it is surprising that even as late as 1980 about one-half of the respondents thought Japan's food situation would worsen in the next 10 years. There is a big difference between the two polls. In 1980 only about 10 percent of those polled expected a drastic change in the food situation, either for the worse or better, whereas in 1975, 20 percent expected major change. The percentage of people who anticipated a little improvement increased from zero to 6 percent during those 5 years, whereas the percentage of those who expected a little deterioration changed very little, except among people living in towns and villages. It should be added

Table 4.    Opinions About What Japan's Food Situation Will Be Like in About
10 Years

|       |                        | Will Worsen | Will Worsen Much | Will Worsen Little | Will Not Differ from Now | Will Improve | Will Improve Much | Will Improve Little | Don't Know |
|-------|------------------------|-------------|------------------|--------------------|--------------------------|--------------|-------------------|---------------------|------------|
|       | Total                  | 58          | 16               | 42                 | 19                       | 5            | 5                 | 0                   | 18         |
| 1975[a] | By Region            |             |                  |                    |                          |              |                   |                     |            |
|       | Ten large cities       | 58          | 17               | 41                 | 20                       | 5            | 5                 | 0                   | 17         |
|       | Other cities           | 58          | 16               | 42                 | 20                       | 4            | 4                 | 0                   | 18         |
|       | Town and villages      | 57          | 15               | 42                 | 19                       | 5            | 5                 | 1                   | 18         |
|       | Total                  | 48          | 8                | 39                 | 30                       | 7            | 1                 | 6                   | 15         |
| 1980  | By Region              |             |                  |                    |                          |              |                   |                     |            |
|       | Eleven large cities    | 53          | 10               | 43                 | 29                       | 7            | 0                 | 7                   | 11         |
|       | Other cities           | 49          | 9                | 41                 | 30                       | 6            | 1                 | 5                   | 15         |
|       | Town and villages      | 39          | 6                | 34                 | 33                       | 9            | 0                 | 8                   | 20         |

[a] 1975 Opinions are from Emery N. Castle and Kenzo Hemmi, 1982, pp. 250–251.

that the percentage of those who responded "Will Not Differ from Now"
increased greatly.

Finally, we must emphasize that judging from Table 4 city people have
been more pessimistic about the future food supply than those living in towns
and villages. We can conclude the following: First, most Japanese people think
that their country's food situation will stay the same or will worsen a little in
the next 10 years. A small number of people think the situation will greatly
worsen or improve slightly. Second, the percentage of these who see a drastic
change in the food situation in the coming 10 years decreased significantly
from 1975 to 1980.

In Japan it is assumed that the decline in food self-sufficiency, coupled with
an increasing reliance on the United States, is a potential source of vulnerability
which may lead to a disruption of the food supply. In recent years self-sufficiency
ratios for wheat, soybean, sugar, and feed grains have been extremely low. Of
course, some Japanese are worried about the country's high dependence on
petroleum, which is almost entirely imported, for domestic food production.
Therefore the generally high ratio of food self-sufficiency does not necessarily
mean a secure supply of food for Japan. However, to the average Japanese,
petroleum is less essential than food. Anyway, Japanese people are extremely
worried about their high dependence on imports for domestic food consumption.
This worry is indicated by the data in Table 5. More than 70 percent of
those who responded thought that, in principle, domestic production should
be increased whenever possible. They are not very conscious of the cost of
increasing domestic food production.

Table 5.  Opinions in 1975 and 1980 About Whether Japan's Food Should Be Imported or Domestic Production Be Increased

| | In Principle, Domestic Production Should Be Increased Whenever Possible | It Is Better to Consume Imported Products if They are Less Expensive | Don't Know |
|---|---|---|---|
| 1975*a* | 71 | 17 | 12 |
| 1980 | 75 | 16 | 9 |

*a* 1975 figures are from Emery N. Castle and Kenzo Hemmi, 1982, p. 253.


## GOVERNMENT REPORT ON FOOD AND AGRICULTURE IN THE YEAR 2000

The view that domestic food production should be increased is also held by Japanese government officials. In projecting the Japanese food and agriculture situation to the year 2000, these government officials assumed that, unlike fossil energy resources, food and agricultural resources are reproducible and can be used forever. Therefore, although it will be necessary to depend on not only domestic production but also on foreign supplies to meet the diversified demands of consumers, it is extremely important to make full use of domestic food and agricultural resources. This is a basic requirement for the stable economic life of the Japanese people. I interpret this assumption to mean that it is essential to maintain domestic food and agricultural production at its full capacity without paying any serious attention to the cost of doing it.

Moreover, government officials assume that Japanese food consumption will be greater in the year 2000. Daily per capita food consumption is expected to rise from an average 2512 calories in 1980 to as much as 2600 calories in the year 2000. Annual per capita rice consumption will decrease from 79 kilograms to 63 to 66 kilograms in 1990 and to around 50 kilograms in the year 2000. Per capita annual consumption of milk and milk products will increase from 62 kilograms in 1980 to more than 80 kilograms in the year 2000. Correspondingly, the consumption of livestock will increase from 22 kilograms in 1980 to more than 30 kilograms by the year 2000. It is assumed that the consumption of vegetables and eggs will not increase and the consumption of fruits will increase moderately.

The rice consumption figures show that annual per capita consumption will decrease 14.5 kilograms in each of the next two decades. I interpret this to mean that in projecting future food consumption for the Japanese people government officials have failed to consider the changing sociodemographic characteristics of Japanese households (the size of households, age structure of the population, the rural–urban distribution of the population, etc.). It seems to me that they view the individual members of the family as the consumption unit rather than the total family. Thus they are not taking into consideration possible impacts of changes in the size of family on food consumption.

I am not prepared to show a possible rice consumption figure for the year 2000, but the projected consumption figures are too high in comparison to both the past trend and the projected changes in the age structure of the Japanese population and the expected change in family size. This means that the predicted increases in livestock and milk consumption are underestimated. Moreover, it is assumed that there will be no change in the relative price of each food category.

There is also a set of assumptions about the world food situation in the year 2000 which is extracted from two works, one by the U.S. government and the other by the Food and Agriculture Organization. It is expected that with a significant increase in yield per hectare there will be a sufficient supply of food in the year 2000 to meet the increased food requirements of the world population. However, the shortage of food in developing countries and in centrally planned economies will increase, and the volume of food trade between these countries and the developed countries, especially the United States, will increase. The bargaining power of the United States as the biggest grain supplier will become more prominent. Since there is no real possibility of expanding cultivated acreage, future food production will depend more and more on increasing inputs such as energy, and the real costs of food production will increase. Moreover, soil erosion, deforestation, desertification, and so on will increase, all of which affect food production unfavorably. The need for the Japanese people to make full use of domestic food and agricultural resources becomes even more compelling.

A government summary of the projected food and agriculture situation for the year 2000 is found in Table 6. It is clear from the table that domestic food and agricultural production will be operating at full capacity while annual per capita consumption of both livestock and milk will be far below the international standard. This situation will not satisfy the market needs of food exporting countries. The following quote from Keizai Shingikai Choki Tembo Iinkai may illuminate the government's position on this question:

> It is without question that in order to satisfy the demand for food for more than 100 million Japanese, who have developed luxurious and diversified tastes in recent years, our food consumption has to depend not only on domestic production but also on imports from other countries. However, in order to secure a stable economic life for the Japanese people, it is important to make full use of domestic food and agricultural resources, while paying due consideration to the necessity of maintaining friendly relations with other countries.

Table 6.   Food and Agriculture Situation in the Year 2000

|  | 1980 | 2000[a] |
|---|---|---|
| *Food Consumption per Capita per Day* | | |
| Energy supply (kilogram calories) | 2512 | 2500–2600 |
| Domestically produced energy supply (K.C.) | 1360 | about 1360 |
| Protein supply (grams) | 80.7 | 80–90 |
| Fat and oil supplies (grams) | 70.1 | 90–93 |
| *Food Supply of Major Food Items* | | |
| Grains | | |
| Net per capita annual food supply (kg) | 113.9 | 88–94 |
| Total supply (10,000 tons) | 3695 | 4000–4100 |
| Domestic production (10,000 tons) | 1075 | 1100–1200 |
| Domestic rice production (10,000 tons) | 1112[b] | 850–900 |
| Livestock Meats | | |
| Net per capita annual food supply (kg) | 22.0 | 32–35 |
| Total supply (10,000 tons) | 369 | 600–640 |
| Domestic production (10,000 tons) | 298 | about 490 |
| Milk and Milk Products | | |
| Net per capita annual food supply (kg) | 62.2 | 84–89 |
| Total supply (10,000 tons) | 756 | 1100–1200 |
| Domestic production (10,000 tons) | 650 | 950–1000 |
| *Land Use* | | |
| Total planted acreage (10,000 hectares) | 564 | 630–640 |
| Acreage under paddy rice (10,000 hectares) | 235 | 150–160 |
| Acreage under fodder crops (10,000 hectares) | 100 | 180–200 |
| Acreage needed to divert from rice (10,000 hectares) | 58 | about 100 |
| Cultivated acreage (10,000 hectares) | 546 | 550–560 |
| *Farm Families and Laborers* | | |
| Farm household population (10,000) | 2137 | about 1420 |
| Farm houshold population 60 years old and older (%) | 21.2 | about 36 |
| Male farm workers working 150 days or more annually on farm (10,000) | 204 | about 115 |
| Male farm workers working 150 days or more on farm (%) annually 60 years old and older | 35.1 | about 53 |
| Number of farm households (10,000) | 466 | about 350 |
| Number of those farm households which have at least one male farm (10,000) worker working 150 days or more on farm annually and younger than 60 years old | 103 | about 40 |

*Source:*   Keizai Shingikai Choki Tembo Iinkai, Hokoku, Keizai Kikakucho, *Nisennen no Nihon— Tojutekina Keizaishakai no Anzen o Motomete*, p. 208.

[a] Projected figures are calculated on various assumption, and be interpreted as subject to modification.
[b] Actual production was 9,750,000 tons due to very bad weather.

This view corresponds to the opinion of two-thirds of the Japanese polled by the Ymoiui Shinbun, one of Japan's major newspapers. Sixty-five percent of those interviewed agreed with the statement that "import liberalization of agricultural commodities must be promoted if it is not harmful to domestic producers." In this connection we have to pay attention to the last section of Table 6 (farm families and labor), which indicates that, although the number of farm households and farm household population will both decrease significantly and the farm population will become older in the coming 20 years, the number of farm households which have at least one male farm worker younger than 60 years old and who works 150 days or more annually on a farm will decrease most drastically.

The government officials calculated that about 1.8 million hectares of agricultural land has to be transferred from those farmers who are retiring to those farmers who want to expand their operations. The transfer will not occur exclusively through changes in ownership: A major part of the transfer will be accomplished by lending. It can easily be assumed that operators of those farm households which employ a young male farm worker or workers working almost year-round are most anxious to expand their operations. Government officials expect that with appropriate policy measures about one-half of the land transfer will be to those farm operators who are most anxious to expand their operations. They expect that as a result of these land transfers, there can be a fairly large number of 10 to 15 hectare grain (rice) farms and 20 to 35 hectares dairy farms. In their opinion these farms can be as productive as farms in the European Community countries, and they conclude that it is therefore possible to reduce the present levels of price supports in Japan to the levels of the EC supports in the coming 20 years.

There are two questions about this scenario. First, there is a question about land use in the year 2000. There must be a substantial amount of acreage diverted from rice cultivation. Rice consumption must decline, and rice yield per hectare acreage has to increase. In their philosophy of making full use of food and agricultural resources, the government officials have to find nonrice crops to be planted on the diverted acreage. Likely candidates are wheat, barley, soybean, and fodder crops. However, past experiences have shown poor results for the first three.

It seems to me that the government officials have decided to increase the acreage under fodder crops by almost 100 percent. In their judgment the number of cattle has to increase from 4.4 million in 1980 to 8.0 million in the year 2000. Moreover, a change in feeding practice is needed. Japanese cattlemen feed their animals too much grain and too little grass. This is another way to increase self-sufficiency in foods. However, there is no technical or economic base for this fodder production program.

Second, the idea of playing catch-up with the EC program seems to assume that EC agriculture will remain stagnant. However, the age distribution of French farm workers, for example, is very uneven. Those who were born around 1925 are very large in number. They will be around 75 years old in 2000 and unable to work on their farms. It is projected that the number of both French farm operators and family farm workers will be reduced by half in the 20 years between 1980 and 2000. However, it is assumed even so that French yields per acre will increase and productivity will rise in that same period. On the other hand, it is very questionable whether Japanese grain farms of 10 to 15 hectares and dairy farms of 20 to 25 hectares can be as productive as farms in the EC countries in the year 2000.

In 1961 the Japanese government redirected its farm program away from increasing grain (rice) production to increasing labor productivity by either improving the basic farming structure or by selectively expanding production. As a result, importation of those crops which have no comparative advantage—such as wheat, feed grains, and soybean—have decreased, and production of those items which have good comparative advantage, such as eggs, chicken, and some vegetables, have increased. The former crops require extensive land use, while the latter are factory-type products. At the same time Japanese food consumption has grown greatly.

As I stated at the beginning of this chapter, the Japanese people believe that in the coming 18 years their country should shift from its present isolation to becoming a full member of the international community. The government agricultural scenario described above clearly contradicts the general national program toward internationalization. Solving this contradiction is one of the biggest problems we Japanese are facing. In this connection I would like to quote from a 1982 publication of the U.S. Congress, Office of Technology Assessment:

> . . . the modelers (of global models) generally agree that the world system is going to change in the near future, and that a continuation of current trends and policies will lead to a change for the worse. They also agree that changes for the better are possible, although they disagree sharply on what those changes should be.

# 19

# A VIEWPOINT ON AGRICULTURE IN DEVELOPING COUNTRIES

RICHARD O. WHEELER

One need not look back very far to find a great many expert viewpoints suggesting that by the 1980s we would be facing widespread famine, our soils would be in the ocean, the real price of food would be so high as to dampen all hope of economic growth, and so on. Perhaps we've been lucky.

But perhaps it hasn't been luck. Maybe it has simply been that the dire predictions for the 1980s were never realistic. And perhaps the even more dire predictions for the twenty-first century are even less to be trusted. Certainly the potentials for significantly increasing food supplies through such technologies as genetic engineering will start influencing the system before the turn of the century. These technologies may not have the impact that many are anticipating, or hoping for, but they are even now becoming available. I join the crowd of those being optimistic about the technology-drive trend toward decreasing real food prices relative to real economic growth well into the twenty-first century.

However, I am less optimistic about whether the developing block of nations will share fully in this more efficient growth in the food supply. My concerns are for the following: (1) the potential for serious sociopolitical disruptions because of temporary tightening of food supplies and wide oscillation of prices;

and (2) foregoing potential gains in growth as nations formulate food and agricultural policies designed to protect against worst-case scenarios and based on short-term political fears and pressures. This lack of stability and the potential consequences have been a topic of concern for many scholars. In my opinion their apprehensions need to be seriously considered when looking toward the twenty-first century.

What can the developing countries do to meet their food needs in the twenty-first century? Obviously, they must take stock of where they are and how they arrived at their current situation. A striking thing about developing countries is the similarity of their problems rather than their differences. They have some very common ailments that must be remedied if the sharing of world progress is to be achieved. However, these ailments are much more obvious than the remedies for their cure.

A common problem is the developing countries' dependence on a narrow base of agricultural export commodities for foreign exchange earnings. These commodities include coffee, sugar, rubber, fruits, palm oil, and so on. Furthermore, there is a growing dependency on external supplies of feed and food grains. Sharp changes in relative world commodity prices can cripple the developing country's treasury and progress.

What are these countries being told to do? One notion is to move toward a more diversified production base but with careful consideration for commodities in which they have a comparative advantage. The political leadership is being reminded that a policy of food security should not mean food self-sufficiency. This is all good advice, since stated policies that suggest total food self-sufficiency are probably the most expensive and among the worst policies for the optional allocation of resources. Most developing countries are willing to consider good advice, but when they ask for a comprehensive strategy to carry it out the expert advice starts getting less and less clear.

Take, for example, a country in the Caribbean or Central American region that is heavily dependent upon sugar for export earnings. In addition to paying for the labor and fixed investment necessary, how can the country pay for its ambassadors, send people to conferences to learn how to solve the food problem, or service the external national debt while trying to shift resources away from sugar toward a more diversified and even more profitable production mix? If donor nations are serious in their advice about diversification, they can help by providing a form of bridge funding during the transition period to allow the shift from one production system to another.

Many other constraints must be faced. For instance, there exists in the Caribbean and Central American regions the potential to produce a wide array of fruit and vegetable crops for which the market prices during certain periods of the year are quite favorable for production and export. U.S. consumers would certainly also benefit if countries in this region produced these crops. However, these commodities require refrigerated sealine shipping. Out of the Caribbean and Central American regions refrigerated banana cargo regularly comes into Miami and Houston. But U.S. import rules and regulations require that most fruit and vegetable commodities be unloaded north of the 39th parallel. Therefore, in order to get shipping lines established, a critical mass of product supply must be developed to attract shipping companies to new and profitable routings.

So, while the U.S. State Department through USAID struggles to bring about a more realistic, diversified agricultural base, the U.S. Department of Agriculture is busy enforcing import rules and regulations that handicap the developing countries' capabilities to diversify. And if subsidies, tariffs, and quotas are not enough, we back them up with import rules and regulations.

Another factor constraining the improved production of food supplies in developing countries is a hodgepodge of government interventions in the marketplace. Egypt is one of the more classic examples. Most Egyptian analysts and policymakers are well aware of this problem. However, this does not stop external groups—from Presidential Missions to occasional visiting scholars—from continually pointing out the obvious. To suggest that it is simply a matter of phased withdrawal by the government begs the issue. For instance, it was not too many years ago that the government attempted to reduce subsidies on bread. Riots broke out in the streets. Peasants started throwing rocks at the palace gates and threatened to hang the political leaders in the courtyard. *That* is political reality. The point of the story might be that we know a great deal about the potential misallocation of resources resulting from government price and subsidy intervention, but we have little in the way of a practical strategy that will allow the developing country to disentangle itself from these policies. We need to encourage our scholars to look more realistically for solutions and less for new ways to state the problems.

Another problem most developing countries currently face started during the colonial period, escalated during the 1960s, and continued through the 1970s. This is the creation of semiautonomous and parastatal government organizations to implement the development process. These organizations take many forms, from government-owned national agricultural development corporations to government organizations involved in the production, processing, and marketing of specific commodities. The formation of many of these, to be sure, resulted from the political philosophy of the developing country that centralized planning was the efficient road to agricultural development. The idea of free-market forces creating a private sector for development simply was not a part of the major development agenda.

Many times these organizations are involved not only in the processing, marketing, and storage of food but also in production agriculture. The intent was for them to engage in commercial activities and generate income for further expansion and to provide revenue for the national treasury. In general, these activities have been an economic development disaster.

What is even worse is that many of these semiautonomous government agencies have been given authority that has resulted in external debt that is driving several developing nations toward bankruptcy. Many, if not most, have the authority to borrow and to obligate the central bank of the developing country. Not only do they run at a loss, which results in a drain on the national treasury, but the governments also are faced with staggering foreign exchange problems as they try to repay the interest on the debts incurred.

There is a growing willingness among many developing countries to look at ways to get out from under the burden of semiautonomous agencies. Willingness to explore is one thing; accomplishing change is another. There are several barriers to changing such structures, some of which are rather nasty.

Among the nastier barriers to progress in the developing countries is downright "corruption." Corruption takes many forms and is generally not brought to light by individuals looking at problems in the developing world. Semiautonomous government organizations are many times the centerpiece of such activity. A tragic part of this problem is that frequently corruption is fed by outside donors, not willingly or overtly, but as a consequence of who they deal with in making the donation. Almost all multilateral and most bilateral government assistance agencies deal only with recipient government agencies. It is further the case that most external assistance agencies would be downright shocked and deeply offended if one accused them of helping to generate corruption—rightfully so, since they are made up of people of high integrity and compassion. They are not corrupt, but one must have some rather large blinders on not to see the slippage as their resources are worked through the local recipient agencies.

In addition to encouraging corruption, assistance by developed countries to semiautonomous government organizations has created some bad investments that the receiving country can ill afford. There are many examples but one case will suffice. A European country donor agrees to provide assistance to the semiautonomous agency of a developing country to build a super modern, ultra-high-temperature milk pasteurization plant. Naturally, the plant had to be installed with equipment made by a manufacturer from the donor country. Since the day the plant was turned on it has run at a loss, and the deficit is a drain on the national treasury. Today it runs much below capacity. But what is interesting, if not tragic, is that 95 percent of the pasteurization process is used to blend imported nonfat dry milk into containers that cost about as much as the contents. The other 5 percent, which the plant management looks at as a nuisance, is local farmers' milk that needs the process.

Problems though there be, I am very much for continued economic assistance from the developed to the developing world. I further subscribe to the argument that it is in the self-interest of the developed countries to continue to provide this assistance even aside from humanitarian reasons, which are also very important.

We must recognize that much of the foreign assistance investment in the developing world has gone into public sector projects—infrastructure and human capital development, institution building, and technology creation and transfer. Essential as these things are, they alone do not generate self-sustaining economic development. The total process is critically limited by the absence of other elements, notably entrepreneurship, management skills, and private capital to stimulate small businesses. The lack of these elements seriously diminishes the payback of investments by the public sector.

If the developing countries are going to contribute to progress in the twenty-first century, they must take steps toward strengthening their private sectors. Those in the development business must look at profits as an important force for generating capital formation and development. Those of us in the development business have for too long allowed the relative importance of the private sector to be placed too low on the development agenda.

Many developing countries are willing and eager to divest themselves of government-run production, marketing, processing, and storage activities. The developed countries have the entrepreneurship and management skills to help. Joint-venture private partnerships must be encouraged, and even subsidized with capital, to break the stagnation in economic development in the hinterlands of developing countries.

One way to help identify the needs and opportunities for improving the system of food and agriculture of a developing nation is to divide the constraints into three broad interrelated types: (1) those behind the farm gate (production oriented); (2) those between the farm gate and the consumer (agroindustry—processing, marketing, distribution, and storage—both inputs and outputs); and (3) public policies that are needed or negatively affect the other two.

Traditional assistance to remove behind-the-farm-gate constraints has focused heavily on technology creation and on attempts to package and transfer that technology to the producer. A barrier to the adoption of technology is very much related to the lack of appropriate price signals to the production system.

In terms of agroindustry, there are many large companies doing business in developing countries. Many are involved in specialty crops that lend themselves to plantation-type production systems. This provides the investor with the opportunity to integrate production and marketing. If profitable opportunities are available, these private companies are well-equipped to move in and make investments. Furthermore, they well understand when government policies or actions make investment too risky or unprofitable. As important as these larger private companies are, they cannot be expected to generate the needed business activity in the hinterlands and at the grass-roots levels. This need for grass-roots agroindustry will have to be met with management skills and with sufficient entrepreneurship to sense opportunities to mobilize local resources into developing profitable, generally small enterprises. However, there is a serious shortage of people with these necessary skills in the developing countries, and, frankly, I don't know how you would go about formally training for entrepreneurship. In terms of management skills for these kinds of enterprises probably the worst strategy is a high-quality business education in a developed country. One way these talents could be developed is through joint-venture, private partnerships where one of the partners brings these skills to the business. The local partner can learn while working.

I believe there are numerous individuals from the developed countries who are ready and willing to invest their talents along with developing country partners to fill the rural agroindustry gap. Furthermore, there is growing willingness by governments in several developing countries to divest government enterprises that could be turned into profit centers. The activities of these enterprises are also needed to advance agriculture. The most constraining resource in all of these cases is venture capital.

How might we go about relaxing this constraint? There are some organizations that do try to fill some of this need. Perhaps one of the oldest is the Commonwealth Development Corporation (CDC), which was established by the British Parliament in 1948. This financial institution's charter calls for it to do the right thing by its dependencies but to do good without losing its own or others' money. CDC takes both equity and loan positions in developing countries around the world. The United States has no counterpart development agency. The closest thing to it is the Latin American Agribusiness Development Corporation (LAAD) in which several large agribusiness firms hold equity positions in the corporation. If the United States is serious about stimulating private sector development as a means of advancing food and agriculture in the developing world, it will require more commitment to advance the private sector than exists today.

A new U.S. agricultural development finance corporation could serve as a catalyst to private sector development. Although the current private enterprise bureau of USAID is attempting to advance private sector investment, the level of funding and the normal government restrictions make it difficult to fill much of the need.

Before suggesting more about what such an organization might do, perhaps one should sort out what it should *not* do. It should not make loans to governments. The international banking system is already in business to service this need. It should not be forced to focus on infrastructure development or on nonprofit activities that other development agencies are set up to do. It should have the freedom to pick among those countries that are willing to make the necessary policy changes to attract for-profit venture capital investors, part of the argument being that financial incentives are likely to be more effective than simply good advice.

If the U.S. private sector is serious about joint private/public sector involvement, they might be called upon to provide as their contribution management and administrative support. This would represent a potential role for both the not-for-profit foundations and the for-profit private firms. The private sector would provide the management of a "trust fund" for the general public. Over time, with payback results, the fund would not need to be recharged from public sources.

I would not claim that an improved private sector involvement in agriculture in developing countries is the total answer for meeting the food and agriculture needs of the twenty-first century. However, in heading an institution dedicated to development, I continually find our efforts thwarted by developing country government policies and the lack of a reasonable agroindustry sector. If experience is of any value, we should know by now that centrally planned agriculture and government-operated industry do not work very well.

There are, however, several developing country governments that are willing to make moves toward more private sector involvement. If we believe the private sector is an important driving force for the twenty-first century, we must have the imagination and willingness to look for alternative ways of channeling our assistance. A venture capital supply to attract and transfer the developed countries' entrepreneurial and management skills will pay very high dividends to the investment in development.

This grain storage, drying, mixing, and transfer center is rapidly adapting to robotic controls. (Courtesy of American Farm Bureau Federation)

# EPILOGUE

# THE NEW AGRICULTURE: A VIEW OF THE TWENTY-FIRST CENTURY

SYLVAN H. WITTWER

The planet earth is a wonder to behold. So it has been recorded by those who have viewed our home from space. Of all the heavenly bodies observed in any detail it is the only one that has the endowments of land, water, light, atmosphere, energy, and temperature to adequately support human habitation. However, many changes—most of them subtle in nature and many of them caused by humans—are occurring on the land, in the oceans, and within the atmosphere. These changes are affecting light, energy, and temperature ranges, which in turn are affecting the biological productivity of the planet. The rates of change in many of these resources, especially during the past decade, may exceed the rates of our understanding of the changes and our ability to cope with them.

The productivity of the land is vital to the future habitability of the earth, especially to the production of food and other renewable resources upon which people depend for food, shelter, fuel, and clothing. It is the land which makes the predominant contribution. We live on the land and gain our primary sustenance from the land and from the adjoining coastal waters. As far as food is concerned, most calories and 95% of the protein consumed by people are derived from the land. The 20 crops that stand between people and starvation are products of the soil.

Agriculture is the world's oldest and largest industry and its first and most basic enterprise. Food is chief among our needs. It is our most important renewable resource. The greatest challenge facing us in the twenty-first century is to produce adequate food to meet the demands of improved diets for an expanding and increasingly affluent global population.

Dependable production is just as important as the magnitude of output. The most sought after goal of humanity should be more stable yields of the crops which directly or indirectly provide over 95% of the food consumed.

Climate is the most critical agricultural production resource. It is both a hazard to be dealt with and a resource to be harnessed. During the 1970s world food production went through one full cycle—from surplus to shortage and back to surplus. Two phenomena—total agricultural output and dependability of production, both primarily crop-dependent—are critical to the future of all people of the earth.

The agricultural system of the United States has five advantages not fully enjoyed by the rest of the world. First, we have a climate–soil–water resource in a vast corn and grain belt adapted for stable high production. It is without equal in any other country. Second, we have a free enterprise system that fosters the profit motive and provides incentives to produce. Third is a unique land-grant university system and the philosophy of a federal–state partnership of more than 100 years duration. They link teaching, research, and extension in each of the 50 states, plus the Tuskegee Institute and the "13 colleges of 1890." Fourth, we have a vibrant, privately supported agricultural research and development sector, equal to or surpassing in importance the public support of food and agricultural research. This private research and development system also provides a vast infrastructure for food processing, mechanization, chemical supplies, credit, and trade. Finally, we have the asset of the English language. English is rapidly becoming the universal communication vehicle for business and science. Through it American scientists and other U.S. ambassadors for food and agricultural technology have an advantage over virtually all nations.

## GLOBAL PERSPECTIVES

Global agricultural productivity set new records in 1982. Never before has so much food been produced, in total and per capita. A world record grain crop of 1.64 billion tons exceeded that of the previous year's 1.62 billion tons. The increase came primarily from U.S. record production of 8.4 billion bushels of corn, 2.8 billion bushels of wheat, and 2.3 billion bushels of soybeans. These increases came primarily from higher yields rather than from expanded acreage. Record production in the European Community and Canada and near-records in Argentina more than offset the drought-reduced crop in Australia.

In the spring of 1983 farmers and governments in the United States and most other industrialized nations were plagued with unprecedented overproduction, surpluses, and low prices of food and many other commodities. It is estimated that by the end of 1983, American farmers, responding to the Payment-in-Kind (PIK) program, will have idled some 83 million acres, or about one-fourth the acreage of all the nation's major field crops in 1981 and 1982. This massive land retirement program is unparalleled in our history. Coupled with adverse summer weather, it could within one season spell near disaster and result in a complete reversal of food supplies, surpluses, and prices. Some nations, particularly the Soviet Union and those in the Warsaw Pact, now have serious food shortages. The Soviet's estimate of 180 million metric tons of grain for 1982 was the fourth year in a row when production fell far below the planned level.

The United States, Canada, Argentina, and Brazil are the grain and oil-seed exporting countries of the world upon whom many nations rely for vast quantities of food. They are the breadbaskets of last resort for the rest of the world. These nations are now accounting for 100 percent of the exports of soybeans, more than 90 percent of the wheat, and more than 75 percent of feed grains. World grain trade approximated 230 million metric tons for 1982 or about 14 percent of total production. Fifty to 60 percent of the grain in international trade has its origin in the United States.

Cereal grains directly provided 60 percent of the calories and 50 percent of the protein consumed by people. Grains consumed directly plus those fed to food animals account for almost 80 percent of the world's caloric intake.

However, America cannot indefinitely serve as the breadbasket for the world. Food production and delivery will become increasingly expensive. Ultimately, food will be have to produced closer to the people who consume it.

## Crops

Globally, 20 crops stand between people and starvation. In approximate order of importance they are wheat, rice, corn, potatoes, barley, sweet potatoes, cassava, soybeans, oats, sorghum, millet, sugarcane, sugar beets, rye, peanuts, field beans, chick-peas, pigeon peas, bananas, and coconuts. New technologies, resource inputs, economic incentives, and government decisions affecting future agriculture or food policy must focus on these crops as well as on the eight forest trees (eucalyptus, radiata pine, Douglas fir, loblolly pine, black locust, teak, Scotch pine, aspen) that provide most of the world's wood and timber.

## Livestock

Animals produce high-quality protein to supplement large quantities of moderate quality, economically produced staples in the human diet. They also work in agriculture and provide manure for fertilizer, solid fuel, and biogas, and serve as a means of capital generation, insurance against risk, and medium of exchange. The livestock reserves of the world exceed those of grain, and they are mobile and better distributed. Worldwide, animal products contribute more than 56 million tons of edible protein and more than 1 billion megacalories of energy annually. This high-value protein is equivalent to more than 50 percent of the protein produced from all cereals.

The products of animal agriculture are becoming increasingly important in meeting world food requirements. Three-fourths of the protein, one-third of the energy, and most of the calcium and phosphorus in the American diet come from animal products. Agriculturally developed and developing nations alike are seeking improved diets and are striving toward the consumption of increased quantities of meat, milk, and eggs; and the grain exports from the United States to both developing and industrialized nations, and to those with centrally controlled economies, are used primarily to feed ever-expanding livestock populations, which will provide the larger quantities of meat.

## Farm Size

Agriculture in the developing world is small in scale compared to America. It is not likely to change in the foreseeable future. Farms are small in land, credit, and capital but relatively plentiful in labor. Most of them are in the tropics or semitropics. Four-fifths of the world's farms are 5 hectares or less and nearly half are a single hectare. In Taiwan there are approximately 1 million farms with an average size of 1 hectare. One of the research challenges for the future will be for science and technology to promote and increase the productivity of small-scale agricultural enterprises.

There is also an ever-increasing number of part-time farmers, both in agriculturally developed and developing nations. In Japan and the United States more than half the income of farm families is derived from off-farm earnings. The small farm can be an economically viable unit: In fact, the outputs per unit of land area and per unit of capital input are often higher on small farms.

They are also usually more labor intensive. With current trends, farming will increasingly become a part-time endeavor, and research and educational programs should be adjusted to this changing structure in American agriculture.

## THE WORLD FOOD SITUATION

The short-term perspective of 2 to 5 years for American agriculture is distinct from that of the longer-term view of 20 years or more. Today there is more food per person on earth than at any time in history. Adequate agricultural production has gone a long way toward alleviating hunger and malnutrition, but it alone is not enough. Today's immediate problem is one of distribution of both food and income. The distribution and delivery of food and the income to buy it pose critical national and global policy issues. Food surpluses, malnutrition, and hunger now exist side-by-side in many nations, including the United States. Only poor people go hungry. The challenge is to get food to people who need it. For hungry people, the food production policy of an agriculturally developing country is far more important than the food aid it may get.

The long-term perspective and challenge is one of sustained growth of productivity. Instability of agriculture in both production and marketing is becoming a major problem in the United States and throughout the world. It would be of great strategic value to know or to be able to estimate the future productive capacity of the U.S. agricultural system. This would include our food supplies and specific changes in the resource base that support primary productivity and other renewable resources.

Significant progress has been made in feeding, clothing, and housing an ever-increasing population. There is no worldwide shortage of food. Even in low income nations, life expectancy at birth has significantly increased during the last 35 years. While there have been a few local occurrences, there have been no major famines in 35 years; nor are any foreseen. Famines resulting from crop failures or other natural causes have been virtually eliminated, and American agriculture can be proud of contributing to this achievement.

## FUTURE TECHNOLOGIES FOR
## AGRICULTURAL PRODUCTIVITY

We project new directions for agricultural and food research in the United States. Here, the three determinants of agricultural productivity come sharply into focus: new technologies, resource inputs, and economic incentives. All three must be simultaneously present and in the appropriate balance. The resource base and new technologies are inseparable. Technologies that add to rather than deplete the earth's resources and that are nonpolluting, environmentally benign, applicable to all sizes of farms, and sparing of capital, management, and nonrenewable resources must be sought after. Likewise, we must search for technologies that will result in stable production at high levels. Finally, there must be economic incentives to produce. Only under such conditions will the food needs of an ever-increasing population be met, a growing food insecurity alleviated, expanding export markets satisfied, the hopes of centrally controlled economies met, and the demands of affluent people of oil-rich countries realized. There is a global desire for improved diets in the form of more meat, milk, eggs, fruits, and vegetables. No single, new production technology will meet all these criteria or expectations, but many can fulfill some of them. There are now many such examples of successful food production technology packages.

The difference between what food production is now and what it could be constitutes the world's greatest technology reserve. The record yields of 20 years ago are only average yields today. Record yields of today should not be considered abnormal occurrences (Table 1). The increases in global food produc-

Table 1.    Average, Best, and World Record Yields in Bushels per Acre

| Food Crop | Average 1981 (U.S.) | Best Farmers (U.S.) | World Record | Ratio Record/Average |
|---|---|---|---|---|
| Maize | 13 | 234 | 354 | 3.1 |
| Wheat | 38 | 103 | 216 | 6.0 |
| Rice (crop/112 days) | | | | |
|   Polished | 88 | 145 | 229 | 2.6 |
|   Unpolished | 109 | 181 | 286 | |
| Sorghum | 59 | 284 | 343 | 5.8 |
| Barley | 56 | 167 | 212 | 3.8 |
| Oats | 59 | 156 | 296 | 5.0 |
| Soybeans | 33 | 54 | 83 | 2.5 |
| Potatoes | 461 | 1011 | 1429 | 3.1 |
| Sugarcane[a] | 98 | 154 | 275 | 2.8 |
| Sugar beets[a] | 20 | 35 | 52 | 2.6 |
| Alfalfa[a] (not irrigated) | 3.3 | 6.7 | 11.0 | 3.3 |
| Milk/cow/1000 lb | 12.1 | 35 | 50 | 4.2 |
| Eggs/hen/year | 235 | 275 | 365 | 1.5 |

[a] Tons per acre.

tion during the past 2 years have not come from an increase in acreage devoted to crop production but from increased output per unit of land area.

## Maximum Yield Trials

Maximum yield trials for crops and performance records for livestock would represent the ultimate in use of technology and modeling for optimizing crop and livestock productivity. They are as important in practice as they are in theory. Studies of the comparative productivity of previously successful agricultural systems could be most rewarding. A systems approach in combining the most desirable technologies to constantly test the limits of available technology and to achieve maximum productivity of the major agricultural, range, and forest crops and performance of livestock should be a continuing effort. This will involve the total agricultural production system. World yield and performance records have for the most part been achieved by successful growers and producers, rather than at agricultural experiment stations or research centers or institutes.

Projections of the future effects of currently emerging and yet to be identified technologies on the productivity of agricultural, forest, and rangelands and performance of livestock are called for. An example is that of corn. The world-record corn yield now stands at 354 bushels per acre. The theoretical maximum is 1066 bushels per acre. Economists at the University of Minnesota have predicted that conventional plant breeding will continue to provide an increase of 1 bushel per acre per year. Improved management will add 0.2 to 0.3 bushels per acre per year. Genetic engineering, on the other hand, will have no effect on the yield in 1990, but by the year 2000 will give increases of 1.7 bushels per acre per year. There is extensive evidence that projected increases in agricultural productivity have consistently fallen short of actual accomplishments. It can be projected that continuing increases in productivity for both crops and livestock will occur in the decades ahead, as they have in the past. Biological limits have neither been defined nor achieved by advances of the past.

## Contrasting Food Production Technologies

Two general types of food production technologies characterize present world agriculture. The one is highly mechanized and makes intensive use of land, water, and energy. The other is more biologically based and scientifically oriented and is sparing of land, energy, and water resources. The first characterizes much of the current U.S. agricultural production system and results in output for farm workers that is the highest in the world. Similar systems exist or are emerging in Canada, Australia, New Zealand, Brazil, and Argentina.

The second type of technology is not as productive per farm worker but produces higher yields per unit of land area, often with a higher cropping index. This system characterizes the current Japanese, Taiwanese, Western European, and Chinese systems. Because of resource constraints—cost and availability—an inevitable shift will occur in U.S. and worldwide agricultural systems to a more scientifically- and biologically-based agriculture. Our world is moving from a demand-driven economy with perceived unlimited resources to a resource-limited economy. Research priorities must be adjusted accordingly. This has already occurred during the first part of the twentieth century in Japan and certain European countries. We are now seeing a transformation in U.S. agriculture from a resource-based to a science-oriented industry; and from a traditional to a high technology sector. U.S. agriculture must maintain its technological leadership.

The new American agriculture will achieve almost all future increases in production as a result of increases in yield (output per unit of land area per unit of time) and from growing additional crops during a given year on the same land. There are really no other viable options. The technologies that will make this possible must be developed today.

Appropriate mechanization and improved crop varieties, that are pest-resistant and environmentally adapted, will play predominant roles in this new agriculture. The intent will be to increase the productivity of both land and labor. Genetically engineered vaccines for disease control in livestock, new highly potent pesticides, monoclonal antibodies, growth hormones, and interferons for improved performance and health of food animals will be used; and explants from super-plant selections will be clonally propagated by new techniques for culturing tissue (see Figure 1). More fertilizer and pesticides of natural and synthetic origin will have to be used but used more efficiently and scientifically. More cultivated land will be irrigated but with greater efficiency of water use and with increased supplemental applications in subhumid crop production areas. Conservation tillage and drip irrigation will continue to receive wide acceptance based both on economics and resource conservation. The full benefits from improved water management (irrigation, drainage) will not be realized without additional fertilizer. Production practices, management procedures, and genetic improvements for both crops and livestock will make them more climate proof. We will see an expanded use of plastics as soil mulches and for covers in protected cultivation. Essential farm operations will be programmed and computerized. Computers in agriculture at the farm level will become commonplace for management decisions, for improved communications, and for instrumentation control. The resource base will change with time and technology.

Figure 1. Stages in the tissue culture propagation of asparagus. Left to right, sequence begins with a single bud or meristem, the stems of which are progressively subdivided to form hundreds of clonal explants.

## Aquaculture

This will become one of the food growth industries of the future in both fresh and salt water, especially in the tropics and semitropics. Larger food creatures eat more and take longer to mature than fish. With fish, there is an approach to a one-to-one feed ratio: 1 pound of feed produces 1 pound of fish. For the United States, aquaculture has lagged and will remain a summertime activity concentrated in the southern states with a continued growth in catfish farming. One possibility will be to raise the temperature of a pond by covering it with single or double layers of plastic sheeting. Specialty luxury species—shrimp, lobster, oyster, and salmon—will continue to predominate in the United States. Integrated farming systems will gain in prominence in countries such as Taiwan, Japan, and China, where fish culture will continue to be combined with the raising of pigs and ducks. Currently, integrated systems of pigs, ducks, and tropical fish that produce 250 pigs or 2500 ducks per hectare will provide sufficient waste for the production of 6 tons per hectare per year of fish of such species as Chinese carp and tilapia. Almost half of the world's cultivated fish—more than 20 million tons annually—are produced in the People's Republic of China. Aquaculture can become a star in agriculture's future. Witness the recent phenomenal increase in catfish farming in the United States and the 11 percent annual growth rate in Taiwan. The "forests and the grass-lands" of the sea have not yet been fully exploited as food sources for humans or beasts.

## Productivity of Forests and Rangelands

Forestry may be the branch of agriculture with the greatest potential for increases in productivity. It has been conservatively estimated that the biological productivity (net realizable growth) of the commercial forestlands of the United States could be more than doubled within a half-century by the immediate and widespread application of proven silviculture practices, provided economic and social conditions permit. These include improved regeneration, tissue culture, species composition, harvesting practices, fertilization, and weed control. It is further projected that with the use of technologies yet to be conceived productivity of forests could be tripled in the same period, and with intensive management forest and rangelands could support three times the present level of grazing.

Significant genetic improvements of forest tree species and range grasses have occurred during the past 30 years. Other developments are provenance testing, hybridization, and tissue culture methods for mass clonal propagation of forest and range plants, including the adaptation of genetic engineering, protoplast fusion, and recombinant DNA methods.

Special consideration will be given in the new agriculture to the physical limitations of growth. These include the availability of nutrients from atmospheric and soil sources. Optimal nutritional requirements of various species in stands of forest and range plants should be identified, along with the role of mycorrhizae and other root symbiotic and pathogenic relationships. Among forest and range plants, soil bacteria, fungi, and algae and their contributions to nutrient availability constitute a remarkable microbiological frontier that has scarcely been explored. In forests and rangelands there is intense competition among plants, animals, and microorganisms. The mechanisms of these multicomponent competitive systems will be pursued to achieve optimal productivity since they now constitute limits to growth.

## CHANGES IN THE RESOURCE BASE

During the first half of the twentieth century human beings have been changing the nature of the earth's land, atmosphere, and oceans. The impact of modern technology on the biosphere is evident worldwide. Human activities and natural events are affecting the biosphere through changes in land cover and biological productivity, soil moisture, groundwater reserves, atmospheric $CO_2$, and trace compounds including pollutants and toxic substances. As already suggested, these rates of change are exceeding our understanding of the implications and our ability to cope with them.

## Land Surface Characteristics

Several global problems are associated with changes in land surface. These include deforestation, desertification, soil erosion, overgrazing, excess tillage, problem soils from salinization, a lack of drainage, and accumulations of toxic metals. Included are changes in tillage practices, marginal soils, conversion of noncropland, and management of rangeland resources. Neither the magnitude nor the rate of these changes are known, yet volumes have been published on the future hazards of some of the above practices and phenomena on life-supporting systems. One estimate of soil erosion in the U.S. corn belt is that 2 bushels of topsoil are lost for each bushel of corn harvested on the land. It is further reported that 90 percent of the land used for row crops and small grains in the United States is tilled largely without soil conservation practices and that erosion today is worse than it was during the Dust Bowl of the 1930s.

But with time and technology the land resource base and ground cover can also be improved, as well as depleted, by cropping. Conservation tillage will be increasingly important for the future. It is becoming a major cultural practice driven by energy costs and soil and water conservation. It stands as the most significant technology yet developed for producing crops and simultane-

Table 2.   Crops and Their Residues Having Allelopathic Properties

| Vegetables | Field Crops (Grain) | Fruit Crops |
|---|---|---|
| Asparagus | Barley | Apple |
| Bean | Corn | Apricot |
| Chinese cabbage | Oats | Citrus |
| Cucumber | Rye | Peach |
| Mustard | Sorghum | |
| Pea | | |
| Potato | | |
| Tomato | | |

*Source:* A. R. Putnam, Professor of Horticulture, Michigan State University.

ously controlling soil erosion. Reduced tillage and no-till systems can maximize crop residue cover and conserve energy, labor, water, soil, fertilizer, and organic matter. Hence the term "conservation tillage." We will continue to see increases in its use for food production. Various forms of conservation tillage, including no-till, and the use of living mulches as cover crops were used by farmers on an estimated 100 million acres of cropland in 1982. This compares with 4 million in 1962. By the year 2000 conservation tillage will double.

The new agriculture in the United States will also see the use of crop residues and plant mulches with allelopathic properties in new conservation tillage practices (Table 2). Allelopathy is chemical warfare between plants. When properly managed, it may reduce or eliminate the need for chemical herbicides and the hazards now associated with herbicide resistance. "Allelochemicals," nature's own herbicides, will come into prominence.

## Soil Moisture and Groundwater Resources

The future magnitude and degree of dependability of agricultural production will rely on the availability of water for crop irrigation in arid and semiarid lands and supplemental irrigation in subhumid areas. Soil moisture in the spring is the most determinant factor for agricultural productivity in the U.S. corn and wheat belts, and July rainfall is the most important variable. The current overdraft of groundwater in the United States, mostly for crop irrigation, is estimated at 20 to 25 million acre-feet per year. This is a largely irreversible use of a nonrenewable resource. It not only will affect future agricultural productivity but is currently causing land subsidence, which is seriously affecting human habitation in some of the valleys of the western United States. The problem of the depletion of freshwater and its importance as a resource for future human habitation is not restricted to the United States. Agriculture is by far the major consumer (approximately 83 percent) of water, and water will likely become the most limited and one of the most expensive of all resources in America for future food production. Water resources cannot be separated from land resources in irrigated agriculture.

Irrigation is becoming increasingly prominent in the subhumid areas of the central and eastern United States. It is the one important option the United States and other nations have for increasing agricultural output resources and assuring dependability of production. Twelve percent of the cultivated farmland in the United States now accounts for 37 percent of crop production.

Currently and traditionally, our use of water resources in the United States has been exploitative. The efficiency of conventionally applied irrigation water varies between 20 and 40 percent in the United States. That compares with

80 to 85 percent efficiency for Israel. Moreover, irrigation efficiency in the United States has changed little over the past 30 years. The projected increases in water use for agricultural irrigation will compound the need for improved efficiency. Consumption of fresh water by production agriculture in the overall U.S. water budget amounts to between 80 and 85 percent of the total. Water for irrigation requires energy for lifting, transporting, and pressurizing. The two resources are inseparable. The water resources for the new agriculture will be much more critical than those for energy.

Drip or trickle irrigation will play an increasingly important role in agricultural food production in the future. As a result of increased efficiencies in the use of water and fertilizer and reductions in labor, we will see significant expansion in the use of drip systems not only for high-value crops but for the main food crops of the world as well. Sixty-three percent of the orchard land and 52 percent of the vegetable crop areas in the United States are now under irrigation. This will increase substantially as we approach the twenty-first century.

## Atmospheric Carbon Dioxide

Inadvertently, human society is now conducting a great biological and environmental experiment with an uncertain outcome. Atmospheric $CO_2$ is increasing at the rate of 2 parts per million per year. This is attributed by most authorities to be the result of the burning of fossil fuels and of deforestation. Most attention to this phenomenon has so far been directed to climatic change and particularly to elevated global temperatures. According to projections, sea levels will rise and ice caps will melt, and some agricultural dislocations and disruptions in food supplies will also result from rising temperatures and shifts in precipitation patterns.

Conversely, the rising level of atmospheric carbon dioxide may have profound beneficial effects on food crop production in the twenty-first century and beyond. The favorable effects of elevated levels of atmospheric $CO_2$ have long been demonstrated in the greenhouse culture of crops. Virtually all crops respond favorably to levels of $CO_2$ higher than the normal, current atmospheric level of 340 parts per million. Extensive reviews of direct effects of elevated levels of atmospheric $CO_2$ on plants indicate the potential for improved photosynthesis and yield, greater biological nitrogen fixation and mycorrhizal activity, enhanced water use efficiency, greater resilience to stresses of water, light, and temperature, and increased protection against air pollutants.

This rising level of atmospheric $CO_2$ may, by the beginning or even before the twenty-first century, have profound effects on total global food production, some of which may be negative and some of which may be positive. The change in this potentially important resource in global food production should be carefully monitored and its biological effects on crop productivity and biological transformations carefully assessed during the coming decades. Carbon dioxide may not be a typical air pollutant.

## Atmospheric Pollutants

There is mounting evidence of the effects of air pollutants, both positive and negative, on food producing systems. These pollutants include sulfur and nitrogen compounds and ozone in the atmosphere as well as carbon monoxide, methane, and acid deposition, all of which have confirmed effects on the productivity of agricultural and rangelands and forests. It will be increasingly important for the agriculture of tomorrow to identify the sources of air pollutants, to monitor changes, to assess their effects on total renewable resource productivity, and to seek means of alleviating the sensitivity of agricultural crops to them. Some progress is being made in identifying chemicals that reduce the damage.

Four points are important with respect to agricultural productivity and air pollutants. First, the effects are regional. Different regions of the United States and the world may show dissimilar effects in response to pollutants. A good example is sulfur gases in the atmosphere, which may either prove to be a beneficial nutrient or a toxic disaster. Second is the subtle magnitude of the effects. Acid rainfall is an example where effects can only be observed over decades. Third is the issue of multiple pollutants, which is the situation under real world conditions. Seldom can we address the effects of a single substance. The fourth relates to the interaction of air pollutants with man-made and environmental stresses such as drought and biological stresses found in natural systems.

## Remote Sensing

Important for tomorrow's agriculture today will be the monitoring through advanced remote sensing techniques of changes in land cover, water and energy, mineral resources and reserves, and the biogeochemical cycles in the atmosphere, on the land, and in the water that affect life-supporting systems. The development of predictive models of natural and man-made physical, chemical, and biological changes concerned with the production of food and other renewable resources will be of great significance.

## THE EFFECTS OF SCIENCE AND NEW TECHNOLOGIES

Four major scientific revolutions have thus far characterized the twentieth century:

1. Genetic engineering—cellular and molecular biology and tissue culture.
2. Atomic energy—the unlocking of the atom.
3. The space age—escaping the earth's gravity.
4. The computer revolution—electronic communications.

The first and fourth revolutions relate directly to science and technologies for agricultural research and the future productivity of American agriculture.

### Computers

Computerized information systems are creating a technological revolution for production agriculture and for agricultural research at many levels. These include colleges, universities, national laboratories, state agricultural experiment stations, extension services, farm organizations, and farms. Computers and "chips" are becoming smaller, cheaper, more versatile, and of greater capacity. All this is parallel to equally new technological developments in satellite and other types of telecommunications and in fiber optics. It is now possible to have immediate access to worldwide market information, prices, storage reserves, production forecasts, and food and feedstocks. Much progress is being made in the storage and retrieval of data and delivery of research findings. Substantial progress can be expected in reducing the time from discovery to adoption of new technologies.

Computers in the future will have enormous impacts at the farm level. They will affect management decisions, programming of operations, improved communications, and instrumentation. Management decisions will include least-cost ratios for feeding livestock; programming of irrigation for crop production; the development of models for control of insects, diseases, and weeds; animal disease control; record keeping, domestic planning, and grain drying.

In the area of communications at the farm level, there will be real-time weather information, electronic fund transfers, and educational networks. Instrumentation control at the farm level will include the programming of growth factors in controlled-environment agriculture and greenhouse culture, robotics in agricultural mechanization, and the monitoring of production dairy, swine, beef cattle, and poultry operations.

One of the most sophisticated developments using computer technologies will be in the control of robotics in agriculture. Japan had 100,000 industrial robots, or 80 percent of the world's total, by the end of 1981. Such machines can be operated 24 hours a day. The use of robotics in agriculture could mean a great leap forward in technology. In the future one operator could control many machines, machines that don't tire, that operate continuously, that perform undesirable jobs, that need no breaks, that require no fringe benefits, and will not strike.

## Genetic Engineering

We are in the midst of the greatest biological revolution of all time and are witnesses to a renaissance in cellular and molecular biology and tissue culture. Globally, no less than 350 firms ranging from large multinationals to small venture capital firms have entered the biotechnology field in less than 5 years. There are an estimated 175 of these companies in the United States alone. Most major seed companies in the United States have merged with biotechnology corporations or chemical, petroleum, or pharmaceutical companies.

The above effects are both far-reaching and challenging. A brain drain on publicly supported institutions for cellular and molecular biologists began in 1979 but is now abating. This migration of human resources from the public sector into the private sector poses the question of who will train the scientists and conduct and support the graduate training programs for the future. Will biological research remain in the public sector, given the current swing toward the private sector that is fueled by tax write-offs and hopes of early profits from the sale of potential seeds, crop varieties, microorganisms, and vaccines?

Through genetic engineering the gap between basic and applied research is being bridged. Some results of basic research are finding immediate application. No longer in biology can we separate the fundamental sciences from practical achievements. Biologists in the academic arena are now active in industrialization and commercialization of their own research and the profit from it. Where will their loyalty lie in the future? As research gains in complexity, goals will be more for profit than for public interest. Acquisitions by chemical companies already provide in-place breeding, reproduction, crop production, and plant protection and distribution networks.

Approximately 10 percent of the currently active biotechnology corporations are engaged in the production of products or the control of systems of potential importance to agriculture (Table 3). The first agricultural applications of genetic engineering will likely be for animal and plant disease control, in the production

Table 3.    Some Biotechnology Companies with Agricultural Interests[a]

| Name | Specialty | Location |
|------|-----------|----------|
| Advanced Genetic Sciences, Inc. | Potato, asparagus, strawberry | Greenwich, CT |
| Agrigenetics Corp. | Cereals, legumes, disease resistance | Denver, CO Madison, WI |
| ARCO Plant Cell Research Institute | Broad interests | Dublin, CA |
| Calegene | Stress–salt tolerance | Davis, CA |
| Cetus Corp. | Nitrogen fixation, inoculants | Berkeley, CA |
| DeKalb–Pfizer Genetics | Corn, sorghum | DeKalb, IL |
| DNA Plant Technology Corp. | Tomato improvement | Cinnaminson, NJ |
| Frito-Lay Inc. | Potato | Dallas, TX |
| International Genetic Engineering, Inc. | Broad interests | Santa Monica, CA |
| International Plant Research Institute (IPRI) | Disease and stress resistance—wheat | San Carlos, CA |
| Life Sciences Inc. | Bulbs, seeds | St. Petersburg, FL |
| Molecular Genetics, Inc. (MGI) | Cereals, sorghum | Minnetonka, MN |
| Native Plants Inc. | Stress tolerance | Salt Lake City, UT |
| Neogen Corp. | Animal disease control, broad interests | East Lansing, MI |
| Phytogen Inc. | Nutrition, disease resistance, photosynthetics | Pasadena, CA |
| Plant Genetics, Inc. | Plant cell biology | Davis, CA |
| R and A Plant/Soils Inc. | Microbial soil inoculants | Pasco, WA |
| Sungene Technologies Corp. | Crop varieties | San Francisco, CA |
| Zoecon Corp. | Pest control, growth enhancement | Palo Alto, CA |

[a] In addition to companies in this listing are the investments of several other major established companies in genetic engineering of potential applications in agriculture. These include but are not limited to the following: Battelle, Celanese, Dow, Du Pont, Eli Lily, Exxon, International Minerals (IMC), Merck, Monsanto, Occidental Petroleum, Pfizer, Stauffer, Upjohn, and Sohio.

of protective vaccines, in food technology, and in the production of new pesticides.

The immediate applications of genetic engineering will likely occur first as a result of microbiological productions, mostly from *Escherichia coli,* and secondly with tissue cultures of specific crops (Table 4).

Table 4.  Crops for Which Tissue Culture
Techniques Are Now Being Used or
Show Promise for Commercial
Propagation of Superior Stocks

| | |
|---|---|
| African oil palm | Onion |
| Asparagus | Papaya |
| Banana | Pineapple |
| Boston fern | Potato |
| Broccoli | Rootstocks (tree fruits) |
| Brussels sprouts | Spinach |
| Carrots | Strawberry |
| Cauliflower | Taro |
| Citrus | Tomato |
| Garlic | Welsh onion |
| Genetically superior trees | Yam |

## Tissue Culture

Cell, embryo, meristem, haploid, and tissue cultures provide a convenient, efficient, and rapid method of crop propagation and for the establishment of experimental crossing lines and genetically superior selections for commercial production. These are the gateways through which the developments of genetic engineering must pass to be useful in the field. Immediate opportunities for genetic improvement lie with such crops as asparagus, potatoes, bananas, the African oil palm, propagation of rootstocks for fruit trees, and a multitude of ornamental and forest trees. Tissue cultures also greatly aid the development of many strains of plants with toxin resistance to diseases of corn, such as the southern corn leaf blight, and tolerance to herbicides. Masses of vegetative cells of corn respond to herbicides or plant toxins as mature plants would respond. *In vitro* selection of cells can be expected to greatly accelerate genetic improvement and disease resistance in crop plants.

Opportunities for tissue culture exist also with both food and forest crops and ornamentals. Silviculture and genetic manipulation of Douglas fir and loblolly pine plantations have increased productivity 70 to 300 percent, respectively, over natural forests. Tissue culture proliferation of super trees will shorten the time required to produce marketable trees and double the productivity of the now intensively managed tree plantations. Economic feasibility will dictate the future of micropropagation and tissue culture. Tissue culture will become an extremely important resource tool and will be the path through which all forms of genetic engineering for crops and plants must go if transitions are to occur from the laboratory to the field.

Meanwhile, classical plant breeding procedures will neither become obsolete nor be replaced. The emerging methodologies from molecular biology will provide powerful new tools that will raise the effectiveness and efficiency of those procedures that already have launched us into the greatest biological revolution of all time.

## Food Handling and Processing

There will be many new developments in the handling and packaging of food. Genetic engineering will have an ever-increasing impact. Rennin for cheese-making that is derived from microbial synthesis rather than from organs of livestock from slaughterhouses is now a reality. Microbial rennets are now used extensively in U.S. cheese-making, as is aspartame for an artificial sweetener.

We will see low ethylene, low oxygen storage for fruits and vegetables. Controlled-atmosphere packages for bread now permit long-term storage without refrigeration or preservatives. Quick, ultra-high-temperature, and aseptic packaging provides a similar option for milk and fruit juices. Hot water (115 to 122°F) sterilization will replace fungicides for the prevention of storage rots in fruits and vegetables.

There will be increased storage, processing, and packaging near the point of production. Packaging, at retail and on the farm, will be in smaller units for convenience and safety, and flexible packages will gain in importance. While the world market for fresh fruits and vegetables will continue to expand for years, there will be less emphasis on air transport and more on storage technology. Post-harvest losses will decrease in the fields and within the marketing chain. Foods containing smaller amounts of moisture will attain more prominence. Fabricated foods and restructured meat products, while not acceptable to the current older generation, will find an increasing market based on price among the younger generation. The proportion of convenience-type foods will increase.

Diets, food sources, and the relationships between food and disease will receive more attention. There will be emphasis on reduced-salt and low-sodium products and a trend away from the use of nitrates and nitrites in meat products. New health and government food items will be introduced. For the human diet, there will be a decrease in fat and calories and sugar, the balance being made up by starch and fiber foods. Many consumers will avoid sugar, salt, caffeine, and preservatives. They will become further and further removed from producers. The current outpouring of books on human nutrition and diets further documents the need to attract new scientific talent to agriculture.

## Reproductive Efficiency in Livestock

The future is now for animal agriculture. Parallel with the great advancements already mentioned in genetic engineering of new vaccines and hybridomas to produce monoclonal antibodies for animal disease control and diagnosis is the potential for rapid genetic improvements in livestock. Improved fertility is now a reality using estrus synchronization, or control of the reproductive cycle. Semen preservation, pregnancy detection, multiple births, superovulation, and nonsurgical embryo transfer and implantation are also realities. All can be maximized to increase the number of offspring from genetically superior parents. Nonsurgical embryo collection is now helped by extremely sensitive microscopic techniques for embryo sexing, freezing, and implantation.

## Improvements in Livestock Metabolism and Nutrition

New frontiers are emerging through improved feeding, management, and genetics to encourage more lean meat and less deposition of fat in food animals. The first option involves intervention with several hormones or anabolic steroids. Some are now available. They consist of progesterone, estrogene, testosterone, and zearolone as feed additives. The second will be administration of the growth hormones (bovine, porcine, etc.), which by the twenty-first century will become available for regulation of the growth and productivity of dairy cattle, beef animals, and swine. These hormones will accelerate the growth of all farm animals and fish, correct genetic diseases, and provide a means of farming genetically valuable animal products.

Forage crop production provides a direct nutritional input for 2.5 billion ruminant animals on earth that are useful for people. Here the opportunity for genetic improvements and management inputs are exceeded only by those for forestry. The most important forage crop in the United States for livestock is alfalfa. The United States produces 30 million acres comprising almost half the world's total. Significant technological inputs have recently boosted nonirrigated annual crop yields beyond the 10-ton-per-acre mark (Table 1). The future will see mechanical harvesting, conditioning, and chemical treatments that will greatly reduce losses and reduce the time for field curing.

The most important feed constituent in forages is cellulose. It is the world's most abundant organic compound. Worldwide production of cellulose is estimated at 100 billion tons per year. This is equivalent to almost 150 pounds of cellulose produced daily for each of the world's 4.5 billion inhabitants. Its conversion into food on an economical basis is accomplished only by ruminants. A remaining challenge for the twenty-first century will be the economical bioconversion of lignin, the world's second most abundant organic compound, to products useful for people.

### Microbiological Transformations and Plant Nutrition— Biological Nitrogen Fixation

Nitrogen fertilizer is the single most important industrial input into agricultural production. Increasing amounts of nitrogen fertilizer, either from organic or inorganic sources, will be required to provide the future food needs of people. Additional renewable and nonrenewable resources must be sought after. Up to now research on biological nitrogen fixation has been outstanding in producing sophisticated basic information, but little practical knowledge has been developed to increase crop production under field conditions.

Both basic and applied research related to enhancement of crop productivity from biologically fixed nitrogen is one of the most neglected of all plant sciences. There are four major nitrogen-fixing biological systems that relate to agriculture, forestry, and rangeland: the *Rhizobia*–legume, the *Azolla–Anabeana*, particularly valuable for rice production in tropical and subtropical areas, the *Actinomycetes*–angiosperm, and the *Spirillum*–grass symbioses (see Figure 2). Advances

Figure 2.    Nitrogen fixation. These soybean root nodules contain rhizobium bacteria in a symbiotic association, enabling the conversion of atmospheric nitrogen into a form useable as a nutrient. This process frees plants from depending on soil and costly fertilizers for their nitrogen nutrient. Experiments are under way to transfer bacterial nitrogen-fixing genes to plant cells, so they can fix nitrogen by themselves. (*Source:*   E.I. du Pont de Nemours & Company.)

Table 5.    Some Nitrogen-Fixing Rhizobial Species
of Significance in Food Production

| Microorganism | Host |
|---|---|
| R. japonicum | Soybeans |
| R. phaseoli | Common beans (navy, kidney) |
| R. sp. (cowpea) | Cowpeas, mung beans, chick-peas, pigeon peas, peanuts, winged bean |
| R. leguminosarum | Broad beans, peas, lentils, vetch |
| R. meliloti | Alfalfa, sweet clover |
| R. trifolii | Clover |
| R. lutini | Lupin |

in biological nitrogen fixation will become increasingly prominent for crop productivity in the future. Major nitrogen fixers in the *Rhizobia*–legume and *Azolla*–*Anabeana* partnerships are listed in Tables 5 and 6. In addition, there are about 160 species of nonleguminous angiosperms that have actinomycetes in their root nodules capable of fixing nitrogen. Only a few of these are now agriculturally important.

## Mycorrhizae

Of equal importance are the mycorrhizae (soil–root–fungus relationships). Mycorrhizae constitute a class of biological organisms with great importance for future crop production. These fungi colonize roots of practically all

Table 6.    Species of Azolla

| Species | Origin |
|---|---|
| Imbricata[a] | China |
| Filiculoides[a] | Hawaii |
| Caroliniana[a] | Ohio |
| Mexicana | Florida |
| Microphylla[a] | Equador |
| Pinnata | Ivory Coast |
| Rubra | Japan |
| Nilotica | Sudan |

[a] Most efficient nitrogen fixers.

food crops. Some 80 species of fungi live in and around root surfaces. Root absorbing surfaces affected by water and nutrient uptake may be increased tenfold by their presence. Phosphorus and many micronutrients are made more available to plants on soils otherwise that are poor in phosphorus and nutrients. They are converted to more soluble forms and transported to the roots of the plants. Mycorrhizae also favor nitrogen-fixing bacteria. They can transport to plants water which is collectively beyond the normal reach of roots, thereby helping the plants withstand drought. Mycorrhizae do not add nutrients or soil moisture to soils, but they may greatly influence the efficiency of crop removal of these essential growth factors. Mycorrhizae may account for only 1 percent of the total weight of plants, but they can give a 150 percent increase in growth. One of the most exciting frontiers for the future will be further studies of the physiology and biochemistry of rhizospheres.

## Root-Colonizing Bacteria

These bacteria are of the genus *Pseudomonas* and suppress many plant diseases caused by other bacteria. The increased plant growth and yield are closely associated with the capacity of the root-colonizing bacteria to produce iron-binding compounds called siderophores. Research suggests that the greatest possibility for increasing crop yields and changes in agricultural practices may involve beneficial rhizobacteria that determine plant health by protecting plant roots from the harmful microorganisms that occur in all agricultural soils. The development of microbial products that are cost-effective and adapted to fit the technology of modern agriculture is a challenge for the future.

## Nitrification and Denitrification

The increased efficiency and utilization of fertilizers applied to crops will be important for the future. Two microbiologically powered processes—nitrification and denitrification—are involved in loss of soil nitrogen. Nitrogen stabilizers or chemical nitrification inhibitors and inhibitors of denitrification are now available. Losses from denitrification can also be reduced by good drainage and management.

Another approach to more efficient use of nitrogen is the introduction of new fertilizer sources such as sulfur-coated urea and supergranules of urea and deep placement for certain crops. The increased cost and limited effectiveness of these materials and problems of handling and application have thus far detracted from their widespread use. Still another means of greatly increasing efficiency of fertilizer usage is application through trickle irrigation systems. Here the value of the fertilizer applied may in some instances be doubled.

## Foliar Applications of Fertilizer

Finally, increased efficiency in the use of chemical fertilizers for crop production may be realized through foliar applications. The scientific merits of leaf feeding (foliar applications of fertilizer) in farm practices is controversial. Increased yields, however, may be possible by using the absorptive capacities of aerial parts as well as the roots, especially when applying nutrients during the early stages of flowering and fruiting. Nutrients sprayed on aerial plant parts are absorbed but the application technology is still lacking. Worldwide, and especially in many developing countries, there is an extensive use of foliar feeding of crops.

The rising costs of nitrogen and other fertilizers and fossil fuel resource requirements relating to their application are providing a continuing stimulus for investigation of the potential for foliar feeding. No comprehensive assessment of foliar application of fertilizer has appeared since our classical report of 1969. This is an example of a future technology adopted by growers and is being extensively applied to food crop production, particularly for micronutrients, with little laboratory or field research to support the practice.

## Alternative Agricultural Production Systems

A view of the new agriculture suggests that attention will be given to new systems of production and marketing. They are described variously as closed-system agricultural enterprises, regenerative agriculture, sustainable agriculture, organic farming and gardening, agroecosystem management, and alternative agricultural systems. The challenge, if they are adopted, will be to maintain current high-producing, conventional systems while seeking alternative production technologies. The issues of resource inputs—energy, fertilizers, pesticides, land, water, mechanization, and human capital—are under review. Traditional production systems are considered wasteful, exploitive of natural resources, and environmentally dangerous through the excessive use of chemical fertilizers and pesticides and in the loss of organic matter and topsoil. There is a need to design flexible, energy-efficient equipment for tillage, culture, harvest, and storage. The challenge for the future will be to develop integrated, resource-efficient production and marketing systems that maximize the stability of production over a wide marketing period while maintaining optimal output and quality with minimal costs. Such challenges pose dimensions for agricultural and food research not now being addressed.

## Integrated Pest Management

We must seek new approaches to pest control. Integrated pest management systems—the use of natural enemies (parasites), resistant varieties, cultural practices, and chemicals—will become essential for the eventual survival of most agricultural systems. A concern will be to ensure the health and safety not only of pesticide applicators but the people in nearby communities and all consumers of agricultural products. Of equal concern will be to contain the increases in the number of pests becoming resistant to chemical pesticides. There are now approximately 430 insects resistant to insecticides, 100 diseases resistant to fungicides and bacteriocides, and 36 weeds resistant to herbicides. The future will see chemicals used for pest control in concert with the use of computer programming for volume and cost reduction. There will be precise targeting of treatments and the use of safer, more powerful, and selective materials. There will be novel approaches that will be sparing of chemicals.

Integrated pest management, in the broad sense and even with individual crops, is still only a concept. There is little chance it will become a reality in this generation. No system has yet been accepted by growers that does not use chemicals. Integration of disciplines at the institutional level has not and is not now occurring. Pest management systems thus far described for cotton, apples, corn, alfalfa, wheat, peanuts, strawberries, and sorghum have addressed only insects, diseases, or weeds. They do not represent a total program. The only exception is the soybean, where consideration has been given to four groups of pests—insects, weeds, diseases, and nematodes.

Institutional arrangements, organizational prerequisites, and funding strategies will have to change before pest management, in the broad sense, becomes a reality. Truly functional integrated pest management research and education programs are interdisciplinary. Scientists engaged in such programs lose professional identity; they must become one of multiple authors of scientific reports, and they are frequently bypassed by current award systems.

The subdisciplines of plant protection (entomology, pathology, nematology, weed science) working independently cannot solve the massive pest problems of crops. Thus far, efforts to achieve truly cooperative research ventures in the United States have bogged down in administrative detail, problems of organization, personnel, prestige, control, benefits, awards, and credits. The problems most difficult to solve are not technical but human. This is the great challenge as we view the new agriculture.

## Plant Growth Regulators

No area in food research or new technology is more promising than that posed by the use of plant growth regulators. The effects are numerous. They range from enhancement of overall productivity, yields, and quality to improved storage life of such crops as potatoes and onions. There is now a new generation of chemicals that produce shorter plants, thicker stems, and result in better filling for heads of cereal grains with less lodging. Ripeners are used on sugarcane with a 10 percent increase in the yield of sugar. The oil content and quality of oilseeds and cereals may be enhanced. With corn, there are increases in yield, earlier pollination, and a longer grain-filling period with better tip fill, larger leaf areas, and heavier kernels. Significant yield increases in soybeans have been achieved and greater resistance to environmental stresses have followed the use of growth regulators on some major food crops (Table 7). Some plant growth regulators such as triacontanol have a wide spectrum of responses on many different species of economically important plants.

Table 7.   Plant Growth Regulators: Some Major Current and Potential Applications

| Product | Source | Plants Treated | Effects |
|---|---|---|---|
| Cycocel (Chloromequat, CCC) | American Cyanamid | Wheat, barley | Shorter plants, thicker stems, better filling of heads, less lodging |
| Ethephon | Several sources | Rubber | Improves latex production |
| Terpal | BASF Wyandott Corp. | Winter rye, barley | Reduces lodging |
| Ethephon + PIX | BASF Wyandott Corp. | Winter rye, barley | Reduces lodging |
| Brassinosteroids | Many sources | Vegetable crops | Yield increases of 25 to 60 percent |
| Gibberellins | Several sources | Valencia oranges, grapes, peaches, tomatoes | Reduces creasing, cell enlargement |
| Ripeners | Several sources | Sugarcane | 10 percent increase in yield of sugar |
| ALAR | U.S. Rubber | Apples, peaches, cherries | Delays harvest, accelerates maturity and color |
| PIX | BASF Wyandott Corp. | Cotton | Shorter and more open plants and whiter cotton |
| Premerge-3 | Dow Chemical | Corn | Yield increases, earlier pollination, longer grain filling period, better tip fill |
| ACA | AMOCA | Corn | Increased yield, bigger root system, larger leaf area, heavier kernels |
| Plant respiration inhibitor | Du Pont | Soybeans | Yield increases up to 100 percent |
| Triacontanol | Several sources | Many crops | Yield increases from 5 to 20 percent |

## Photosynthetic Enhancement

Photosynthesis is the most important biochemical process on earth. Each day green plants store on this planet 17 times as much energy as is presently consumed worldwide. No area is more important for agriculture in the twenty-first century than greater photosynthetic efficiency and partitioning of photosynthetic material to economically valuable structures. Yet most photosynthesis research thus far has been directed toward further understanding the nature of the process rather than relating the results of improvements of crop productivity under field conditions.

Research and technology initiatives relative to enhancement of crop productivity through greater photosynthetic efficiency include fast initial growth for a quick ground cover, a high leaf area index, improved plant architecture with more erect leaves, delayed or slow senescence, the use of plant growth regulators, genetic selections for photosynthetically superior lines, and reductions in photo and maintenance respiration. Specifically, ribulose diphosphate carboxylase and herbicide binding proteins may undergo genetic changes that alter the photosynthetic behavior of plants. Both photosynthetic rates and dry matter production have dramatically increased. Similarly, genetic modification of herbicide binding sites will permit the engineering of plants with altered susceptibility to herbicides. There are many promising areas for future research on photosynthesis, both fundamental and problem solving, which could be translated into improved crop productivity.

## Greater Resilience to Environmental Stresses

Heretofore, little agricultural research has been directed toward alleviation of climatic stresses on crops and livestock and the instability these stresses cause in production and marketing. Harvest shortfalls and surpluses and frequent price disruptions are the result. Such research holds top priority for achieving stable production at high levels but has not received support among the priorities in the competitive grant programs administered by the U.S. Department of Agriculture. Improved resistance to environmental and climatic stresses— drought, heat, cold, problem soils, salinity, and air pollutants—will be achieved through genetic improvements of crops and breeds of livestock, chemical treatments, and better management. The effects of unfavorable climates on agricultural productivity are as pervasive as those of unfavorable soils. Productivity will be enhanced in the now marginal growing areas, resulting in greater stability of supplies and improvements in quality. If there is to be climate change from the rising level of atmospheric carbon dioxide coupled with the present reality of interannual variations in climate which cause current disruptions in supplies and markets, there are valid reasons for seeking means of alteration of climatic stress. A second alternative is controlled-environment agriculture. The agriculture of tomorrow will thus become that of today.

Improved housing for livestock will be prompted by interests both in animal comfort, better waste disposal, and increases in efficiency in feed utilization and productivity. Controlled photoperiod environments for food animals will regulate reproduction, stimulate body growth, and increase the output of meat, milk, and eggs in several domestic species. The complete industrialization of hog production (confinement from birth to market) prompted by economics and biology will become a reality as we approach the twenty-first century.

## Genetic Diversity in Plants

Plants provide directly or indirectly up to 95 percent of the world's total food. Of the 350,000 species of plants on earth only about 0.1 percent (less than 300) are currently used worldwide for food. Large numbers of individual selections of major food crops, however, have been assembled by the International Agricultural Research Centers. These include more than 40,000 for rice, 26,000 for wheat, 13,000 for corn, 14,000 for sorghum, 10,500 for soybeans, 5000 each for pearl millet and mung beans, 12,000 for potatoes, 2000 for cassava, 11,500 for chick-peas, 5500 for pigeon peas, and 3000 each for field beans and peanuts. Added to these are 3000 genetic stocks for barley, 5000 for peas, 5400 for tomatoes, 1025 for sweet potatoes, and 800 for Chinese cabbage.

The establishment in the United States of the National Plant and Germplasm System is a landmark for the future of American agriculture and for the salvation of genetic resources for food production. More than 400,000 selections of seed and vegetatively propagated stocks have been collected. Within this so-called world collection, however, is a significant deficiency in wild races.

Major efforts are now in progress to seek out crop selections and wild races that are resistant to or will tolerate salinity, other soil problems, and environmental stresses. Included are both food and forage crops. More than 3.8 million square miles of soils are too salty to grow crops in the world of today. Some occur in the United States. Likely candidates for improved tolerance to salinity are the wild relatives of barley, wheat, sorghum, rice, millet, sugar beets, tomato, and date palm. Also included are alfalfa, Ladino clover, creeping bent, and Bermuda grasses. Genetic selections of barley and tomato have already been identified that can be grown in seawater once the seeds have germinated.

Genetic vulnerability to pests and environmental limitations stem from genetic uniformity. Some major food crops in the United States, on this basis, are highly vulnerable. Witness the southern corn leaf blight of 1970, which reduced corn yields by 15 percent nationwide. The answer lies in greater genetic diversity, the best insurance against vulnerability to climate and pests. Greater diversity of genes will do two important things for the future—overcome the current vulnerability to pests and extend boundaries of production now constrained by unfavorable environments. The continued and improved productivity of American agriculture for the twenty-first century and beyond will be conditioned by genetic pools available for future generations. The magnitude of this pool of genetic resources will depend upon preservation of what we now have, seeking out the wild races which still exist, and our forthcoming successes in engineering new additions. This stockpile will include not only the food crops but organisms at the microbial level that will serve useful functions or produce products for the improvement of human welfare.

### New Crops

They are continuously being sought after as new sources for food, feed, fiber, and as renewable energy resources. Some crops such as jojoba, buffalo gourd, and most *Euphorbia* species are desert plants. Their productivity will continue to be limited because deserts without irrigation are notoriously low-producing, and water will become an increasingly more limited and expensive resource. No major shift from food-producing to energy-producing crops is anticipated.

The sunflower was the most rapidly expanding crop in the United States during the 1970s but has now plateaued at about 3 million acres. The oil is highly nutritious for human consumption and may have some utility as a substitute for diesel fuel in powering motorized vehicles. Amaranth, as a food crop, has been given great publicity for its seed as well as its foliage. Leucaena is a promising forage legume and forest tree specie but is very specific as to its climatic and soil requirements. The new cereal grain Triticale, the synthetic hybrid of wheat and rye, has been given high visibility for its superior yields, adaptability, and nutritional values. High lysine corn was discovered 20 years ago but still awaits human acceptance. Perennial grains have been declared as the ultimate prevention for soil erosion but have many limitations as to culture and productivity.

Food habits of people are not changed easily or quickly. What people consume is based on tradition and more on appearance, taste, color, and texture rather than on nutritional values. Any major impact on food habits of people for future generations will likely have its origin in school lunch programs.

## Protected Cultivation: The Plastic Revolution

No technological development has so much modified the course of controlled-environment agriculture as the development of plastic films. They are now used extensively for constructing new greenhouses, covering old glass greenhouses, for air-inflated, self-contained, or supported bubble houses, solar greenhouses, solar stills, and in containing atmospheres for carbon dioxide enrichment of greenhouse-grown crops. Plastic films are used for covering row crops and seed beds and for soil mulches. Plastic tubes, laterals, and drip lines are important components of drip irrigation systems. They are also used for the packaging of many perishable food crops. Recent introduction of peat modules, rock wool, and nutrient film techniques for plant growing in greenhouses involve plastic films for precise control of all aspects of the root environment.

The modern-day plastic revolution is manifest in Japan and the countries of western Europe. Little recognition has been given to what has been happening and is continuing to happen in the most populated nation on earth, the People's Republic of China. For China in 1981 there were 55,000 acres of plastic mulches and 30,000 acres of plastic greenhouses, all constructed since 1979 and used almost exclusively for food crop production. Each successive year in China since 1979 has shown a tenfold increase in use of plastic row covers and mulches for improving yields and extending the season for many crops such as tomatoes, cucumbers, peppers, eggplant, cabbage, melons, cotton, sugarcane, and peanuts.

The new agriculture will see additional developments in the use of plastics for crop production in the decades ahead. They will include the use of films with white or silver stripes to repel insects. Biodegradable plastic soil mulches are becoming available. Plastic films and covers will extend and push crop productivity into new geographical frontiers in both agriculturally developed and less-developed nations. Their use promises much for agriculture in the twenty-first century.

## CONCLUSIONS

As we project the new agriculture, we will witness no dramatic changes in the food habits of people. Some 20 food crops will continue to meet the primary needs. With the appropriate technologies, resource inputs, and economic incentives there will be no worldwide shortage of food. Because of resource constraints, costs, and availability, the future will see an inevitable shift from our now highly mechanized, labor-saving, single-crop or livestock production systems to more science and biological opportunities and a more diversified, resource-conserving set of agricultural production technologies. The new agriculture for America will see almost all future advances in productivity as a result of increases in yield, and alternative agricultural systems that use resources sparingly will come into prominence.

A wide variety of new crop production technologies resulting in a collective system of inputs will have a significant impact on increasing food productivity in the year 2030 (Table 8). The future will see even more production of technology inputs than the past.

Table 8.    Increase in Crop Productivity by 2030 A.D.
Anticipated from Various Crop Technologies and Changes in the Resource Base (1983 = 100)

| Technologies | Productivity Increases (%) |
|---|---|
| Plant breeding | 135 |
| Irrigation and crops to conserve water | 133 |
| Genetic engineering | 125 |
| Growth regulators | 124 |
| Rising level of atmospheric $CO_2$ | 120 |
| Biological nitrogen fixation | 118 |
| Multiple cropping and polyculture | 115 |
| Improved photosynthetic efficiency | 117 |
| Temperature acclimation | 113 |
| Forage nutritional quality | 112 |
| Crop maturity | 111 |
| Transpiration suppressants | 110 |
| Intercropping | 108 |
| Protected cultivation | 105 |

*Source:*    Adapted from G. H. Heichel (1982).

When we consider changes in the resource base, new technologies, and the impacts of science, we are mindful of the creativity, innovations, and vision of people—the human resource—which is the greatest of all resources. It has been called human capital. No limitations can be ascribed to the creativity of the human mind, and it will be the unpredictable spark of creativity that will be the most important contributor to the new agriculture.

# BIBLIOGRAPHY

## CHAPTER 2 PRIORITY ISSUES IN LAND, SOIL, AND FORESTRY

Priority Issues in Soil Conservation

*Sandra S. Batie*

Batie, Sandra, 1983, *Soil Erosion: Crisis in America's Croplands?*, The Conservation Foundation, Washington, D.C.

Clayton, K. and C. Ogg, 1982, "Soil Conservation Under More Integrated Farm Programs," unpublished paper, U.S. Department of Agriculture, Washington, D.C.

Crosson, P. R. and S. Brubaker, 1982, *Resource and Environmental Effects of U.S. Agriculture*, The Johns Hopkins University Press (for Resources for the Future), Baltimore.

Libby, L. W., 1980, "Who Should Pay for Soil Conservation?," *Journal of Soil and Water Conservation*, 35(4), pp. 155–157.

Ogg, C. and A. Miller, 1981, "Minimizing Erosion on Cultivated Land: Concentration of Erosion Problems and the Effectiveness of Conservation Practices," *Policy Research Notes*, U.S. Department of Agriculture, Washington, D.C.

U.S. Department of Agriculture, 1981, "Agricultural Stabilization and Conservation Service," *National Summary Evaluation of the Agricultural Conservation Program Phase I*, U.S. Government Printing Office, Washington, D.C.

U.S. Department of Agriculture, 1982, *A National Program for Soil and Water Conservation: 1982 Final Report and Environmental Impact Statement*, U.S. Government Printing Office, Washington, D.C.

## Issues in U.S. Forestland Use

### W. David Klemperer

Clawson, M., 1976, "The Economics of National Forest Management," Working Paper EN-6, Resources for the Future, The Johns Hopkins University Press, Baltimore.

Clawson, M., 1977, "Economic Timber Production Characteristics of Nonindustrial Private Forests in the United States," Discussion Paper, Resources for the Future, Washington, D.C., p. 77.

Clawson, M., 1979, "Forests in the Long Sweep of American History," *Science*, **204**, pp. 1168–1174.

Cordell, H. K. and J. C. Hendee, 1982, "Renewable Resources Recreation in the United States: Supply, Demand, and Critical Policy Issues," American Forestry Association, Washington, D.C.

Hyde, W. F., 1981, "National Forest Logs Red Ink for Treasury," *The Wharton Magazine*, Fall issue.

Sedjo, R. A. and D. M. Ostermeier, 1978, *Policy Alternatives for Nonindustrial Private Forests*, Society of American Foresters and Resources for the Future.

Siegel, W. C., 1974, "Long-Term Timber Contracts in the South," *Forest Farmer*, **34**(1), pp. 8–9.

U.S. Department of Justice, 1981, *Redwoods Litigation*, Lands and Natural Resources Division, DOJ-1982-01.

U.S. Forest Service, 1980, *A Recommended Renewable Resources Program—1980 Update*, Publication FS-346, Government Printing Office, Washington, D.C., p. 540 plus appendices.

U.S. Forest Service, 1981, *An Assessment of the Forest and Range Land Situation in the United States*, Forest Resource Report No. 22, Washington, D.C.

U.S. Forest Service, 1982, *An Analysis of the Timber Situation in the United States: 1952–2030*, Forest Resource Report No. 23, Washington, D.C.

## CHAPTER 3   ENERGY: STRATEGIES FOR THE FUTURE

### Energy Use in the Food Sector

### John N. Walker

Council for Agricultural Science and Technology, August 1977, *Energy Use in Agriculture, Now and for the Future*, Report No. 68.

Council for Agricultural Science and Technology, 1983 (Draft), "Energy Use and Production in Agriculture," revision of Report No. 68.

Doering, O. C., 1980, "Energy Dependence and the Future of American Agriculture," in S. S. Batie and R. G. Healy, Eds., *The Future of American Agriculture as a Strategic Resource*, The Conservation Foundation, Washington, D.C., Chapter 5, pp. 191–223.

Doering, O. C., E. Gavett, T. VanArsdell, E. Rall, P. Devlin, June 23, 1977, "Current Energy Use in the Food and Fiber System," a briefing paper prepared by the Economic Research Service, U.S. Department of Agriculture.

McFate, K. L., 1982, "Food, Energy and Your Future," National Food and Energy Council.

Pimental, D., 1980, *1980 Handbook of Energy Utilization in Agriculture*, CRC Press, Inc., Boca Raton.

Pimental, D., September 22–24, 1982, "Energy Inputs and U.S. Food Security," Workshop on Food Security in the United States, College of Agriculture, University of Kentucky.

Stout, B. A., December 19, 1978, *Energy—A Vital Resource for the U.S. Food System*, a Public Issues Study Report, American Society of Agricultural Engineers.

Walker, J. N., October 1–3, 1975, "Energy Usage in Crops Production," Proceedings of Southern Regional Education Board's Energy in Agriculture Conference.

## Energy from Biomass: A New Commodity

### Wayne H. Smith

Alexander, A. G., 1981, "Management of Tropical Grasses as a Year-Round Alternative Energy Source," in D. L. Klass, Ed., *Symposium Papers, Energy from Biomass and Wastes V*, Institute of Gas Technology, Chicago, pp. 87–103.

Barr, T. N., 1981, "The World Food Situation and Global Grain Prospects," *Science*, **214**, pp. 1087–1095.

Boyer, J. S., 1982, "Plant Productivity and Environment," *Science*, **218**, pp. 443–448.

Brady, N. C., 1982, "Chemistry and World Food Supplies," *Science*, **218**, pp. 847–853.

Calvin, M., 1983, "New Sources for Fuel and Materials," *Science*, **219**, pp. 24–26.

Erdman, J. G. and D. G. Petty, 1980, "The Future of Petroleum as an Energy Resource," *Revue de l'Institut Francais de Petrole*, **35**(2), pp. 198–214.

Hall, D. O., 1982, "Food Versus Fuel, A World Problem?," Proceedings from the Second European Communities Conference, in A. Strub, P. Chartier, and G. Schleser, Eds., *Energy from Biomass*, Applied Science Publishers, New York, pp. 43–62.

International Energy Agency, 1982, *Energy Outlook*, Organization for Economic Cooperation and Development, Paris.

Menz, K. M. and C. F. Neumeyer, 1982, "Evaluation of Five Emerging Technologies for Maize," *BioScience*, **32**, pp. 675–77.

Resources and Technology Management Corporation, 1981, *Alternative Energy Data Summary for the United States*, Vol. 1, Arlington.

Sheppard, W. J. and E. S. Lupinsky, 1982, "Chemicals from Biomass," Proceedings from the Second European Communities Conference, in A. Strub, P. Chartier, and G. Schleser, Eds., *Energy from Biomass*, Applied Science Publishers, New York, pp. 63–71.

Smith, W. H. and M. L. Dowd, 1981, "Biomass Production in Florida," *Journal of Forestry*, **79**, pp. 508–511.

Smith, W. H., P. H. Smith, and J. R. Frank, 1982, "Biomass Feedstocks for Methane Production," Proceedings from the Second European Communities Conference, in A. Strub, P. Chartier, and G. Schleser, Eds., *Energy from Biomass*, Applied Science Publishers, New York, pp. 122–126.

Tyner, W. E., 1980, "Our Energy Transition: The Next Twenty Years," *American Journal of Agricultural Economics*, **62**, pp. 957–964.

Zellner, J. A. and R. M. Lamm, 1982, "Agriculture's Vital Role for Us All," *1982 Yearbook of Agriculture*, U.S. Department of Agriculture, Washington, D.C., pp. 2–9.

## CHAPTER 4 PUBLIC POLICY: PRESENT AND FUTURE

## Milestones of U.S. Agricultural Policy

### Harold M. Harris, Jr.

Benedict, Murray R., 1953, *Farm Policies of the United States, 1790–1950*, The Twentieth Century Fund, New York.

Brandow, G. E., 1977, "Policy for Commercial Agriculture" in Lee R. Martin, Ed., *A Survey of Agricultural Economics Literature*, University of Minnesota Press, Minneapolis, Vol. 1.

Cochrane, Willard W. and Mary E. Ryan, 1976, *American Farm Policy 1948–1973*, University of Minnesota Press, Minneapolis.

Dubov, Irving and E. L. Rawls, 1974, *American Farm Price and Income Policies: Main Lines of Development, 1920–1973*, University of Tennessee Agriculture Experiment Station Bulletin 539.

Economic Perspectives, Inc., 1982, *An Analysis of Proposed Changes to the Nation's Dairy Policies— 1982*, Report prepared for the Milk Industry Foundation.

O'Rourke, A. Desmond, 1978, *The Changing Dimensions of U.S. Agricultural Policy*, Prentice-Hall, Englewood Cliffs.

Rasmussen, Wayne D. and Gladys L. Baker, 1979, *Price-Support and Adjustment Programs from 1933 through 1978: A Short History*, U.S. Department of Agriculture, Economic Statistics and Cooperative Service, Information Bulletin No. 424.

U.S. Department of Agriculture, Agricultural Stabilization and Conservation Service, 1976, *Farm Commodity and Related Programs*, Agriculture Handbook No. 345.

Wilcox, Walter W., Willard W. Cochrane, and Robert W. Herdt, 1974, *Economics of American Agriculture*, Prentice-Hall, Englewood Cliffs.

## Resource Policy and American Agriculture

### James C. Hite

Batov, Francis M., August 1958, "The Anatomy of Market Failure," *Quarterly Journal of Economics*, 72, pp. 351–379.

Dales, J. H., 1968, *Pollution, Property and Prices*, University of Toronto Press, Toronto.

Haveman, Robert, May 1973, "Common Property, Congestion and Environmental Pollution," *Quarterly Journal of Economics*, 87, pp. 278–287.

Schmid, A. Allen, 1978, *Property, Power and Public Choice*, Praeger, New York.

U.S. Bureau of the Census, November 1982, *Farm Population of the United States: 1981*, U.S. Department of Commerce and U.S. Department of Agriculture, Current Population Reports, Series P-27, No. 55.

## Food and Agricultural Policy in the Twenty-First Century

### B. H. Robinson

Duncan, Marvin and C. Edward Harshbarger, September–October 1977, "A Primer on Agricultural Policy," *Monthly Review*, Federal Reserve Bank of Kansas City.

Duncan, Marvin and C. Edward Harshbarger, November 1977, "Agricultural Policy: Evolution and Goals," *Monthly Review*, Federal Reserve Bank of Kansas City.

Knutson, Ronald D., 1979, "Summit Review," *Consensus and Conflict in U.S. Agriculture: Perspectives from the National Farm Summit*, Texas A&M University Press, College Station.

Knutson, Ronald D., 1982, "Agricultural Policy at a Decision Point," *Increasing Understanding of Public Problems and Policies—1982*, The Farm Foundation, Oak Brook, Ill.

Mayer, Leo, September 1980, "Agriculture, Food and Nutrition—Policy Issues for the 1980s," Speech at the Southern Regional Agricultural Outlook Conference, Atlanta.

## CHAPTER 5    THE OUTLOOK FOR AGRICULTURAL RESEARCH AND TECHNOLOGY

### Ralph W. F. Hardy

Barton, J. H., 1982, "The International Breeder's Rights System and Crop Plant Innovation," *Science*, 216, pp. 1071–1075.

Bell, E. A., Ed., 1983, *Better Crops for Food*, Ciba Foundation Symposium 97, Putnam Books, London.

Black, C. A., Ed., 1983, *Science of Food and Agriculture*, a new magazine for high school science teachers by CAST, Ames.

Brown, A. W. A., T. C. Byerly, M. Gibbs, and A. San Pietro, Eds., 1975, *Crop Productivity—Research Imperatives*, Michigan State Agricultural Experiment Station, East Lansing, pp. 1–399.

Brown, W., J. Tavares, and A. Hollaender, 1983, *Genetic Engineering of Plants*, National Academy of Sciences, Washington, D.C.

Evenson, R. E., P. E. Waggoner, and V. W. Ruttan, 1979, "Economic Benefits from Research: An Example from Agriculture," *Science*, 205, pp. 1101–1107.

Hardy, R. W. F., 1979, "Chemical Plant Growth Regulators in World Agriculture," in T. Scott, Ed., *Plant Regulation and World Agriculture*, Plenum Press, New York, pp. 165–206.

The International Rice Research Institute, 1982, *International Conference on Chemistry and World Food Supplies: The New Frontiers. CHEMRAWN II Conference Handbook and Abstracts*, CHEMRAWN II Secretariat, Manila, The Philippines.

Kornberg, W., Ed., 1982, "The Plant Sciences," *Mosaic*, 13(3), pp. 1–52.

Krieger, J. H., December 20, 1982, "Chemistry Confronts Global Food Crisis," *Chemical and Engineering News*, pp. 9–23.

Lemon, E., 1983, *Increasing Atmospheric $CO_2$ and Plant Productivity*, American Association for the Advancement of Science, Washington, D.C.

National Academy of Sciences, 1983, *Research Briefings, Report of the Briefing Panel on Agricultural Research*, Washington, D.C., pp. 57–79.

National Research Council, *Priorities in Biotechnology Research for International Development*, 1982, Proceedings of a Workshop, Board on Science and Technology for International Development, National Academy Press, Washington, D.C.

Pino, J. A. and D. J. Prager, Eds., 1982, *Science for Agriculture*, Report of a Workshop on Critical Issues in American Agricultural Research, Rockefeller Foundation, New York.

Pond, W. G., R. A. Merkel, L. D. McGilliard, and V. J. Rhodes, Eds., 1980, *Animal Agriculture: Research to Meet Human Needs in the 21st Century*, Westview Press, Boulder.

Rachie, K. O. and J. M. Lyman, Eds., 1981, *Genetic Engineering for Crop Improvement,* The Rockefeller Foundation, New York.

Swaminathan, M. S., 1982, "Biotechnology Research and Third World Agriculture," *Science,* **218,** pp. 967–972.

U.S. Government Printing Office, 1981, *An Assessment of the United States Food and Agricultural Research System,* Washington, D.C.

Verne, R. V., Ed., 1982, "Genetic Engineering of Plants," *California Agriculture,* **36**(8), pp. 1–36.

Zelitch, I., 1982, "The Close Relationship between Net Photosynthesis and Crop Yield," *BioScience,* **32,** pp. 796–802.

## CHAPTER 6  NEW DIRECTIONS IN THE PLANT AND ANIMAL SCIENCES

### Plant Research and Technology

### *W. Ronnie Coffman*

Barton, K. A. and W. J. Brill, 1983, "Prospects in Plant Genetic Engineering," *Science,* **219,** pp. 671–676.

Borlaug, N. E., 1983, "Contributions of Conventional Plant Breeding to Food Production," *Science,* **219,** pp. 689–693.

Boyer, J. S., 1982, "Plant Productivity and Environment," *Science,* **218,** pp. 443–448.

Chaleff, R. S., 1983, "Isolation of Agronomically Useful Mutants from Plant Cell Cultures," *Science,* **219,** pp. 676–682.

Flavell, R. B., 1982, "Recognition and Modification of Crop Plant Genotypes Using Techniques of Molecular Biology," in I. K. Vasil, W. R. Scowcroft, and K. J. Frey, Eds., *Plant Improvement and Somatic Cell Genetics,* Academic Press, New York.

Galun, E., 1982, "Somatic Cell Fusion for Inducing Cytoplasmic Exchange: A New Biological System for Cytoplasmic Genetics in Higher Plants," in I. K. Vasil, W. R. Scowcroft, and K. J. Frey, Eds., *Plant Improvement and Somatic Cell Genetics,* Academic Press, New York.

Menz, K. M. and C. F. Neumeyer, 1982, "Evaluation of Five Emerging Bio-technologies for Maize," *BioScience,* **32,** pp. 675–676.

Office of Technology Assessment, 1982, *Genetic Technology, A New Frontier,* Westview Press, Boulder.

Sheppard, J. F., D. Bidney, T. Barsby, and R. Kemble, 1983, "Genetic Transfer in Plants Through Interspecific Protoplast Fusion," *Science,* **219,** pp. 683–688.

Wittwer, S. H., 1975, "Food Production: Technology and the Resource Base," *Science,* **188,** pp. 579–584.

### Research and Technology in Animal Science

### *J. M. Elliot*

Abelson, P. H., 1983, "Biotechnology: An Overview," *Science,* **219,** p. 611.

Bauman, D. E. and W. B. Currie, 1980, "Partitioning of Nutrients during Pregnancy and Lactation:

A Review of Mechanisms Involving Homeostasis and Homeorhesis," *Journal of Dairy Science,* 63, p. 1514.

Brackett, B. G., G. E. Seidel, and S. M. Seidel, Eds., 1981, *New Technologies in Animal Breeding,* Academic Press, New York.

Fronk, T. J., D. E. Bauman, and C. J. Peel, 1981, "Effect of Growth Hormone on Performance of High Producing Dairy Cows," Proceedings from Cornell Nutrition Conference for Feed Manufacturers, p. 42.

National Research Council, 1982, *Priorities in Biotechnology Research for International Development,* Proceedings of a Workshop, Board on Science and Technology for International Development, National Academy Press, Washington, D.C.

Palmiter, R. D., R. L. Brinster, R. E. Hammer, M. E. Trumbauer, M. G. Rosenfeld, N. C. Birnberg, and R. M. Evans, 1982, "Dramatic Growth of Mice that Develop from Eggs Microinjected with Metallothionein-Growth Hormone Fusion Genes," *Nature,* 300, p. 611.

Peel, C. J., D. E. Bauman, R. C. Gorewit, and C. J. Sniffen, 1981, "Effect of Exogenous Growth Hormone on Lactational Performance in High Yielding Dairy Cows," *Journal of Nutrition,* III, p. 1662.

Pond, W. G., R. A. Merkel, L. D. McGilliard, and V. J. Rhodes, 1980, *Animal Agriculture: Research to Meet Human Needs in the 21st Century,* Westview Press, Boulder.

Trevis, J. and A. Bertelsen, February 1, 1982, "Genetic Engineering: Promise for Agricultural Industries," *Feedstuffs,* p. 32.

## CHAPTER 7    FUTURE APPROACHES FOR MEETING NUTRITIONAL NEEDS

## Nutritional Needs for the Twenty-First Century

### Ernest J. Briskey

Abelson, P. H., 1983, "Biotechnology: An Overview," *Science,* 219, p. 611.

Barr, T. N., 1981, "The World Food Situation and Global Grain Prospects," *Science,* 214, p. 1087.

Batie, S. S. and R. G. Healy, 1983, "The Future of American Agriculture," *Scientific American,* 248(2), p. 45.

Brady, N. C., 1981, "Food Resource Needs of the World," in J. T. Manassah and E. J. Briskey, Eds., *Advances in Food Producing Systems for Arid and Semi-Arid Lands,* Academic Press, New York, Vol. A.

Brady, N. C., 1982, "Chemistry and World Food Supplies," *Science,* 218, p. 847.

Briskey, E. J., 1981, "Overview of Recent Developments and Prospects for Our Future Food Supply," the Sixth William Underwood Technical Conference.

Briskey, E. J., 1983, "Universities and the Global Hunger Problem: Opportunities and Challenges in the Third World," Board for International Food and Agricultural Development Regional Meeting, Corvallis, February 3–4, 1983.

California Agricultural Experiment Station, 1982, "California Agriculture," in special issue: *Genetic Engineering of Plants,* 36(8), p. 36.

Castle, E. N., 1980, *Agricultural Education and Research,* The 1980 Kellogg Foundation Lecture.

Crosson, P. R. and K. D. Frederick, 1977, *The World Food Situation*, Resources for the Future, Washington, D.C.

Drucker, P. F., 1979, *Science and Industry, Challenges of Antagonistic Interdependence.*

Hulse, J. H., 1981, "Research and Postproduction Systems," in J. T. Manassah and E. J. Briskey, Eds., *Advances in Food Producing Systems for Arid and Semi-Arid Lands*, Academic Press, New York, Vol. B.

Kifer, P. E., 1981, "Appropriate and Advanced Processing Technologies," in J. T. Manassah and E. J. Briskey, Eds., *Advances in Food Producing Systems for Arid and Semi-Arid Lands*, Academic Press, New York, Vol. B.

Krieger, J. H., December 20, 1982, "Chemistry Confronts Global Food Crisis," *Chemistry and Engineering News*, pp. 9–23.

McMahon, Walter W., 1982, "Externalities in Education," Faculty Working Paper 877, College of Commerce and Business Administration, University of Illinois.

McMahon, Walter W., 1982, *The Integration of R and D and Human Capital in Medium Term Macroeconomic Models*, Division of Science Resource Studies, National Science Foundation, Washington, D.C.

McMahon, W. W., 1983, "Sources of the Slowdown in Productivity Growth: A Structural Interpretation," in John Kendrick, Ed., *International Comparisons of Productivity and Causes of the Slowdown*, American Enterprise Institute, Washington, D.C.

Manassah, J. T. and E. J. Briskey, 1981, *Advances in Food Producing Systems for Arid and Semi-Arid Lands*, Academic Press, New York.

Population Reference Bureau, 1982, "The 1982 Data Sheet Now and 38 Years from Now," Vol. 10, No. 4.

Seidel, G. E., Jr., 1981, "Perspectives on Animal Breeding," in *New Technologies in Animal Breeding*, Academic Press, New York.

Thomas, G. W., 1983, "The International Dimension: Unrealized Potential," 1983 Agricultural Conference Days Proceedings, Oregon State University.

Thomas, W. D., October 28, 1982, "U.S. Development Assistance Policy—Middle Income Countries," Board for International Food and Agricultural Development.

Wharton, Jr., Clifton R., 1981, "Statement before the Sub-Committee for Foreign Operations House Appropriations Committee."

Williams, Simon, September 1, 1979, "Confronting World Hunger Anew—A Wider Role for Agribusiness," *Agribusiness World Wide.*

Wittwer, S., 1982, "The Dawning Future—World Food Supplies," *The Professional Nutritionist.*

Wittwer, S., 1983, "The Future is Now in American Agriculture—A View from the 21st Century," 1983 Agricultural Conference Days Proceedings, Oregon State University.

Wortman, S. and R. W. Cummings, Jr., 1978, *To Feed This World—The Challenge and the Strategy*, The Johns Hopkins University Press, Baltimore.

# Implications for Animal Agriculture

## Virgil W. Hays

Bartley, E. E., E. L. Herod, R. M. Bechtle, D. A. Sapienza, and B. E. Brent, 1979, "Effect of Monensin or Lasalocid, with and without Niacin or Amicloral, on Rumen Fermentation and Feed Efficiency," *Journal of Animal Science*, **49**, pp. 1066–1075.

Broughman, R. W., 1982, "Practical Livestock-Forage Systems: Model to Managers," in J. A. Smith and V. W. Hays, Eds., *Proceedings, Fourteenth International Grassland Congress*, Westview Press, Boulder, pp. 48–53.

Brumby, P. J. and J. Hancock, 1955, "The Galactopoietic Role of Growth Hormone in Dairy Cattle," *New Zealand Journal of Science and Technology*, A36, p. 417.

Byerly, T. C., 1966, "The Role of Livestock in Food Production," *Journal of Animal Science*, 25, pp. 552–566.

Ely, D. G., T. W. Robb, and K. P. Coffey, 1982, "Potential for Increasing Productivity of Ruminants by Treating Forages with a Plant Growth Regulator, Mefluidide," private communication.

Hansel, W., December 5–9, 1982, "Future Agricultural Technology and Resource Conservation: Beef and Dairy Cattle, Sheep, Goats, Swine, and Poultry," Resource Conservation Act Symposium, U.S. Department of Agriculture, Washington, D.C.

Hays, V. W., 1980, "Effectiveness of Feed Additive Usage of Antibacterial Agents in Swine and Poultry Production," *The Hays Report*, Rachelle Laboratories, Inc., Long Beach.

Hays, V. W., R. L. Bard, R. F. Brokken, S. A. Edgar, W. T. Elliott, M. Finland, W. D. Heffernan, C. A. Lassiter, H. W. Moon, S. A. Naqi, A. A. Paulsen, J. V. Shutze, H. S. Teague, R. Vinson, W. J. Visek, G. Vocke, H. D. Wallace, R. H. White-Stevens, and D. Wolf, 1981, "Antibiotics in Animal Feeds," *Council for Agricultural Science and Technology, Report 88*, Ames.

Heersche, G. E., 1982, "Artificial Insemination of Dairy Heifers," Proceedings of the National Dairy Cattle Workshop, U.S. Department of Agriculture Economic Service, Washington, D.C.

Johnson, A. O., 1978, "Food Consumption, Prices and Expenditures," *Agricultural Economic Report* 138, 1977 Supplement.

Klopfenstein, T., 1978, "Chemical Treatment of Crop Residues," *Journal of Animal Science*, 46, pp. 841–848.

Machlin, L. J., 1973, "Effect of Growth Hormone on Milk Production and Feed Utilization in Dairy Cows," *Journal of Dairy Science*, 56, pp. 575–580.

Overfield, Jr., J. R. and E. E. Hatfield, 1975, *Anabolic Agents for Beef Cattle*, University of Illinois Beef Cattle Report AS-669, Urbana, pp. 48–51.

Palmiter, R. D., R. L. Brinster, R. E. Hammer, M. E. Trumbauer, M. G. Rosenfeld, N. C. Birnberg, and R. M. Evans, 1982, "Dramatic Growth of Mice that Develop from Eggs Microinjected with Metallothionein-Growth Hormone Fusion Genes," *Nature*, 300, p. 615.

Pursel, V. G., D. O. Elliot, R. B. Staigmiller, and C. W. Newman, 1982, "Fertility of Gilts Fed Altrenogest with and without Hormone Injections to Control Ovulation," *Journal of Animal Science*, 55, Supplement 1, p. 379.

Reid, J. T., 1977, "Potential for Increased Use of Forage in Dairy and Beef Rations," *Proceedings of the Tenth Research-Industry Conference*, American Forage Grassland Council, Lexington, p. 165.

Seidel, Jr., G. E., 1981, "Management of Reproduction of Cattle in the 1990s," Carl B. and Florence E. King Visiting Scholar Lectures, Special Report 77, University of Arkansas.

Wagner, D., 1982, "Ionophore Comparisons for Feedlot Cattle in the Future of the Beef Industry," National Beef Symposium and Oklahoma Cattle Feeders Seminar, Oklahoma State University, Stillwater, pp. N1–N9.

## Implications for Plant Agriculture
### Roger L. Mitchell

Borlaug, N. E., 1983, "Contributions of Conventional Plant Breeding to Food Production," *Science*, 219, pp. 689–693.

Boyer, J. S., 1982, "Plant Productivity and Environment," *Science*, 218, pp. 443–448.

Feldman, Moshe and E. R. Sears, 1981, "The Wild Gene Resources of Wheat," *Scientific American*, 244, pp. 102–112.

Gale, M. D. and C. S. Law, 1976, *The Identification and Exploitation of Norin 10 Semi-Dwarfing Genes*, Annual Report of the Plant Breeding Institute, Cambridge, pp. 21–35.

Heichel, G. H., 1982, "Anticipating Productivity Responses to Crop Technologies," Resources Conservation Act Symposium, Future Agricultural Technology and Resource Conservation, December 5–9, 1982, Washington, D.C.

Jensen, N. F., 1978, "Limits to Growth in World Food Production," *Science*, 201, pp. 317–320.

Larson, W. E., F. J. Pierce, and R. H. Dowdy, 1983, "The Threat of Soil Erosion to Long-Term Crop Production," *Science*, 219, pp. 458–465.

"Science for Agriculture," 1982, Report of a workshop on critical issues in American Agricultural Research, jointly sponsored by the Rockefeller Foundation and the Office of Science and Technology Policy.

Stangel, P. J., 1976, "World Fertilizer Reserves in Relation to Future Demand," in M. J. Wright, Ed., *Plant Adaptation to Mineral Stress*, Proceedings of Workshop, Cornell Univ. Experiment Station, Beltsville.

U.S. Department of Agriculture and Council on Environmental Quality, 1981, *National Agricultural Lands-Study*, Final Report, U.S. Government Printing Office, Washington, D.C.

## CHAPTER 8   CROPS FROM THE DESERT, SEA, AND SPACE
## Jojoba: A Source of a Superior, Nonpetrochemical Lubricant
### Demetrios M. Yermanos

Miwa, T. K. and J. W. Hageman, 1976, "Physical and Chemical Properties of Jojoba Liquid and Solid Waxes," in T. K. Miwa and X. Murrieta, Eds., *Proceedings of the Second International Conference on Jojoba*, Ensenada, Mexico.

## New Crops from the Sea
### Michael Neushul

Cameron, F. K., 1915, *Potash from Kelp*, U.S. Department of Agriculture, Report No. 100.

Coon, D. A., 1981, *Measurements of Harvested and Unharvested Populations of the Marine Crop Plant Macrocystis*, Proceedings from the International Seaweed Symposium, Science Press, New Jersey, Vol. 7, pp. 678–687.

Crowder, B., Ed., 1982, *Sea Grant Aquaculture Plan*, Texas A&M University Press, College Station.

Cushing, D. H., 1980, "The Biology of Fish Populations," in D. S. Simonett, Ed., *Marine Sciences and Ocean Policy Symposium*, University of California Press, Santa Barbara.

Food and Agriculture Organization, 1981, *Agriculture: Toward 2000,* Food and Agriculture Organization of the United Nations, Rome.

Harger, B. W. W. and M. Neushul, 1983, "Test-Farming of the Giant Kelp, *Macrocystis,* as a Marine Biomass Producer," *Journal of World Mariculture Association,* 14.

Lasker, R., 1978, "The Relationship between Oceanographic Conditions and Larval Anchovy Food in the California Current: Identification of Factors Contributing to Recruitment Failure," *Report from the International Council for the Exploration of the Sea* (France), Vol. 173, pp. 212–230.

Lasker, R., 1979, "Factors Contributing to Variable Recruitment of the Northern Anchovy (*Engraulis mordax*) in the California Current: Contrasting Years 1975 through 1978," *Symposium on the Early Life History of Fish,* International Council for the Exploration of the Sea, Woods Hole, April 1979.

Lasker, R., J. Pelaez, and R. M. Laurs, 1981, "The Use of Satellite Infrared Imagery for Describing Ocean Processes in Relation to Spawning of the Northern Anchovy (*Engraulis mordax*), *Remote Sensing of Environment,* 11, pp. 439–453.

Lehman, J. T. and D. Scavia, 1982, "Microscale Patchyness of Nutrients," *Science,* 216, pp. 729–730.

Mitchell, B. and R. Sandbrook, 1980, *The Management of the Southern Ocean,* Temtex Ltd., London.

Neushul, M., B. W. W. Harger, and J. W. Woessner, 1981, *Laboratory and Nearshore Field Studies of the Giant California Kelp as an Energy Crop Plant,* Proceedings of the International Gas Research Conference, Government Institutes, Rockville, pp. 401–410.

Ryther, J. H., 1979, "Aquaculture in China," *Oceanus 22,* Woods Hole, pp. 21–28.

Smayda, T. J., 1970, "The Suspension and Sinking of Phytoplankton in the Sea," *Annual Review of Oceanography and Marine Biology,* 8, pp. 352–414.

## Farming in Space

### Robert D. MacElroy

Moore, B., R. Kaufmann, and C. Reinhold, 1982, "Design and Control Strategies for CELSS; Integrating Mechanistic Paradigms and Biological Complexities," American Society of Mechanical Engineers, 82-ENAs-43.

Schwartzkipf, S. and P. M. Stofan, 1981, "A Chamber Design for Closed Ecological Systems Research," American Society of Mechanical Engineers, 81-ENAs-37.

### CELSS Research Reports

Aroeste, H., January 1982, *Application of Guided Inquiry System Technique (GIST) to Controlled Ecological Life Support Systems (CELSS),* NASA CR-166312.

Auslander, D. M., R. C. Spear, and G. E. Young, August 1982, *Application of Control Theory to Dynamic Systems Simulation,* NASA CR-166383.

Averner, M., August 1981, *An Approach to Mathematical Modeling of a Controlled Ecological Life Support System,* NASA CR-166331.

Ballou, E. V., March 1982, *Mineral Separation and Recycle in a Controlled Ecological Life Support System (CELSS),* NASA CR-166388.

Carden, J. L. and R. Browner, August 1982, *Preparation and Analysis of Standardized Waste Samples for Controlled Ecological Life Support Systems (CELSS),* NASA CR-166392.

Fong, F. and E. A. Funkhouser, August 1982, *Air Pollutant Production by Algal Cell Cultures,* NASA CR-166384.

Gustan, E. and T. Vinopal, November 1982, *Controlled Ecological Life Support Systems: Transportation Analysis*, NASA CR-166420.

Hoff, J. E., J. M. Howe, and C. A. Mitchell, March 1982, *Nutritional and Cultural Aspects of Plants Species Selection for a Controlled Ecological Life Support System*, NASA CR-1663234.

Hornberger, G. M. and E. B. Rastetter, March 1982, *Sensitivity Analysis as an Aid in Modeling and Control of (Poorly-Defined) Ecological Systems*, NASA CR-166308.

Howe, J. M. and J. E. Hoff, June 1982, *Plant Diversity to Support Humans in a CELSS Ground-Based Demonstrator*, NASA CR-166357.

Huffaker, R. C., D. W. Rains, and C. O. Qualset, October 1982, *Utilization of Urea, Ammonia, Nitrite, and Nitrate by Crop Plants in a Controlled Ecological Life Support System (CELSS)*, NASA CR-166417.

Johnson, Emmett J., March 1982, *Genetic Engineering Possibilities for CELSS: A Bibliography and Summary of Techniques*, NASA CR-166306.

Karel, M., June 1982, *Evaluation of Engineering Foods for Controlled Ecological Life Support Systems (CELSS)*, NASA CR-166359.

Maguire, B., August 1980, *Bibliography of Human Carried Microbes' Interaction with Plants*, NASA CR-16630.

Mason, R. M., April 1980, *CELSS Scenario Analysis: Breakeven Calculation*, NASA CR-166319.

Mason, R. M. and J. L. Carden, 1982, *Controlled Ecological Life Support System: Research and Development Guidelines*, NASA CP-2232.

Moore, B. and R. D. MacElroy, 1982, *Controlled Ecological Life Support System: Biological Problems*, NASA CP-2233.

Moore, B., III, R. A. Wharton, Jr., and R. D. MacElroy, 1982, *Controlled Ecological Life Support System: First Principal Investigators Meeting*, NASA CP-2247.

Radmer, R., O. Ollinger, A. Venables, and E. Fernandez, July 1982, *Algal Culture Studies Related to a Closed Ecological Life Support System (CELSS)*, NASA CR-166375.

Raper, Jr., C. David, November 1982, *Plant Growth in Controlled Environments in Response to Characteristics of Nutrient Solutions*, NASA CR-166431.

Stahr, J. D., D. M. Auslander, R. C. Spear, and G. E. Young, July 1982, *An Approach to the Preliminary Evaluation of Closed-Ecological Life Support System (CELSS) Scenarios and Control Strategies*, NASA CR-166368.

Tibbits, T. W. and D. K. Alford, 1982, *Controlled Ecological Life Support System: Use of Higher Plants*, NASA CP-2231.

Young, G., May 1982, *A Design Methodology for Nonlinear Systems Containing Parameter Uncertainty: Application to Nonlinear Controller Design*, NASA CR-166358.

## CHAPTER 9   THE MISSOURI PARTNERSHIP FOR ECONOMIC DEVELOPMENT

### An Industry Perspective

*John T. Marvel*

*Agricultural Statistics*, 1982, U.S. Department of Agriculture, U.S. Government Printing Office, Washington, D.C.

*An Analysis of Federal R&D Funding by Function,* 1963–1973, National Science Foundation, NSF 72-313.

*An Analysis of Federal R&D Funding by Function,* 1969–1979, National Science Foundation, NSF 78-320.

*An Assessment of the United States Food and Agricultural Research System,* December 1981, Office of Technology Assessment, Washington, D.C.

Borlaug, N., 1983, "Contributions of Conventional Plant Breeding to Food Production," *Science,* 219, p. 689.

Chaleff, R. S. and M. F. Parsons, 1978, *Proceedings of the National Academy of Science USA,* 75, pp. 5104–5107.

*Federal Funds for Research and Development,* 1978–1980, National Science Foundation, Final Report, NSF 80-315.

*Federal Funds for Research and Development,* 1979–1981, National Science Foundation, Final Report, NSF 81-306.

*Federal R&D Funding by Budget Function,* 1981–1983, National Science Foundation, April 1982.

Frey, N., June 9, 1982, "Potential Application of Recombinant DNA and Genetics on Agricultural Sciences," Hearings before the Committee on Science and Technology, U.S. House of Representatives, Ninety-seventh Congress.

Fujiwara, A., 1982, *Plant Tissue Culture 1982,* The Japanese Association for Plant Tissue Culture, Tokyo.

Reinert, J. and Y. P. S. Bajaj, 1977, *Plant Cell, Tissue and Organ Culture,* Springer-Verlag, New York.

Vasil, Indra, W. R. Scowcroft, and K. J. Frey, 1982, *Plant Improvement and Somatic Cell Genetics,* Academic Press, New York.

## CHAPTER 11   NEW ROLES FOR THE AMERICAN FARMER

### New Approaches to Farm and Ranch Management

#### *Steven T. Sonka*

Arthur Andersen and Company, 1982, *The Management Difference: Information Needs of Farmers and Ranchers,* special study report, Chicago.

Toffler, Alvin, 1981, *The Third Wave,* Bantam Books, New York.

## CHAPTER 12   AGRICULTURAL INFORMATION DELIVERY SYSTEMS

### On-Farm Information Systems

#### *John R. Schmidt*

Ad Hoc Computer Task Force of the Extension Committee on Organization and Policy, 1982, "The Computer: Management Power for Modern Agriculture," Indiana Cooperative Extension Service.

Arthur Andersen and Company, 1982, *The Management Difference: Future Information Needs of Commercial Farmers and Ranchers*, special study report, Chicago.

"Farm Computer News," December 1982, *Successful Farming*, **2**, No. 12.

Fuller, Earl I., 1983, "The Future in On-farm Computing: 1983 and Beyond," unpublished mimeograph contribution to Minnesota Agricultural Experiment Station Project 14–036, prepared for presentation at a Farm Microcomputer Conference, Oklahoma State University, February 23, 1983.

Harsh, Stephen B., July 1981, "Management on a Chip—Version II," Agricultural Economics Staff Paper No. 81–51, Michigan State University.

Huggins, L. F., 1982, "Computers in Agriculture: Will Agriculture Engineering Departments Provide the Leadership?," *Computer Applications in Agricultural Engineering: Present and Future*, Conference Proceedings, North Central Computer Institute.

Knoblauch, Wayne A., January 1983, "The Cornell Minicomputer Dairy Management System," unpublished mimeograph prepared for presentation at the New York DHI Farm Computing Seminar, Syracuse, New York.

Kramer, Robert C., September 1980, "The Future of Computers in American Agriculture," *FACTS Computer System: An Update*, Conference Proceedings, Purdue University.

Schmidt, John R., November 1981, "Computer Based Information Services for Agriculture," Staff Paper No. 1, North Central Computer Institute.

Schmidt, John R., March 1982, "On-Farm Computer Usage: Today Vs. 1990," Staff Paper No. 3, North Central Computer Institute.

Schmidt, John R. and W. Terry Howard, 1981, "Design Considerations: Least Cost Dairy Ration Computer Program," Research Bulletins R3140 and R3140–1 (Supplement), Wisconsin Agricultural Experiment Station.

## Agricultural Information Utilities

### Larry R. Whiting

Priddy, Tom, 1982, "Evaluation of a Videotex System in Kentucky," Department of Agriculture Engineering, University of Kentucky, Lexington.

Warner, Paul D. and Frank Clearfield, January 1983, "An Evaluation of a Computer-based Videotext Information Delivery System for Farmers: The Green Thumb Project," Department of Sociology, Universitiy of Kentucky, Lexington.

Whiting, Larry R., 1981, "Communications Technology in the Land-Grant University Setting: A Focus on Computer-based Innovations for Information Dissemination to External Audiences," unpublished Ph.D. dissertation, Iowa State University, Ames.

## CHAPTER 13   MANAGING AGRICULTURAL TECHNOLOGY

### Science and American Agriculture

### John T. Caldwell

Bury, J. B., 1932, 1955, *The Idea of Progress*, Dover Press, New York.

Daniels, George H., 1971, *Science in American Society*, Alfred A. Knopf, New York.

Dupree, A. Hunter, 1957, *Science in the Federal Government*, Harvard University Press, Cambridge.

*Encyclopedia Britannica*, 1960.

Nevins, Allan, 1962, "The Historical Background of the Land-Grant Institutions," in *Charting the Second Century Under the Land-Grant Philosophy of Education*, Cornell University Press, Ithaca.

Rockefeller Foundation, 1983, *RF Illustrated*.

Van Tassell, David D. and Michael G. Hall, 1966, *Science and Society in the U.S.*, Dorsey Press, Homewood.

## Generating Technology

### Kenneth R. Keller

National Academy of Sciences, 1977, *World Food and Nutrition Study. The Potential Contributions of Research*, National Academy of Sciences, Washington, D.C.

Wortman, S. and R. W. Cummings, Jr., 1978, *To Feed This World—The Challenge and The Strategy*, The Johns Hopkins University Press, Baltimore.

## Transferring Technology

### George Hyatt, Jr.

Carpenter, William L., 1980, *Serving Personal and Community Needs through Adult Education*, Jossey-Bass, San Francisco.

Department of State and the President's Council on Environmental Quality, 1980, *Global 2000 Report to the President*, U.S. Government Printing Office, Washington, D.C.

Hildreth, R. J., 1976, "Heritage Horizons Extension's Commitment to People," *Journal of Extension*, Special Centennial volume.

Leagans, Paul J., 1979, *Adoption of Modern Agricultural Technology by Small Farm Operators*, Cornell University Press, Ithaca.

Rhoades, Robert, E., 1982, "Understanding Small Farmers Sociocultural Perspectives on Experimental Farm Trials," International Potato Center, Lima, Peru.

Simon, Julian L., 1981, *The Ultimate Resource*, Princeton University Press, Princeton.

Wortman, S. and R. W. Cummings, Jr., 1978, *To Feed This World—The Challenge and the Strategy*, The Johns Hopkins University Press, Baltimore.

## CHAPTER 15    FINANCING THE STRUCTURE OF AGRICULTURE

## Preparing for the Twenty-First Century

### Marvin R. Duncan and Dean W. Hughes

Adair, Ann Laing, John B. Penson, Jr., and Marvin Duncan, June 1981, "Monitoring Lease-Financing in Agriculture," *Economic Review*, Federal Reserve Bank of Kansas City, pp. 16–27.

Barry, Peter J., July 1981, "Agricultural Lending by Commercial Banks," *Agricultural Finance Review*, **41**, pp. 28–40.

Barry, Peter J., December 1981, "Impacts of Regulatory Change on Financial Markets for Agriculture," *American Journal of Agricultural Economics*, **63**, pp. 905–912.

Battelle Memorial Institute, 1983, *Agriculture 2000: A Look at the Future*, Battelle Press, Columbus.

Boehlje, Michael, July 1981, "Noninstitutional Lenders in the Agricultural Credit Market," *Agricultural Finance Review*, **41**, pp. 50–57.

Breimyer, Harold F., "Agricultural Policy Issues: Historical Perspective and Current Setting," Proceedings from the National Agricultural Policy Symposium, March 27–29, 1983, Kansas City, sponsored by the University of Missouri–Columbia, Dept. of Agric. Ecs.

Duncan, Marvin, "U.S. Agricultural Supply Setting and Future Agricultural Adjustments: A Discussion," Proceedings from the National Agricultural Policy Symposium, March 27–29, 1983, Kansas City, sponsored by the University of Missouri–Columbia, Dept. of Agric. Ecs.

Fenwick, Richard S., C. Edward Harshbarger, and Gene L. Swackhamer, July 1981, "The Role of the Farm Credit System in Financing Agriculture," *Agricultural Finance Review*, **41**, pp. 20–27.

Frederick, Roy A. L., "Symposium Overview and Summary," Proceedings from the National Agricultural Policy Symposium, March 27–29, 1983, Kansas City, sponsored by the University of Missouri–Columbia, Dept. of Agric. Ecs.

Herr, William McD. and Eddy LaDue, July 1981, "The Farmers Home Administration's Changing Role and Mission," *Agricultural Finance Review*, **41**, pp. 58–72.

Hottel, Bruce, July 1981, "The Commodity Credit Corporation and Agricultural Lending," *Agricultural Finance Review*, **41**, pp. 73–82.

Hughes, Dean W., July 1981, "An Overview of Farm Sector Capital and Credit Needs," *Agricultural Finance Review*, **41**, pp. 1–19.

Hughes, Dean W., December 1981, "Impacts of Regulatory Change on Financial Markets for Agriculture: Discussion," *American Journal of Agricultural Economics*, **63**, pp. 921–922.

Hughes, Dean W., Stephen C. Gabriel, Ronald L. Meekhof, Michael D. Boehlje, and George R. Amols, February 1982, "Financing the Farm Sector in the 1980s: Aggregate Needs and the Roles of Public and Private Institutions," Economic Research Service, Staff Report No. AGES820128, U.S. Department of Agriculture, National Economics Division.

Kaufman, George G., Larry R. Mote, and Harvey Rosenblum, "The Future of Commercial Banks in the Financial Services Industry," in George J. Benston, Ed., *Financial Services: The Changing Institutions and Government Policy*, Prentice-Hall, Englewood Cliffs.

Lins, David A., July 1981, "Life Insurance Company Lending to Agriculture," *Agricultural Finance Review*, **41**, pp. 41–49.

Penson, Jr., John B. and Marvin Duncan, July 1981, "Farmers' Alternatives to Debt Financing," *Agricultural Finance Review*, **41**, pp. 83–91.

U.S. Department of Agriculture, January 1981, *A Time to Choose: Summary Report on the Structure of Agriculture*, Washington, D.C.

U.S. Department of the Treasury, 1981, *Geographic Restrictions of Banking in the United States*, Washington, D.C.

# Financial Needs of Agriculture in the Next Century
## Michael Boehlje

Boehlje, Michael, July 1981, "Noninstitutional Lenders in the Agricultural Credit Market," *Agricultural Finance Review*, **41**, pp. 50–57.

Boehlje, Michael and Ron Durst, 1982, "Getting Started in Farming is Tough," in *Food—From Farm to Table*, 1982 Yearbook of Agriculture, U.S. Department of Agriculture, U.S. Government Printing Office, Washington, D.C.

Davenport, Charles, Michael Boehlje, and David B. H. Martin, February 1982, "The Effects of Tax Policy on American Agriculture," AER-480, U.S. Department of Agriculture, Economic Research Service.

Hughes, Dean W., July 1981, "An Overview of Farm Sector Capital and Credit Needs," *Agricultural Finance Review*, **41**, pp. 1–19.

Penson, Jr., John B. and Marvin Duncan, July 1981, "Farmers' Alternatives to Debt Financing," *Agricultural Finance Review*, **41**, pp. 83–91.

Schertz, Lyle P., and others, December 1979, "Another Revolution in U.S. Farming?," AER-441, U.S. Department of Agriculture, Economic Statistics and Cooperative Service.

U.S. Department of Agriculture, January 1981, *A Time to Choose: Summary Report on the Structure of Agriculture*, Washington, D.C.

## Financing the Small Farmer

### *Richard D. Robbins*

Carlin, Thomas A., 1979, *Small Farm Component of the U.S. Farm Structure, Structure Issues of American Agriculture*, U.S. Department of Agricultural Economics, Statistics, and Cooperative Services, U.S. Government Printing Office, Washington, D.C.

Coley, Basil G. and David Richard William, 1982, *The Feasibility of Establishing Marketing Cooperatives among Low Income Farmers in Six North Carolina Counties*, a series of six reports published in 1982, The Research Bulletin Series, North Carolina A&T State University, Greensboro.

Federal Reserve Bank of Atlanta, 1969, "Bank Financing of the Southern Agricultural Revolution," Reading in Southern Finances, No. IX, Atlanta.

Federal Reserve Bank of Chicago, March 1983, *Agricultural Letter*, No. 1599.

Horne, James E., September 1979, *Small Farms: A Review of Characteristics, Constraints and Policy Implications*, Southern Rural Development Center Series Publication No. 33, Mississippi State University.

Khan, Anwar S., Summer 1982, *An Economic Analysis of Small Farms: What Lies Ahead in North Carolina*, The Research Bulletin Series, North Carolina A&T State University, Greensboro.

Londhe, Suresh R., Ashok K. Singh, and Emily M. Crawford, August 1981, *The Potential for Increasing Incomes on Limited Resource Farms in South Carolina*, Research Bulletin No. 21, South Carolina State College, Orangeburg.

Madden, Patrick J., 1978, *Small Farms in America—The Diversity of Problems and Solutions, Increasing Understanding of Public Problems and Policies*, Farm Foundation, Oak Brook.

U.S. Department of Commerce, Bureau of the Census, 1978, *Census of Agriculture*, Vol. 51, Summary and State Data, 1981, U.S. Government Printing Office, Washington, D.C.

## CHAPTER 16   AGRICULTURE AND U.S. TRADE POLICY
### D. Gale Johnson

American Enterprise Institute, 1974, *The Sugar Program: Large Costs and Small Benefits,* Washington, D.C.

Buckwell, Allan E., David R. Harvey, Kenneth J. Thomson, and Kevin A. Parton, 1982, *The Costs of the Common Agricultural Policy,* Croom Helm Ltd., London.

Council on International Economic Policy, 1973, *Agricultural Trade and the Proposed Round of Multilateral Negotiations,* The Flanigan Report, U.S. Government Printing Office, Washington, D.C.

Johnson, D. Gale, August 1967, "Agricultural Trade and Foreign Economic Policy," in *Foreign Trade and Agricultural Policy,* National Advisory Commission on Food and Fiber, Technical Papers, Vol. VI, U.S. Government Printing Office, Washington, D.C., pp. 3–34.

Mangum, Jr. Fred A. and George E. Rossmiller, 1982, "Trade Protection Versus Trade Liberalization: A Dilemma for the United States," paper presented at First International Food and Finance Conference, September 12–14, 1982, Minneapolis.

Smith, Richard A., 1981, *Testimony before Committee on Agriculture Subcommittee on Livestock, Dairy and Poultry,* U.S. House of Representatives, September 10, 1981, published by U.S. Department of Agriculture Major News Releases and Speeches, September 4–11, 1981.

U.S. Department of Agriculture, Economic Research Service, *Feed: Outlook Situation,* various issues.

Villain, Claude, 1982, "Agriculture and Foreign Policy," paper presented at University of Minnesota, April 30, 1982.

## EPILOGUE   THE NEW AGRICULTURE: A VIEW OF THE TWENTY-FIRST CENTURY
### Sylvan H. Wittwer

Abegglen, J. C. and A. Etori, 1982, "Japanese Technology Today," *Scientific American,* **247**(4), p. 110.

Ames, D., 1980, "Thermal Environment Affects Production Efficiency of Livestock," *BioScience,* **30**(7), pp. 457–460.

Barr, T. N., 1981, "The World Food Situation and Global Grain Prospects," *Science,* **214**, pp. 1087–1095.

Batie, S. S. and R. G. Healy, 1983, "The Future of American Agriculture," *Scientific American,* **248**(2), pp. 45–53.

Bergen, W. G., 1980, "Future Challenges and Expected Direction of Ruminant Nutrition," Proceedings of the Western Nutrition Conference, March 4–5, 1980, Saskatoon, Saskatchewan, Canada, pp. 212–215.

Bergen, W. G. and D. B. Bates, 1983, "Ionophores: Their Effect on Production Efficiency and Mode of Action," Ruminant Nutrition Laboratory, Department of Animal Science, Michigan State University, East Lansing.

Bingham, C. W., 1980, "Forest Plantations in the 21st Century. Weyerhaeuser Company's Needs and Expectations," in *Forest Plantation, the Shape of the Future,* a Weyerhaeuser Science Symposium, Tacoma, April 30–May 3, 1979.

Blaxter, K., 1982, "Animal Agriculture in a Global Context," paper presented at a joint meeting of the American Society of Animal Science and the Canadian Society of Animal Science, University of Guelph, Ontario, Canada, August 9, 1982.

Bonnen, J. T., 1963, "Analysis of Long Run Economic Projections for American Agriculture," Proceedings of the Agricultural Research Institute, Twelfth Annual Meeting, Washington, D.C., pp. 69–114.

Boyeldieu, J., 1982, "Organic Farming and Its Prospects, Compared with Conventional Farming," *Phosphorus in Agriculture*, No. 82, pp. 31–38.

Boyer, J. S., 1982, "Plant Productivity and Environment," *Science*, 218, pp. 443–448.

Brady, N. C., 1982, "Chemistry and World Food Supplies," *Science*, 218, pp. 847–853.

Brown, A. W. A., T. C. Byerly, M. Gibbs, and A. San Pietro, Eds., 1975, "Crop Productivity—Research Imperatives," Proceedings of an International Conference, October 20–24, 1975, Boyne Highlands, Michigan Agricultural Experiment Station and Charles F. Kettering Foundation, Yellow Springs.

Brown, L. R., 1980, *Food of Fuel: New Competition for the World's Cropland*, Worldwatch Paper 35, Worldwatch Institute, Washington, D.C.

Browning, J. A., 1982, "Goal for Plant Health in the 'Age of Plants': A National Plant Health System," Department of Plant Sciences, Texas A&M University, College Station.

California Agricultural Experiment Station, 1982, *California Agriculture*, Special Issue, *Genetic Engineering of Plants*, Berkeley, Vol. 36(8).

Cardwell, V. B., 1982, "Fifty Years of Minnesota Corn Production: Sources of Yield Increase," *Agronomy Journal*, 74, pp. 984–990.

Cathey, H. M. and H. E. Heggestad, 1982, "Ozone and Sulfur Dioxide Sensitivity of Petunia; Ozone Sensitivity of Herbaceous Plants; Ozone Sensitivity of Woody Plants: Modification by Ethylenediurea," *Journal of the American Society of Horticultural Science*, 107(6), pp. 1028–1045.

Chaleff, R. S., 1983, "Isolation of Agronomically Useful Mutants from Plant Cell Cultures," *Science*, 219, pp. 676–682.

Chandler, Jr., R. J., 1983, *The Potential for Breeding Heat-Tolerant Vegetables for the Tropics*, 10th Anniversary Monograph Series, Asian Vegetable Research and Development Center, Shanhua, Taiwan.

Clark, W. C., Ed., 1982, *Carbon Dioxide Review 1982*, Clarendon Press, Oxford, and Oxford University Press, New York.

Council for Agricultural Science and Technology, 1982, *Soil Erosion: Its Agricultural, Environmental, and Socio-economic Implications*, Report 92, Ames.

Cunha, T. J., 1980, "Action Programs to Advance Swine Production Efficiency," *Journal of American Science*, 51(6), pp. 1429–1433.

Dalley, E. L., D. E. Eveleigh, B. S. Montenecourt, H. W. Stokes, and R. L. Williams, July 1981, "Recombinant DNA Technology—Food for Thought," *Food Technology*, pp. 26–33.

Devash, Y., S. Biggs, and I. Sela, 1982, "Multiplication of Tobacco Mosaic Virus in Tobacco Leaf Disks is Inhibited by (2'-5')oligoadenylate," *Science*, 216, pp. 1415–1416.

Diamond, B. A., D. E. Yelton, and M. D. Scharff, May 28, 1981, "Monoclonal Antibodies," *The New England Journal of Medicine*, pp. 1344–1349.

D'Itri, F., Ed., 1982, *Acid Precipitation, Effects on Ecological Systems*, Ann Arbor Science Publishers, Ann Arbor.

Dybing, C. D. and C. Lay, 1982, "Growth Regulators that Alter Oil Content and Quality," *Plant Growth Regulator Bulletin*, 10(3), pp. 9–12.

Eckholm, E. P., 1976, *Losing Ground,* W. W. Norton Co., New York.

Edens, T. C. and D. L. Haynes, 1982, "Closed System Agriculture: Resource Constraints, Management Options, and Design Alternatives," *Annual Review of Phytopathology,* **20,** pp. 363–395.

Eisbrenner, G. and H. J. Evans, 1983, "Aspects of Hydrogen Metabolism in Nitrogen-Legumes and Other Plant-Microbe Associations," *Annual Review of Plant Physiology,* **34,** pp. 105–136.

Elsden, R. P., G. E. Seidel, Jr., T. Takeda, and G. D. Farrand, 1982, "Field Experiments with Frozen–Thawed Bovine Embryos Transferred Nonsurgically," *Theriogeneology,* **17**(1), pp. 1–10.

Farnum, P., R. Timmis, and J. L. Kulp, 1983, "Biotechnology of Forest Yield," *Science,* **219,** pp. 694–702.

Fennema, O., January 1983, "The Food Industry: Charting a Course to the Year 2000," *Food Technology,* pp. 46–62.

Gerber, J. F., 1982, "Priority Areas for Horticultural Science," *HortScience,* **17**(5), pp. 706–707.

Gibson, A. H. and W. E. Newton, Eds., 1981, "Current Perspectives in Nitrogen Fixation," *Proceedings of the Fourth International Symposium on Nitrogen Fixation,* Canberra, Australia, December 1–5, 1980, Australian Academy of Science, Canberra.

Gogerty, R., 1982, "Growth Regulators Come of Age," *The Furrow,* **87**(6), pp. 10–13.

Hahn, G. L., 1976, "Rational Environmental Planning for Efficient Livestock Production," *Biometerology,* **6,** pp. 106–114.

Hannah, L. C., 1982, "Nonconventional Approaches to Plant Environment," *HortScience,* **17**(4), pp. 554–555.

Heck, W. W., O. C. Taylor, R. Adams, G. Bingham, J. Miller, E. Preston, and L. Weinstein, 1982, "Assessment of Crop Loss from Ozone," *Journal of the Air Pollution Control Association,* **32**(4), pp. 353–361.

Heichel, G. H., 1982, *Anticipating Productivity Responses to Crop Technologies,* Resource Conservation Act Symposium on Future Agricultural Technology and Resource Conservation, December 5–9, 1982, Washington, D.C.

Holgate, M. W., M. Kassas, and G. F. White, 1982, "World Environmental Trends between 1972 and 1982," *Environmental Conservation,* **9**(1), pp. 11–29.

Idso, S., 1983, *Carbon Dioxide. Friend or Foe? An Inquiry into the Climate and Agricultural Consequences of the Rapidly Rising $CO_2$ Content of Earth's Atmosphere,* IBR Press, Tempe.

Ingle, M. B., 1982, *Impact of Recombinant DNA Technology on Animal and Nutrient Production,* International Minerals and Chemical Corporation, IMC Staff Publication, Northbrook.

Jensen, M. E., 1980, "Irrigation Methods and Efficiencies," *Proceedings of Agricultural Sector Symposium,* January 9–10, The World Bank, Washington, D.C., pp. 170–211.

Jensen, M. E. 1982, "Water Resource Technology and Management," paper presented at the Resource Conservation Symposium, December 5–9, 1982, Washington, D.C.

Jensen, M. H., 1981, "Tomorrow's Agriculture Today," in *The 1981 Longwood Program Seminars with Symposium: Room to Live,* University of Delaware, Vol. 13.

Johansson, T. B., S. Steen, E. Bogren, and R. Fredriksson, 1983, "Sweden Beyond Oil: The Efficient Use of Energy," *Science,* **219,** pp. 355–361.

Johnson, D. G., 1982, "The World's Poor: Can They Hope for a Better Future?" in *Perspective '83. The World Food Situation,* International Minerals and Chemical Corporation, Northbrook, pp. 8–15.

Karnosky, D. F., 1981, "Potential for Forest Tree Improvement via Tissue Culture," *BioScience*, 31(2), pp. 114–120.

Kenney, M., F. H. Buttel, J. T. Cowan, and J. Kloppenburg, Jr., 1982, *Genetic Engineering and Agriculture: Exploring the Impacts of Biotechnology on Industrial Structure, Industry–University Relationships, and the Social Organization of U.S. Agriculture*, Cornell Rural Sociology Bulletin Series, No. 125, Cornell University Press, Ithaca.

Kerr, Jr., H. W. and L. Knutson, Eds., 1982, *Research for Small Farms*, U.S. Department of Agriculture, Agricultural Research Service, Miscellaneous Publication No. 1422, Beltsville.

Larson, W. E., L. M. Walsh, B. A. Stewart, and D. H. Boelter, Eds., 1981, *Soil and Water Resources: Research Priorities for the Nation*, Soil Science Society of America, Madison.

Larson, W. E., F. J. Pierce, and R. H. Dowdy, 1983, "The Threat of Soil Erosion to Long-Term Crop Production," *Science*, 219, pp. 458–465.

Laughlin, R. G., R. L. Munyon, S. K. Ries, and V. F. Wert, 1983, "Growth Enhancement of Plants by Femtomole Doses of Colloidally Dispersed Triacontanol," *Science*, 219, pp. 1219–1221.

Lawson, G. E., 1981, "Plasticulture in Japan," *American Vegetable Grower*, 29(10), pp. 14, 20, 21.

Lemon, E., Ed., 1983, *$CO_2$ and Plants. The Response of Plants to Rising Levels of Atmospheric Carbon Dioxide*, Westview Press, Boulder.

Lumpkin, T. A. and D. L. Plucknett, 1982, *Azolla as a Green Manure: Use and Management in Crop Production*, Westview Tropical Agriculture Series No. 5, Westview Press, Boulder.

McElkannon, W. S. and H. A. Mills, 1981, "Inhibition of Denitrification of Nitrapyrin with Field-Grown Sweet Corn," *Journal of the American Society of Horticultural Science*, 106(5), pp. 673–677.

Morris, C. E., 1981, "Genetic Engineering: Its Impact on the Food Industry," *Food Engineering*, 55(5), pp. 57–67.

Myers, N., 1983, *A Wealth of Wild Species. Storehouse for Human Welfare*, Westview Press, Boulder.

National Academy of Sciences, 1972, "Genetic Vulnerability of Major Crops," Committee on Genetic Vulnerability of Major Crops, Agricultural Board, Division of Biology and Agriculture, National Research Council, National Academy of Sciences, Washington, D.C.

National Academy of Sciences, 1975, *World Food and Nutrition Study. Enhancement of Food Production for the United States*, Report of the Board on Agriculture and Renewable Resources, Commission on Natural Resources, National Research Council, National Academy of Sciences, Washington, D.C.

National Academy of Sciences, 1976, *Climate and Food*, Board on Agriculture and Renewable Resources, National Academy Press, Washington, D.C.

National Academy of Sciences, 1977, *World Food and Nutrition Study. The Potential Contributions of Research*, National Academy of Sciences, Washington, D.C., pp. 77–80.

National Academy of Sciences, 1982, *Impacts of Emerging Agricultural Trends on Fish and Wildlife Habitat*, Committee on Impacts of Emerging Agricultural Trends on Fish and Wildlife Habitat, Board on Agriculture and Renewable Resources, Commission on Natural Resources, National Research Council, National Academy of Sciences, Washington, D.C.

National Academy of Sciences, 1983, *Report of the Briefing Panel on Agricultural Research*, Committee on Science, Engineering and Public Policy, National Academy Press, Washington, D.C.

National Science Foundation, March 19, 1973, *Interactions of Science and Technology in the Innovative Process*, case studies and final report, Battelle Columbus Laboratories.

Newman, J. E., 1982, "Seasonal Weather Impacts as Related to Seasonal Changes in USDA Crop Forecasts," Department of Agronomy, Purdue University, West Lafayette.

Nickel, L. G., 1982, "Plant-Growth Substances," in Kirk-Othmer, *Encyclopedia of Chemical Technology*, 3rd ed., John Wiley & Sons, New York, Vol. 18, pp. 1–23.

Office of Technology Assessment, 1982, *Genetic Technology, a New Frontier*, Westview Press, Boulder.

Okon, Y., 1982, "Recent Progress in Research on Biological Nitrogen Fixation with Non-Leguminous Crops," *Phosphorus in Agriculture* (Paris), No. 82, pp. 3–10.

Palmiter, R. D., R. L. Brinster, R. E. Hammer, M. E. Trumbauer, M. G. Rosenfeld, N. C. Birnberg, and R. M. Evans, 1982, "Dramatic Growth of Mice that Develop from Eggs Microinjected with Metallothionein-Growth Hormone Fusion Genes," *Nature*, 300(5893), pp. 611–615.

Phatak, S. C., D. R. Sumner, H. D. Wells, D. K. Bell, and N. C. Glaze, 1983, "Biological Control of Yellow Nutsedge (*Cypress esculentus Z.*) with the Indigenous Rust Fungus (*Puccinia canaliculta* [Schw.] Lagerh.)," *Science*, 219, pp. 1446–1447.

Phillip, R. E., R. L. Blevins, G. W. Thomas, W. W. Frye, and S. H. Phillips, 1980, "No-Tillage Agriculture," *Science*, 208, pp. 1108–1113.

Pimentel, D., 1981, *Handbook of Best Management in Agriculture*, Vol. 3, CRC Press, Boca Raton.

Plucknett, D. L. and N. J. H. Smith, 1982, "Agricultural Research and Third World Food Production," *Science*, 217, pp. 215–220.

Plucknett, D. L., N. J. H. Smith, J. T. Williams, and N. Murthi Anishetty, 1983, "Crop Germplasm Conservation and Developing Countries," *Science*, 22, pp. 163–169.

Pond, W. G., 1983, "Modern Pork Production," *Scientific American*, 248, pp. 96–103.

Pond, W. G., R. A. Merkel, L. D. McGilliard, and V. J. Rhodes, Eds., 1980, *Animal Agriculture. Research to Meet Human Needs in the 21st Century*, Westview Press, Boulder.

Putnam, A. R., 1983, "Allelopathic Chemicals, Nature's Herbicides in Action," *Chemical and Engineering News*, 61(14), pp. 34–45.

Putnam, A. R., J. DeFrank, and J. P. Barnes, 1983, "Exploitation of Allelopathy for Weed Control in Annual and Perennial Cropping Systems," *Plant Physiology*, in press.

Putter, I., J. G. MacConnell, F. A. Preiser, A. A. Haidri, S. S. Ristich, and R. A. Dybas, 1981, "Avermectins: Novel Insecticides, Acaricides and Nematicides from a Soil Microorganism," *Experientia*, 37, pp. 963–964.

Raper, Jr. C. D. and P. J. Kramer, Eds., 1983, *Crop Reactions to Water and Temperature Stresses in Humid, Temperate Climates*, Westview Press, Boulder.

Rasmussen, W. D., 1982, "The Mechanization of Agriculture," *Scientific American*, 247(3), pp. 77–89.

Resources for the Future, 1982, *Water for Western Agriculture*, Washington, D.C.

Revelle, R., 1982, "Carbon Dioxide and World Climate," *Scientific American*, 247(2), pp. 35–43.

Rhodes, L. H., 1980, *"Mycorrhizae,"* Outlook on Agriculture, 10(6), pp. 275–281.

Ries, S., V. Wert, and J. Biernbaum, 1982, "Factors Altering the Efficacy of Triacontanol Applications to Plants," paper presented at the 9th Annual Plant Growth Regulator Society of America, Asilomar, Monterey, California, July 5–9, 1982.

Rodale, R., 1982, "Regenerative Agriculture, a Search for Low-Cost, Self-Renewing Solutions to Farming's Problems," Rodale Press, Emmaus.

Ruttan, V. W., 1977, "Induced Innovation and Agricultural Development," *Food Policy*, 2(3), pp. 196–216.

Schroth, M. N. and J. G. Hancock, 1982, "Disease-Suppressive Soil and Root-Colonizing Bacteria," *Science*, 216, pp. 1376–1381.

Schultz, T. W., 1981, *Investing in People: the Economics of Population Quality*, University of California Press, Berkeley.

Schultz, T. W., 1982, "The Dynamics of Soil Erosion in the United States," Agricultural Economics Paper 82:8, University of Chicago, presented at a conference on soil conservation, Agricultural Council of America, March 17, 1982.

Seidel, Jr., G. E., 1981, "Superovulation and Embryo Transfer in Cattle," *Science*, 211, pp. 351–358.

Seidel, Jr., G. E., 1982, "Applications of Microsurgery to Mammalian Embryos," *Theriogeneology*, 17(1), pp. 23–24.

Seidel, Jr., G. E. and B. G. Brackett, 1981, "Perspectives on Animal Breeding," in *New Technologies in Animal Breeding*, Academic Press, New York, pp. 3–9.

Sharp, W. R. and D. A. Evans, 1982, "Plant Tissue Culture: the Foundation for Genetic Engineering in Higher Plants," presented at the North Atlantic Region of the American Society of Agricultural Engineers Annual Meeting, Burlington, August 9, 1982.

Shepard, J. F., 1982, "The Regeneration of Potato Plants from Leaf-Cell Protoplasts," *Scientific American*, 246(5), pp. 154–166.

Shoji, K., 1977, "Drip Irrigation," *Scientific American*, 237(5), pp. 62–68.

Sill, Jr., W. H., 1982, *Plant Protection, an Integrated Interdisciplinary Approach*, The Iowa State University Press, Ames.

Sims, W. L., 1982, "Mechanical Harvesting of Vegetable Crops," paper presented at the 21st International Horticultural Congress, August 29–September 4, 1982, Hamburg, Federal Republic of Germany.

Splittstoesser, W. E., 1982, "Mycorrhizae and How They Influence the Growth of Plants," *Journal Korean Society Horticultural Science*, 23(2), pp. 1–11.

Subba Rao, N. S., Ed., 1982, *Advances in Agricultural Microbiology*, Oxford and IBH Publishing Company, New Delhi, India.

Sundquist, W. B., K. M. Menz, and C. F. Neumeyer, 1982, "A Technology Assessment of Commercial Corn Production in the United States," Minnesota Agricultural Experiment Station Bulletin 546.

Swaminathan, M. S., 1982, "Biotechnology Research and Third World Agriculture," *Science*, 218, pp. 967–972.

Thompson, L. M., 1980, "Climate Change and World Grain Production," in D. Gale Johnson, Ed., *The Politics of Food*, The Chicago Council on Foreign Relations, Chicago, pp. 100–123.

Tiedje, J. M., 1982, "Denitrification," in *Methods of Soil Analysis* (rev. ed.), American Society of Agronomy, Madison.

Tucker, H. A. and R. K. Ringer, 1982, "Controlled Photoperiod Environments for Food Animals," *Science*, 216, pp. 1381–1386.

U.S. Department of Agriculture, 1980, *Report and Recommendations on Organic Farming*, Washington, D.C.

U.S. Department of Agriculture, 1981, *Aquaculture*, Economic Research Service AS-2, Washington, D.C.

U.S. Department of Agricultural, November 10, 1982, *Foreign Agricultural Circular—World Crop Production*, Foreign Agricultural Science, Economic Research Service, Washington, D.C.

U.S. Water Resources Commission, 1978, *The Nation's Water Resources 1975–2000*, Vol. 1 Summary, Washington, D.C.

White, R., 1982, "The Ultimate Environmental Dilemma: Is Man Changing the Climate?," *Mazingira*, 6(3), pp. 32–43.

Wortman, S. and J. W. Cummings, Jr., 1978, *To Feed This World: The Challenge and the Strategy*, The Johns Hopkins University Press, Baltimore, pp. 186–230.

# BIOGRAPHICAL NOTES

## THE AUTHORS

RICHARD BARROWS, Ph.D., Professor of Agricultural Economics, University of Wisconsin–Madison. Dr. Barrows' research and extension activities involve all areas of agricultural and natural resource policy. He is a W. F. Kellogg Foundation National Fellow and has served as a policy advisor to the National Agricultural Land Study. Dr. Barrows was the first director of Wisconsin's Farmland Preservation Program.

SANDRA S. BATIE, Ph.D., Associate Professor of Agricultural Economics, Virginia Polytechnic Institute and State University. Dr. Batie has conducted research on the economics of soil conservation, water allocation, alternative management strategies for coastal wetlands, and agricultural land use. In 1980 she was senior associate of the Conservation Foundation where she developed a rural resource management research program.

MICHAEL BOEHLJE, Ph.D., Professor of Economics, Iowa State University. Born on a farm, Dr. Boehlje teaches undergraduate and graduate courses in agricultural finance and conducts extension programs in finance, estate and business planning for farmers and lenders. He is also assistant director of the Iowa Experiment Station, associate director of the Iowa Agricultural Credit School, and vice president and board member of Waldel, Ltd., a family farm corporation in Iowa.

JAMES B. BOILLOT, Director of Agriculture, State of Missouri. Currently serving his second term as director, Mr. Boillot is also a member of the Farm Income Protection Insurance Program Task Force. He is past president of the Midwest Association of State Departments of Agriculture and a former director of the National Association of State Departments of Agriculture. An active farmer since high school, Mr. Boillot received both his B.S. and M.S. from the University of Missouri College of Agriculture.

ERNEST J. BRISKEY, Ph.D., Dean of Agriculture, Oregon State University. Dr. Briskey is in charge of the University's experiment station, extension service, resident instruction, and international agriculture division. Before assuming his current post, he had been a vice president of the Campbell Soup Company and vice president-research, Campbell Institute for Food Research. He presently serves on Oregon's Board of Agriculture and as a senior advisor to the Kuwait Institute for Scientific Research.

JOHN T. CALDWELL, Ph.D., Professor, Department of Political Science, North Carolina State University. In addition to his teaching responsibilities, Dr. Caldwell is a director of the Triangle Universities Center for Advanced Studies. He has also served as president of Alabama College and the University of Arkansas and as chancellor of North Carolina State. Dr. Caldwell is currently a trustee of the National Humanities Center.

W. RONNIE COFFMAN, Ph.D., Professor of Plant Breeding and International Agriculture, New York State College of Agriculture and Life Sciences, Cornell University. Before joining Cornell's Department of Plant Breeding and Biometry in 1981, Dr. Coffman was a plant breeder at the International Rice Research Institute in the Philippines. He has worked on rice improvement and helped establish a network for the international testing of rice. Dr. Coffman has also assisted with the spring wheat improvement project for New York State.

ROBERT B. DELANO, President, American Farm Bureau Federation. Mr. Delano is president of the world's largest general farm organization. A strong advocate of market-oriented agriculture and a self-described "trade optimist," he is a member of President Reagan's Advisory Committee for Trade Negotiations and has led trade missions to many nations which are seeking expansion of agricultural trade. He is also a former president of the Virginia Farm Bureau Federation and the Richmond County Farm Bureau. Mr. Delano still operates a 400-acre wheat, corn, and soybean farm in Richmond County, Virginia.

MARVIN R. DUNCAN, Ph.D., Vice President and Economist, Federal Reserve Bank of Kansas City. As head of the regional economics group within the bank's Economic Research Department, Dr. Duncan directs research concerning agriculture, energy, and regional economics. Before joining the bank in 1975, he had devoted 15 years to farm management and agribusiness as well as serving on the faculties of North Dakota State University and Iowa State University. Dr. Duncan is currently a member of the USDA's National Agricultural Cost of Production Standards Review Board.

J. M. ELLIOT, Ph.D., Professor of Animal Science, New York State College of Agriculture and Life Sciences, Cornell University. Dr. Elliot has been on the faculty at Cornell since 1960. In addition to his teaching responsibilities, he conducts research on dairy cattle nutrition and is studying the effect of diet on the metabolism of nutrients in lactating cows. Dr. Elliot recently received the American Dairy Science Association's 1982 Ralston Purina Teaching Award.

KENNETH R. FARRELL, Ph.D., Senior Fellow and Director, Food and Agricultural Policy Program, Resources for the Future. Dr. Farrell joined Resources for the Future in 1981 to develop its food and agricultural policy research and education program. His current projects include an assessment of world food and agricultural prospects in the twenty-first century and a study of productivity in the U.S. food and agricultural sector. Before joining RFF, Dr. Farrell headed the Economics and Statistics Service of the U.S. Department of Agriculture and also worked for the USDA's Economic Research Service.

RALPH W. F. HARDY, Ph.D., Director of Life Sciences, Central Research and Development Department, E. I. du Pont de Nemours & Company. Dr. Hardy, a specialist in $N_2$ fixation and photosynthesis, is a native of Ontario, Canada, and began his career in 1960 as an assistant professor of biochemistry at that province's University of Guelph. He came to Du Pont as a research biochemist in 1963 and was appointed a research supervisor 4 years later. Dr. Hardy was named associate director of the company's life sciences division in 1974 and has held his current position since 1979. He is a member of the National Research Council's Board of Agriculture and the International Council of Scientific Unions' Committee on Genetic Experimentation.

HAROLD M. HARRIS, JR., Ph.D., Extension Agricultural Economist, College of Agricultural Sciences, Clemson University. Before joining the Agricultural Economics Department at Clemson in 1975, Dr. Harris was on the faculty at VPI. His major activities are in the areas of dairy marketing, grain marketing, economic situation and outlook, producer marketing strategy, and agricultural policy.

VIRGIL W. HAYS, Ph.D., Chairman and Professor, Department of Animal Sciences, University of Kentucky. Dr. Hays is an authority on animal nutrition and swine production and has been professor of animal sciences at Kentucky since 1967. The recipient of the Distinguished Service Award of the Department of Animal Science, University of Oklahoma, in 1981, he has participated in many international animal nutrition symposiums and served on the National Academy of Sciences' Review Team to review animal agriculture in the People's Republic of China.

KENZO HEMMI, Ph.D., Professor, College of Agriculture, University of Tokyo. Dr. Hemmi, a former dean of the Faculty of Agriculture at the University of Tokyo, joined the teaching staff at the University 21 years ago. He currently serves on the boards of trustees of the International Rice Research Institute in the Philippines and the Agricultural Development Council in New York. Dr. Hemmi is a member of Japan's Population Council, Livestock Industry Promotion Council, and the Rice Price Deliberation Council. He has led Japanese government economic cooperation missions to Africa and Thailand.

JAMES C. HITE, Ph.D., Alumni Professor of Agricultural Economics, Clemson University. The author of four books, Dr. Hite teaches undergraduate and graduate courses on natural resource economics and public finance. His research interests focus on natural resource policy and regional economic development. In addition to his teaching responsibilities, Professor Hite is editor of *The Review of Regional Studies*.

DEAN W. HUGHES, Ph.D., Economist, Federal Reserve Bank of Kansas City. Dr. Hughes works principally with agricultural modeling and forecasting. Before joining the bank in 1981 he had been with the Economic Research Service of the USDA, an investment advisor, and had taught economics at Purdue University.

GEORGE HYATT, JR., Ph.D., Emeritus Director and Professor, Department of Adult and Community College Education, North Carolina State University. Following a distinguished 26-year career at N.C. State, Dr. Hyatt retired as associate dean of the University's School of Agriculture and Life Sciences and director of its extension service in 1978. A long-time dairyman, he has written numerous articles for scientific journals and breed and farm magazines.

D. GALE JOHNSON, Ph.D., Eliakim Hastings Moore Distinguished Service Professor of Economics and Chairman, Department of Economics, The University of Chicago. In addition to his nearly 40 years of work at the University, Dr. Johnson has served as a consultant to the National Academy of Sciences, the National Research Council, the Steering Committee for the President's Food and Nutrition Study, and the U.S. Council on International Economic Policy. From 1970 to 1972 he was a member of the National Commission on Population Growth and the American Future.

KENNETH R. KELLER, Ph.D., Managing Director, Bright Belt Warehouse Association. Dr. Keller was director of research and associate dean of North Carolina State University's School of Agriculture and Life Sciences at his retirement in 1979. A research award endowment has been established at the University to recognize his outstanding contributions in the field of tobacco research. Dr. Keller is a former chairman of the Tobacco Science Council, the Burley Tobacco Workers' Conference, and the Tobacco Workers' Conference.

W. DAVID KLEMPERER, Ph.D., Associate Professor of Forest Economics, Virginia Polytechnic Institute and State University. Dr. Klemperer's professional activities include teaching, research, and occasional consulting in his field of forest economics. He has also been on the staff of Associated Oregon Industries, Oregon State University, and the Washington State University extension service.

RICHARD KRUMME, Editor, *Successful Farming* magazine. Before joining the magazine in 1967, Mr. Krumme was an assistant editor for the Department of Agricultural Information at Texas A&M University. The winner of four major agricultural writing awards, he is a past president of the American Agricultural Editors' Association. In addition to his literary activities, Mr. Krumme owns and operates a farm in Iowa and also operates his family farm in Missouri.

MAX LENNON, Ph.D., Vice President for Agricultural Administration, The Ohio State University. Dr. Lennon moved to Ohio State in July 1983 from the University of Missouri–Columbia where he had been Dean of the College of Agriculture. Before going to Missouri, he held administrative and teaching posts at Texas Tech University and North Carolina State University. Dr. Lennon has also worked in the private sector as a swine research specialist.

W. ARTHUR LEWIS, Ph.D., James Madison Professor of Political Economy, Woodrow Wilson School of Public and International Affairs, Princeton University. A 1979 Nobel Prize winner in economics and a world-renowned authority on the economics of developing countries, he was knighted by Queen Elizabeth in 1963 for his work as Vice Chancellor of the University of the West Indies. His 1954 book, *The Theory of Economic Growth*, has become a modern-day classic. Born in St. Lucia, British West Indies, and educated at the University of London, Sir Arthur is a founder and former president of the Caribbean Development Bank.

ROBERT D. MacELROY, Ph.D., Research Scientist, CELSS Program Manager, NASA Ames Research Center. Dr. MacElroy joined the Ames staff in 1969 and currently manages the Controlled Environment Life Support System (CELSS) research program. His research programs include numerical simulation studies of molecular interactions and life support systems based upon recycling materials needed for human sustenance.

JOHN T. MARVEL, Ph.D., General Manager, Research Division, Monsanto Agricultural Products Company. Dr. Marvel's many areas of expertise include pesticide metabolism, the synthesis/biology of herbicides and plant growth regulators, and exploratory biology. He joined Monsanto as a senior research chemist in 1968 and assumed his present position in 1981. He serves on several agricultural committees, including the USDA's National Agricultural Research and Extension Users Advisory Board and Missouri Governor Christopher S. Bond's Task Force on Rural Development.

JOHN L. MERRILL, Director, Ranch Management Program, Texas Christian University. Mr. Merrill's position combines intensive classroom work with 10,000 miles of travel each year to study working ranches. He is a past president of the Society for Range Management and a current executive member of the Governor's Task Force on Agriculture. Mr. Merrill, the fourth generation of a Texas cattle ranching family, operates the XXX Ranch in Crowley, Texas.

WALTER W. MINGER, Senior Vice President, Bank of America N.T. & S.A. Mr. Minger is in charge of the Bank of America's worldwide agribusiness relations and policy in connection with the bank's extensive agricultural lending activities. He is also a member of the Food and Agriculture Subcommittee of the U.S. Chamber of Commerce and a board member of the San Francisco District Export Council for the U.S. Department of Commerce. Mr. Minger serves on the board of the American Society for Agricultural Consultants International. A resident of California, Mr. Minger serves with that state's 4-H Foundation and the California Council for International Trade.

ROGER L. MITCHELL, Ph.D., Dean, College of Agriculture, University of Missouri–Columbia. Dr. Mitchell, who has also been a dean of the University's extension service, was vice president for agriculture at Kansas State University from 1975 to 1981. A former Danforth Fellow, he is a past president of both the Crop Science Society and the American Society of Agronomy. Dr. Mitchell is the author of a textbook on crop physiology entitled *Crop Growth and Culture*.

MICHAEL NEUSHUL, Ph.D., Professor of Marine Botany, University of California, Santa Barbara. Dr. Neushul, an authority on marine phycology, algal cytology, and biological oceanography, has been at the University since 1963. He is also board chairman and president of Neushul Mariculture Incorporated. Dr. Neushul has received the Darbaker Prize from the Botanical Society of America for his studies of microscopic algae.

RICHARD D. ROBBINS, Ph.D., Professor and Chairman, Department of Agricultural Economics and Rural Sociology, North Carolina A&T State University. Dr. Robbins has been at the University since 1971 and in 1981–1982 served as a farm management specialist for USAID in the Science and Technology Bureau. He is presently a member of the Governor's Oversight Committee for Official Labor Market Information.

B. H. ROBINSON, Ph.D., Professor and Head, Department of Agricultural Economics and Rural Sociology, Clemson University. In addition to his distinguished career at Clemson, Dr. Robinson has worked for the USDA's Economics Research Service. He is currently a member of USDA's Cost of Production Review Board. Dr. Robinson has also been chairman of the National Public Policy Education Committee and served on the steering committee for the National Project on International Trade Policy Education.

JOHN R. SCHMIDT, Ph.D., Director, North Central Computer Institute, and Professor of Agricultural Economics, University of Wisconsin. Dr. Schmidt's responsibilities include research, teaching, and providing extension services in farm management and production economics. Dr. Schmidt is working on designs for electronic farm records systems and interactive decision aids for farmers. He has served as a consultant to the World Bank in its monitoring of agricultural credit programs in underdeveloped countries.

WAYNE H. SMITH, Ph.D., Institute of Food and Agricultural Sciences, University of Florida. Dr. Smith is director of the Institute's Center for Environmental and Natural Resources Programs and the Center for Biomass Energy Systems. He is also technical manager for its Gas Research Institute. Dr. Smith, who dubs himself a "serious hobby farmer," was a board member of the Florida Solar Coalition.

STEVEN T. SONKA, Ph.D., Associate Professor of Agricultural Economics, University of Illinois. Twice recognized for distinguished research by the American Association of Agricultural Economics, Dr. Sonka has published extensively in the areas of agricultural finance, policy, and natural resource economics. His book on the use of small computers in agriculture, *Computers in Farming: Selection and Use*, was published last spring.

ARTURO R. TANCO, JR., Minister of Agriculture, Republic of the Philippines. When he was appointed Minister of Agriculture in 1971 at age 38, Minister Tanco was the youngest person ever to assume that position. In 1978 he was elected a member of the Philippine Parliament. Among several accomplishments he has conceived and managed "Masagana 99," the Philippines' highly innovative rice production program. From 1977 to 1981 he served as president of the U.N. World Food Council and played a key role in the creation of that organization's Manila Communique, the Program of Action on Eradication of Hunger and Malnutrition, which the General Assembly adopted in 1977. He is presently vice chairman of the International Rice Research Institute. Minister Tanco has studied at Harvard, Cornell, and Union College in the United States as well as at universities in Manila.

THOMAS L. THOMPSON, Ph.D., Professor of Agricultural Engineering, University of Nebraska. Born and raised on a dairy farm, Dr. Thompson specializes in grain drying and storage and computer simulation. He has been active in the North Center Computer Institute for more than 3 years and has co-developed and led a national agricultural computer network called AGNET, which offers a variety of information services.

JO ANN VOGEL, Cato, Wisconsin. With her family, Mrs. Vogel farms 510 acres and oversees 325 dairy and beef cows. She is actively involved in agricultural issues, focusing on the role of women. Mrs. Vogel serves as legislative chair of Wisconsin's Women for Agriculture and as first vice president for American Agri-Women. She is presently working on a documentary depicting the contemporary farm woman.

JOHN N. WALKER, Ph.D., Associate Dean, College of Agriculture, University of Kentucky. Dr. Walker, a professor of agricultural engineering, is responsible for development programs, physical planning, computing services, and interdisciplinary projects within the College of Agriculture. He was acting director of the University's Institute for Mining and Minerals Research from 1981 to 1982. In addition, he has been director for Education and Research and a board member of the American Society of Agricultural Engineers.

RICHARD O. WHEELER, Ph.D., President and Chief Executive Officer, Winrock International. Before joining Winrock, Dr. Wheeler served as a consultant to Schnittker Associates and was a research economist with the Economic Research Service, USDA. His activities have included research to identify and weigh consequences of policy actions for both producers and consumers and to study intraregional trade in the field of agriculture. Dr. Wheeler has acted as a consultant and technical assistant in East Africa, West Africa, and Southeast Asia.

LARRY R. WHITING, Ph.D., Chairman, Department of Information and Publications, University of Maryland. Dr. Whiting's department is the communications arm of the Maryland Cooperative Extension Service, the Maryland Agriculture Experiment Station, and the Division of Agricultural and Life Sciences. A former farm editor for several newspapers, his current research interests focus on computer-based information dissemination systems.

SYLVAN H. WITTWER, Ph.D., Director Emeritus, Agricultural Experiment Station, College of Agriculture and Natural Resources, Michigan State University. Dr. Wittwer, a pioneer in the use of radioisotopes in studies of foliar nutrition of agricultural crops, is an authority on plant growth regulants in the chemical control of flowering, fruiting, and senescence and factors affecting flowering and fruit setting in horticultural crops. He is a world-renowned expert on the biological limits of crop and livestock productivity and projections for food supply in the twenty-first century. For the past 23 years Dr. Wittwer has traveled extensively to advise and present lectures on agricultural productivity and research in many nations, including the People's Republic of China, India, the U.S.S.R., Egypt, and Mexico. Prior to his recent retirement, Dr. Wittwer had served as a professor at Michigan State for 37 years and as director of the Agricultural Experiment Station for 18 years.

DEMETRIOS M. YERMANOS, Ph.D., Oil Crops Project Leader, University of California, Riverside. Dr. Yermanos, a native of Salonica, Greece, taught quantitative genetics and statistical designs at the University from 1961 to 1982. He is an authority on the genetic and environmental factors which affect the quality of oil and protein in seeds. A former chairman of the U.S. National Committee on Jojoba, Dr. Yermanos serves as a consultant to the United Nations Development Program and to the National Academy of Sciences.

## THE EDITOR

JOHN W. ROSENBLUM, D.B.A., Dean and Professor, The Colgate Darden Graduate School of Business Administration, University of Virginia. Dr. Rosenblum served on the faculty at the Harvard Business School for 10 years before coming to the University of Virginia in 1979. He has participated in many national and international Executive Education Programs and has served as a corporate consultant. Dr. Rosenblum, who received his M.B.A. and D.B.A. from Harvard University, is a director of Dansk Designs, Ltd. He is also a co-author of *Strategy and Organization—Text and Cases in General Management* and *Case Studies in Political Economy: Japan 1854–1977.*

# INDEX

Pages in *italics* refer to tables or figures